NUTRITION RESEARCH REVIEWS 1994

EDITORS
J. W. T. Dickerson
J. M. Forbes
M. I. Gurr
D. J. Millward

Nutrition Research Reviews is published annually by Cambridge University Press for the Nutrition Society which owns it and controls its publication. Information about the Nutrition Society, which has as its object the advancement of the scientific study of nutrition and its application to the maintenance of human and animal health, may be obtained from the Honorary Secretary, Dr J. C. Mathers, The Nutrition Society, 10 Cambridge Court, 210 Shepherds Bush Road, London, W6 7NJ, U.K. Cambridge University Press also publishes the *British Journal of Nutrition* and the *Proceedings of the Nutrition Society* for the Society.

Nutrition Research Reviews (0954-4224) may be purchased separately from booksellers or directly from Cambridge University Press. The subscription including postage (excluding VAT) to volume 7, 1994, is £46 payable in advance to Cambridge University Press, The Edinburgh Building, Shaftesbury Road, Cambridge CB2 2RU. The subscription in USA, Canada and Mexico is US$85, and inquiries there should be addressed to Cambridge University Press, Journals Department, 40 West 20th Street, New York, NY 10011-4211. All orders must be accompanied by payment. EU subscribers (outside the UK) who are not registered for VAT should add VAT at their country's rate. VAT registered subscribers should provide their VAT registration number, Japanese prices for institutions (including ASP delivery) are available from Kinokuniya Company Ltd, P.O. Box 55, Chitose, Tokyo 156, Japan.

Copying. This journal is registered with the Copyright Clearance Center, 222 Rosewood Drive, Danvers, MA 01923. Organizations in the USA who are registered with the C.C.C. may therefore copy material (beyond the limits permitted by sections 107 and 108 of the US copyright law) subject to payment to C.C.C. of the per-copy fee of $5.00. This consent does not extend to multiple copying for promotional or commercial purposes. Code 0954-4224/94 $5.00 + .00. Organizations authorized by the Copyright Licensing Agency may also copy material subject to the usual conditions.

ISI Tear Sheet Service, 3501 Market Street, Philadelphia, Pennsylvania 19104, USA, is authorized to supply single copies of separate articles for private use only.

For all other use, permission should be sought from the Cambridge or New York offices of the Cambridge University Press.

Future issues
Recommendations for topics and authors in any field of nutritional research are invited. Recommendations should be reasonably precise and must be accompanied by a clear statement on why the proposer thinks that a review of the chosen subject would be particularly timely and why the particular author or group of authors is being proposed. We hope to give as much prominence to authors from overseas, as long as they write well in English, as to those from the United Kingdom. Proposals, all of which will be considered carefully, should be sent to: Professor M. I. Gurr, Vale View Cottage, Maypole, St Mary's, Isles of Scilly, TR21 0NU, United Kingdom. Proposals for Volume 9 (1996) should be sent to arrive by June 1995.

NUTRITION RESEARCH REVIEWS 1994

VOLUME 7

GENERAL EDITOR
M. I. GURR

Published by the Press Syndicate of the University of Cambridge
The Pitt Building, Trumpington Street, Cambridge CB2 1RP
40 West 20th Street, New York, N.Y. 10011–4211, U.S.A.
10 Stamford Road, Oakleigh, Melbourne 3166, Australia

© Nutrition Society 1994

ISBN 0521 47767 0
ISSN 0954-4224

Printed in Great Britain by the University Press, Cambridge

CONTENTS

Contributors   vii

Preface   ix

The composition of human milk as a model for the design of infant formulas: recent findings and possible applications   1
ANNEMIEK C. GOEDHART and JACQUES G. BINDELS

Feto-maternal interaction of antibody and antigen transfer, immunity and allergy development   25
JULIE A. LOVEGROVE and JANE B. MORGAN

Physical activity, not diet, should be the focus of measures for the primary prevention of cardiovascular disease   43
ALEXANDER L. MACNAIR

Nutrition, physical activity and bone health in women   67
J. H. WILSON

Human bioavailability of vitamins   93
C. J. BATES and H. HESEKER and members of the EC Flair Concerted Action No. 10

Biology of zinc and biological value of dietary organic zinc complexes and chelates   129
JOHANNES W. G. M. SWINKELS, ERVIN T. KORNEGAY and MARTIN W. A. VERSTEGEN

Zinc nutrition in developing countries   151
ROSALIND S. GIBSON

Anticarcinogenic factors in plant foods: a new class of nutrients?   175
I. T. JOHNSON, G. WILLIAMSON and S. R. R. MUSK

Glucosinolates and glucosinolate derivatives: implications for protection against chemical carcinogenesis   205
LIONELLE NUGON-BAUDON and SYLVIE RABOT

Nutritional influences on interactions between bacteria and the small intestinal mucosa   233
DENISE KELLY, R. BEGBIE and T. P. KING

Subject Index   259

CONTRIBUTORS

Christopher J. Bates *MRC Dunn Nutritional Laboratory, Downham's Lane, Milton Road, Cambridge CB4 1XJ, UK*

R. Begbie *Rowett Research Institute, Bucksburn, Aberdeen, Scotland, AB2 9SB, UK*

Jacques G. Bindels *Nutricia Research, P.O. Box 1, 2700 MA Zoetermeer, The Netherlands*

Rosalind S. Gibson *Division of Applied Human Nutrition, University of Guelph, Guelph, Ontario N1G 2W1, Canada*

Annemiek C. Goedhart *Nutricia Research, P.O. Box 1, 2700 MA Zoetermeer, The Netherlands*

H. Heseker *University of Paderborn, D-33098 Paderborn, Warburgerstrasse 100, Germany*

I. T. Johnson *AFRC Institute of Food Research, Norwich Laboratory, Norwich Research Park, Colney, Norwich NR4 7UA, UK*

D. Kelly *Rowett Research Institute, Bucksburn, Aberdeen, Scotland, AB2 9SB, UK*

T. P. King *Rowett Research Institute, Bucksburn, Aberdeen, Scotland, AB2 9SB, UK*

Ervin T. Kornegay *Department of Animal Science, Virginia Polytechnic Institute and State University, Blacksburg, VA 24061, USA*

Julie A. Lovegrove *Nutritional Metabolism Research Group, School of Biological Sciences, University of Surrey, Guildford GU2 5XH, UK*

Alexander L. Macnair *20 Wimpole Street, London W1M 7AD, UK*

Jane B. Morgan *Nutritional Metabolism Research Group, School of Biological Sciences, University of Surrey, Guildford GU2 5XH, UK*

S. R. R. Musk *AFRC Institute of Food Research, Norwich Laboratory, Norwich Research Park, Colney, Norwich NR4 7UA, UK*

Lionelle Nugon-Baudon *Unité d'Ecologie et de Physiologie du Système Digestif, Centre de Recherches de Jouy, Institut National de la Recherche Agronomique, 78352 Jouy-en-Josas Cédex, France*

Sylvie Rabot *Unité d'Ecologie et de Physiologie du Système Digestif, Centre de Recherches de Jouy, Institut National de la Recherche Agronomique, 78352 Jouy-en-Josas Cédex, France*

Johannes W. G. M. Swinkels *Research Institute for Pig Husbandry, P.O. Box 83, 5240 AB Rosmalen, The Netherlands*

Martin W. A. Verstegen *Agricultural University, Department of Animal Nutrition, Haagsteeg 4, 6708 PM Wageningen, The Netherlands*

G. Williamson *AFRC Institute of Food Research, Norwich Laboratory, Norwich Research Park, Colney, Norwich NR4 7UA, UK*

J. H. Wilson *Royal Oldham Hospital, Rochdale Road, Oldham, OL1 2JH, UK*

PREFACE

A major area of research interest at the present time is the extent to which nutrition in early life can influence later health and disease. In volume 8 we hope to include an article on this topic. But what is the best nutrition for the human infant? *Goedhart and Bindels* begin this volume with an exploration of the composition of human milk and the features that make this food so special. They highlight the important balance between a nutrient profile that is designed to take account of the relative immaturity of the baby's gut, kidneys and brain, and an array of molecules whose function is to protect the baby against infection.

Not all mothers are able to breast feed and these authors write from the point of view of scientists in the infant foods industry who seek to produce a food that, while not aiming to mimic the composition of human milk in every respect, nevertheless achieves the physiological effects expected in breast fed infants. They rightly point out that it will be extremely difficult to demonstrate clear functional advantages of supplementing formulas with specific components of human milk (for example lactoferrin) and that benefits of compositional modifications should be weighed against costs. Guidelines to the industry should provide a reasonable balance between protecting the public against opportunistic supplementations while leaving room for scientific innovation.

The perinatal period is an important time for the development of the infant's immune defence system. As indicated in *Goedhart and Bindels'* article, the newborn receives some immunological assistance from components of human milk. However, as *Lovegrove and Morgan* point out, the infant's immune defence has already been given a start by the transfer of immunoglobulins from mother to fetus across the placenta. These authors are mainly concerned with factors that disrupt the normal functioning of the immune system to produce states of hypersensitivity and allergy. Thus infants, especially those born into families with a history of atopy, may already have been sensitized to allergens transferred *in utero* from the mother *via* the placenta or may be sensitized by allergens in the breast milk derived from foods the mother has eaten during lactation. These authors make a plea for parents with atopy to be aware of the potential for their offspring to develop allergic symptoms and to adopt strategies to minimize the risk, such as avoiding foods known to be allergenic certainly in lactation and possibly also during pregnancy. They do warn however that, such is our state of knowledge, there is as yet no simple means of guaranteeing prevention of infant allergy.

Proponents of the view that diet is a major, if not *the* major factor in both the development and prevention of cardiovascular disease are prone to state that this topic is not now one of scientific controversy. *Macnair's* paper should cause readers to reflect seriously upon, if not to abandon, this simple view. *Macnair* considers four key so-called 'risk factors' for cardiovascular disease: blood lipids; insulin resistance; hypertension and failure of energy balance. The author poses some important questions about the physiological significance of the relatively high concentrations of plasma lipids in human infants. He argues that the human nervous system has an exceptional requirement for cholesterol in the first few months of life. Furthermore he concludes that a key factor in metabolizing circulating lipids in later life is the degree to which muscle tissue is 'trained' by physical activity. It is commonplace now to find 'meta-analyses' of publications relating plasma lipid profiles to dietary fatty acid intakes but it is less widely known that the literature describing physical activity as a key determinant of plasma lipids is quite as large.

Concepts relating physical activity to each of the 'risk factors' and finally to atherosclerosis and ischaemic heart disease are plausibly integrated in a summarizing diagram (page 60).

Continuing the theme of interactions between nutrition and physical activity, *Wilson* addresses their combined effects on women's bone health, taking as particular examples female athletes. While regular physical activity is recognized to have many beneficial effects, not the least of which is the enhancement of bone mineralization and protection against premature osteoporosis, there is a paradox in that intensive exercise in the young may actually reduce bone mineral density. The author identifies menstrual dysfunction as a key factor in this problem. Nutrition is important because of the high prevalence of eating disorders in athletes. Whereas the increased levels of physical activity demanded result in higher energy requirements, many athletes desire to have a low body weight for either functional or aesthetic reasons and may develop a psychology that results in undereating, low body fat reserves, menstrual dysfunction and compromised bone metabolism. The quality as well as the quantity of dietary intake may be important. Greater awareness is needed of the prevalence of eating disorders among athletes, and nutrition should be placed high on the agenda for optimum performance.

Many organizations worldwide are dedicated to the continuing assessment of important practical aspects of human nutrition. One of these is the European Union through its 'FLAIR Concerted Action' Programmes. One sometimes has the impression that the hard work undertaken by the numerous expert committees of scientists in producing reports to these organizations is not sufficiently rewarded in terms of the widest possible dissemination of the results of their efforts. *Nutrition Research Reviews*, while focusing mainly on nutritional concepts, nevertheless believes that there is room in the *Journal* for 'archival material' which might not otherwise be so readily available to the general reader. The scholarly review by *Bates*, *Heseker* and their colleagues on vitamin bioavailability in this issue is an example. The aim has been for comprehensive coverage and this has necessitated rather brief commentaries on each aspect of each vitamin. We hope that the succinct and up to date information contained in this review will be helpful to teachers, nutritional advisers and research workers alike and welcome comments from our readers on the extent to which the *Journal* should devote future space to similarly treated topics.

This issue contains two reviews devoted to dietary minerals, both concerned with zinc. *Swinkels*, *Kornegay and Verstegen* approach the topic from the standpoint of animal production. The authors review the biology of Zn particularly in farm animals, with emphasis on mechanisms of absorption. They then discuss the relative efficiencies of absorption of various Zn complexes. There are clearly opportunities for developing Zn compounds with higher biological value but more research will be needed on basic mechanisms of absorption before the animal feeds industry can benefit fully.

Gibson draws attention to the relative lack of recognition by international agencies of the importance of Zn deficiency in developing countries compared with the attention given to iron, iodine and vitamin A. She reviews requirements for Zn and concludes that widespread Zn deficiency does contribute to high morbidity and mortality. However, probability estimates for risk of Zn inadequacy do not identify actual individuals in the population who are deficient or define the severity of the nutrient inadequacy. Such information must be obtained by combining dietary intake data with laboratory and clinical indices of Zn status. In discussing strategies for prevention of Zn deficiency (supplementation, fortification and dietary modification and diversification based on traditional household principles) the author gives us important insights into ways in which political, economic and logistic factors impinge upon what are basically simple nutritional problems.

Whereas it is probably true that most if not all essential nutrients (absence of which from the diet eventually results in well-defined deficiency states fully reversible by including the specific nutrient in the diet) have now been identified, there are many components of the diet that are now believed to influence general health and wellbeing and even protect against degenerative diseases. While not being 'essential' in the established sense, should these components be called 'nutrients'? *Johnson, Williamson and Musk* explore this question in regard to a range of compounds with diverse chemical structures: isothiocyanates, indoles, flavonoids, glucosinolates, polyphenols, phyto-oestrogens and many other substances present in a variety of plant foods. To argue their case they need to discuss mechanisms of carcinogenesis and the metabolic pathways that are involved in both the activation and the defence against carcinogens (xenobiotic metabolizing enzymes). For readers with minimal biochemical background, these sections will be difficult reading but the effort is well worthwhile since this opens up uncharted areas of nutrition science. An important question is raised as to whether the enzymes that protect us against carcinogens are 'inducible' by substances in the diet. If so, the regular exposure to certain dietary components (for which the authors have coined the term 'dietary phytoprotectants') may be a necessary part of normal nutrition.

Nugon-Baudon and Rabot continue the same theme, focusing specifically on the glucosinolates and their derivatives. These are sulphur-containing molecules produced from amino acids by the secondary metabolism of plants and they occur mainly in such cruciferous vegetables as cabbage, sprouts, turnip and cauliflower. Epidemiological studies seem to suggest that higher consumption of these vegetables is associated with reduced risk of cancers of the gastrointestinal tract. Experiments to investigate the effects of cruciferous vegetables or their isolated glucosinolate derivatives have given inconsistent results. There is suggestive evidence (all in laboratory animals) that glucosinolates may induce or otherwise modify the activities of the 'xenobiotic metabolizing enzymes' (also described by *Johnson and colleagues*). Inconsistencies may arise because, although these enzymes are normally regarded as protecting against carcinogenesis, they may themselves cause toxicity under certain circumstances. There is clearly a great deal to be learned about the strategies adopted by the body to defend itself against toxic compounds and the role of normal dietary constituents in this protection and we can expect this area of nutrition to develop rapidly in the coming years.

This issue concludes with a comprehensive review by *Kelly, Begbie and King* of the ways in which the endogenous microflora attach themselves to the membranes of the cells lining the gastrointestinal tract as a necessary part of colonization and the nutritional implications thereof. It has now been established that bacterial glycoproteins (adhesins) interact with receptors, also glycoproteins, on intestinal membranes and mucus. A combination of molecular, biological and X-ray crystallographic techniques is contributing to an understanding of these interactions and their role in pathogenesis and protection against pathogens. Nutrition may be important in modifying receptor expression. For example, changes in the amounts and types of proteins, fatty acids and possibly other nutrients may affect the activities of glycosyltransferases required to glycosylate receptors and adhesins. Another nutritional approach may be to tailor make oligosaccharides as probiotics for the prophylaxis and therapy of intestinal infections.

The Editorial Board of *Nutrition Research Reviews* has recently been enhanced by the introduction of four new international editors: Noel W. Solomons (Guatemala); John Nolan (Armidale, Australia); Leif Hambraeus (Uppsala, Sweden) and Lindsay Allen (Connecticut, USA). Their role will be to introduce new topics and writers to the *Journal* and to act as points of reference in their areas of the globe to make our publication more widely known. We welcome our overseas colleagues and are confident that the quality, timeliness and liveliness of the *Journal* will be thereby greatly enhanced.

THE COMPOSITION OF HUMAN MILK AS A MODEL FOR THE DESIGN OF INFANT FORMULAS: RECENT FINDINGS AND POSSIBLE APPLICATIONS

ANNEMIEK C. GOEDHART AND JACQUES G. BINDELS

Nutricia Research, P.O. Box 1, 2700 MA Zoetermeer, The Netherlands

CONTENTS

INTRODUCTION	1
ENERGY	2
PROTEIN	3
PROTEIN QUANTITY	3
PROTEIN QUALITY	4
LACTOFERRIN	5
IMMUNOGLOBULINS	6
NON-PROTEIN NITROGEN	7
LIPIDS	7
LONG CHAIN POLYUNSATURATED FATTY ACIDS	7
STRUCTURED LIPIDS	9
CHOLESTEROL	10
CARBOHYDRATES	11
MINERALS AND TRACE ELEMENTS	12
CALCIUM AND PHOSPHORUS	12
IRON	13
SELENIUM, MOLYBDENUM AND CHROMIUM	14
VITAMINS AND CONDITIONALLY ESSENTIAL SUBSTANCES	14
β-CAROTENE	14
INOSITOL	15
TAURINE	15
NUCLEOTIDES	15
CONCLUDING REMARKS	16
REFERENCES	16

INTRODUCTION

Human milk is assumed to be the ideal food for the infant at least up to the age of 5 or 6 months, ensuring optimal growth and development (ESPGAN, 1982). In many respects human milk, the most natural food available, is unique. The nutritional composition of human milk varies from mother to mother, from day to day, during the day and even during a feed, and is generally suited to the individual needs of the infant. There is little doubt that human milk serves a role in infant physiology greater than being a supply of energy and nutrients. For instance, the immunological properties of human milk (immunoglobulins, bacteriostatic proteins, living cells, antiviral lipids) are well documented. In developing countries these established beneficial properties can be translated into demonstrable advantages to the breast fed over the bottle fed infant, in terms of reduced morbidity and mortality (Jason *et al.* 1984; Hanson *et al.* 1994). Controversy persists, however, about whether breast feeding protects infants in Western countries from infectious diseases and, if so, the magnitude of this effect. Four of the five epidemiological studies in industrialized countries published between 1970 and 1984 that met the methodological standards set by Bauchner *et al.* (1986) concluded that breast feeding was not protective against infections. Only one of the studies did show a protective effect of breast feeding against gastrointestinal infections among infants younger than four months. Newer studies continue to suggest only a minimal protective effect of breast feeding in industrialized countries, with the clearest impact on gastroenteritis (Wright *et al.* 1989; Howie *et al.* 1990; Rubin *et al.* 1990).

While human milk is superior for the newborn infant, milk substitutes play a necessary role in infant nutrition when breast feeding is not possible, desirable or sufficient. The search for human milk substitutes has been conducted since the Stone Age, but it was not until sanitation practices developed in the late nineteenth century that feeding these substitutes, mostly based on cows' milk, to infants could be safely accomplished. Since then, there have been progressive attempts to bring the composition of these formulations closer to human milk. Important modifications included reduction of the protein and electrolyte content, addition of vitamins, trace elements and lactose, alteration of the casein:whey protein ratio, and substitution of unsaturated vegetable fats for butterfat. With the addition of taurine some years ago a new phase in the development of infant formulas has commenced. Presently, research is concentrating on those substances in human milk which serve other than traditional nutritional roles. Attempts are in progress to supplement infant formulas with protective and trophic factors so far unique only to human milk. The final aim is not necessarily to mimic the composition of human milk in every respect, but to achieve physiological effects as in breast fed infants.

In this paper, the relevant aspects of the composition of human milk and of the current infant formulas will be reviewed, and an outline of some of the expected developments in the composition of infant formulas will be given.

ENERGY

The energy requirements of infants reflect levels of energy intake that will promote health, adequate growth, optimal body composition, and levels of physical activity appropriate for their developmental age. The average energy density of mature human milk is generally used as a basis for the assessment of the infant's energy requirement, and thus as a guideline for the energy content of infant formulas.

Traditionally, the energy density of human milk has been assessed by analysis of its macronutrient composition, obtained by manual or mechanical expression from the breast. To calculate the energy density of the milk from its macronutrient composition, the standard Atwater conversion factors are used. The mean energy value of mature milk obtained by this method is around 700 kcal/l (Department of Health and Social Security, 1977). This method of estimating energy density is subject to error, however, as milk that has been obtained by whole breast expression may not be identical in composition to suckled milk. In particular, the fat content, a major determinant of energy content, may be higher in expressed milk, which may result in an overestimation of the energy density of suckled milk (Lucas et al. 1990). Another potential problem of this method of estimating energy density is that the Atwater factors used to estimate metabolizable energy from macronutrient composition may be inappropriate for young infants, as nutrient absorption may be less than that accounted for in these conversion factors. The recent development of the doubly labelled water method for determining energy expenditure, energy stored in new tissue, and water (and therefore milk) intake has made it possible to estimate the energy density of suckled breast milk. Using this method, the metabolizable energy content of breast milk has been found to be 530 kcal/l and 580 kcal/l at 6 weeks and 3 months of age respectively (Lucas et al. 1990). The metabolizable energy values obtained for infants fed formula with a calculated energy density of 680 kcal/l at these ages were 600 and 660 kcal/l respectively, suggesting that the use of the Atwater factors may indeed lead to an overestimation of the metabolizable energy intake in the first weeks of life.

These data have important potential consequences for infant nutrition. It may be that current guidelines for energy intake in infancy are too high and thus need revision (Prentice et al. 1988). This may explain why breast fed infants grow less rapidly than formula fed infants, despite their lower total daily energy expenditure (Butte et al. 1990; Heinig et al. 1993).

Further research is needed to confirm the energy values for suckled milk found by Lucas et al. (1990), and to determine whether the energy density of infant formulas should be lowered towards these values.

PROTEIN

PROTEIN QUANTITY

It is now generally accepted that the true protein content of mature milk is only about 8–10 g/l and that the earlier overestimation was due to a large proportion of nitrogen that is not part of human milk protein (Hambraeus et al. 1978; Räihä, 1985; Harzer et al. 1986). This so-called non-protein nitrogen fraction includes about 20–25% in human milk (Hambraeus et al. 1978; Harzer et al. 1986).

Since some of the protective whey proteins of human milk, particularly secretory IgA and lactoferrin, are quite resistant to low pH and the action of proteolytic enzymes and are to a significant extent excreted in the infant's stools, the nutritionally available protein of mature human milk may be as low as 6–8 g/l (Räihä, 1985; Davidson & Lönnerdal, 1987).

Through the last decades, there has been a trend towards lowering the protein levels of infant formulas, in order to make them more similar to human milk. A major reason for this decrease has been the finding that formula fed infants have elevated plasma concentrations of urea and of specific amino acids, which suggests that these infants may be exposed to unnecessary metabolic stress (Järvenpää et al. 1982a, b; Janas et al. 1985). Several studies evaluated the effects of lowering the protein content of formula to values of 11–13 g/l, while varying the casein:whey protein ratio (Räihä et al. 1986a, b; Picone et al. 1989; Lönnerdal & Chen, 1990). Infants receiving such a low protein formula had growth rates and indices of protein nutritional status similar to those of breast fed infants. During the first four weeks of life, however, Räihä et al. (1986a, b) observed that the infants fed the low protein formula (12·5 g/l) had blood urea nitrogen and urine nitrogen concentrations that were significantly lower than those of the breast fed group. The fact that this finding could not be confirmed by other investigators, even when a formula with a protein level of as low as 11 g/l was given, may at least be partly explained by differences in non-protein nitrogen levels between the formulas used (Donovan & Lönnerdal, 1989). In none of the studies could plasma amino acid patterns identical to those of breast fed infants be produced, suggesting that factors other than the amino acid pattern of human milk influence plasma amino acid levels (Picone *et al.* 1989). Reducing the protein content of formula to values of around 13 g/l has been observed to result in low plasma tryptophan values relative to those of breast fed infants, presumably owing to inadequate intake and reduced bioavailability of tryptophan from bovine whey protein (Janas *et al.* 1987). Low plasma tryptophan levels may be of concern as tryptophan is an essential amino acid and a precursor of serotonin and niacin. Addition of free tryptophan to a low protein formula has been found to result in plasma tryptophan levels similar to those of breast fed infants, and to influence the infants' sleep latency (Hanning *et al.* 1992; Fazzolari-Nesci *et al.* 1992; Steinberg *et al.* 1992).

PROTEIN QUALITY

The concentration of whey proteins in human milk decreases from early lactation and continues to fall. These changes result in a casein:whey protein ratio of about 10:90 in the first days of lactation and of about 45:55 in mature milk (Harzer *et al.* 1986; Kunz & Lönnerdal, 1992). The optimal casein:whey protein ratio of infant formulas is still a point of controversy. Growth rates do not differ between infants fed whey predominant formulas and those receiving casein predominant formulas (Harrison *et al.* 1987; Janas *et al.* 1987; Lönnerdal & Chen, 1990). Theoretically, whey predominant formulas may offer some minor advantages for newborns because they form a finer, softer curd than casein predominant formulas leading to higher gastric emptying rates, which are more comparable to those of breast fed infants (Nakai & Li-Chan, 1987; Billeaud *et al.* 1990). Further, during the first two months a whey predominant formula was found to induce a faecal flora somewhat closer to that of breast fed babies than a casein predominant formula (Balmer *et al.* 1989). It should be noted, however, that the casein predominant formula used in this study had higher protein and phosphate contents than the whey predominant one, and therefore presumably a higher buffering capacity. The functional consequences of the observed differences in gastric emptying and intestinal flora still have to be demonstrated. The same holds true for the putative advantages of casein predominant formula being more satisfying or less allergenic.

Since there is no convincing evidence that whey predominant formulas are superior in composition or physiological effect, the Scientific Committee for Food, set up by the European Communities, recommended in their opinion expressed on 17 September 1993 not to adhere any longer to different minimal values for the protein content of infant formula dependent on the casein:whey protein ratio.

Human milk casein and its subunits represent the least understood and characterized class of proteins in human milk. Important differences exist between human and bovine caseins. β-Casein is the predominant casein in human milk, whereas cows' milk contains a large proportion of α-caseins (Eigel *et al.* 1984; Kunz & Lönnerdal, 1990). Additionally, physicochemical differences between human and bovine caseins affect curd formation, which in turn influences gastric emptying and intestinal transit time (Nakai & Li-Chan, 1987; Billeaud *et al.* 1990). Partial enzymic dephosphorylation and/or rennet modification may improve coagulation characteristics and digestibility of bovine caseins for infant feeding (Nakai & Li-Chan, 1987; Li-Chan & Nakai, 1989). The physiological roles of casein in mineral absorption and in providing fragments with immunomodulating and opioid-like activities need further study (Migliore-Samour & Jollès, 1988; Daniel *et al.* 1990).

LACTOFERRIN

Of the protective whey proteins in human milk, the iron binding glycoprotein lactoferrin is the second most abundant one, being present in colostrum and mature milk in concentrations of about 5–7 and 1–2 g/l respectively (Hambraeus et al. 1978; Goldman et al. 1982). Lactoferrin is remarkably resistant to degradation by proteinases (Davidson & Lönnerdal, 1987).

The possible physiological functions of lactoferrin have recently been reviewed by Iyer & Lönnerdal (1993). Lactoferrin has been suggested to have bacteriostatic activity, to enhance iron absorption, and to stimulate mucosal proliferation. The bacteriostatic and bactericidal effects of lactoferrin in vitro have been studied in a wide range of microorganisms. Lactoferrin withholds iron from invading organisms by its high iron affinity as well as its slow rate of change to a conformation in which the iron site is exposed (Chung & Raymond, 1993). Next to iron withholding, other mechanisms may be involved in the antibacterial action of lactoferrin, and a synergic mechanism with lysozyme and/or IgG has been proposed (Iyer & Lönnerdal, 1993). Convincing in vivo data which confirm the in vitro findings are still lacking. Limited hydrolysis of bovine lactoferrin resulted in potent antibacterial activity, suggesting that lactoferrin latently contains at least one antibacterial peptide region that is released when the molecule is hydrolysed. The bactericidal activity of the peptide fragments did not have any relation to iron chelation (Saito et al. 1991). The putative bactericidal domain, lactoferricin, has recently been isolated and described in bovine and human lactoferrin. The region is distinct from the iron binding region. The antimicrobial peptide of bovine lactoferrin was found to be more active than that of human lactoferrin (Bellamy et al. 1992). Further studies are required to determine whether digestion of either human or bovine lactoferrin in vivo generates potent bactericidal peptides in sufficient quantities to be of biological importance for neonates.

The precise role of lactoferrin in iron absorption has not yet been defined. Clinical trials in which infant formulas were supplemented with bovine lactoferrin have failed to demonstrate an improvement in iron absorption (Fairweather-Tait et al. 1987; Schulz-Lell et al. 1991). Possible explanations for these negative results are (1) the species specificity of the human lactoferrin receptor (if this receptor is involved in iron transport), and (2) inappropriateness of the composition of the formulas used (high in citrate and phosphate, low in bicarbonate). Chierici et al. (1992) reported, however, that infants fed a lactoferrin supplemented formula had higher ferritin levels than those fed the control formula, and not significantly different from the breast fed group. Both the experimental and the control formulas contained no added iron. One possible flaw of this study was the fact that the lactoferrin supplemented formula contained more endogenous iron than the non-supplemented control formula. Recently, Davidsson et al. (1994) surprisingly observed that iron absorption in infants was significantly lower from human milk with its native content of lactoferrin than from lactoferrin free human milk, which suggests that human milk lactoferrin may have no direct role in the enhancement of iron absorption. More studies are needed to define the role of lactoferrin in iron transport better. Lactoferrin could well be involved in iron metabolism, acting as a regulator of iron absorption when iron stores are adequate and as an enhancer of iron absorption in deficiency states (Iyer & Lönnerdal, 1993).

The hypothesis that lactoferrin acts as a growth factor for the intestine remains to be confirmed in vivo (Nichols et al. 1987; Iyer & Lönnerdal, 1993).

In the coming years, it will be possible to perform in vivo supplementation trials using human lactoferrin produced in the mammary glands of transgenic dairy animals. Transgenic dairy calves have already been generated carrying the human lactoferrin fusion gene (Krimpenfort, 1993). It is hoped that these studies will lead to a better understanding of the biological functions of human milk lactoferrin.

IMMUNOGLOBULINS

Secretory IgA comprises over 90% of the immunoglobulins in human milk. The highest concentrations of sIgA (~ 9 g/l) are found in colostrum. Mature milk has sIgA levels averaging 1–2 g/l (Harzer & Bindels, 1985; Goldman & Goldblum, 1989). Human milk sIgA is directed against enteric and respiratory immunogens that have triggered the maternal enterobronchial mammary gland pathways (Goldman & Goldblum, 1989). Secretory IgA antibodies in human milk neutralize bacterial toxins and virulence factors and inhibit adherence and proliferation of bacteria on epithelial surfaces by binding to the bacterial adhesins (Goldman & Goldblum, 1989; Davin et al. 1991).

In order to improve the immunological composition of infant formulas, in theory IgA derived from human milk or produced through genetic engineering should be used. It is extremely difficult, however, to harvest large amounts of IgA from human milk for commercial purposes. Additionally, large scale production of sIgA by recombinant DNA technology may be a daunting task since the molecule has four different types of peptide chains, the formation of antigen combining sites requires the rearrangement of four different groups of genes, once the relevant peptides are produced it would be necessary to link them, and many different antibodies would be required to create a broad range of specificities (Goldman, 1989).

As a more feasible alternative to human IgA, bovine antibodies could be used. Whereas sIgA is the predominant class of immunoglobulins in human milk, IgG_1 is predominant in cows' milk. Even though it is structurally different, it appears to have the same function (Facon et al. 1993). Attempts have been made to increase the immunoglobulin concentration of cows' milk and to manipulate its specificity by immunization. In several studies infants were prophylactically supplemented with immunoglobulins isolated from the milk of cows immunized to specific pathogens. Davidson et al. (1989) demonstrated that bovine colostrum with high antibody titre against four human rotavirus serotypes, given as a supplement, was highly effective in protecting hospitalized infants against rotavirus infection. Additionally, administration of colostrum from rotavirus immunized cows prevented rotavirus infections in infants living in an orphanage (Ebina et al. 1985). Turner & Kelsey (1991) reported that administration of bovine milk antibodies to human rotavirus did reduce rotavirus associated illness but not rotavirus infection. Brunser et al. (1992) failed to prevent rotavirus and Escherichia coli infections in infants living in low socioeconomic conditions by providing a formula containing bovine milk antibodies against these pathogens. The investigators postulated that the dose of immunoglobulin given may have been too low and/or the age of the infants (majority 9–12 months) may have had an influence on the digestive enzymes and thereby on the unprotected passage of the antibodies through the intestine.

Larger scale controlled clinical trials are needed to prove the efficacy of passive oral immunization with milk antibodies from immunized cows, to ascertain whether the risks of such antibody supplementation are minimal, and to determine the minimal effective dose (Goldman, 1989; Boesman-Finkelstein & Finkelstein, 1991). It is not inconceivable that in the future immunoglobulin fortified infant formulas will be developed, which can be given to non-breast fed infants susceptible to infections by intestinal pathogens, such as those that are immunodeficient (Shield et al. 1993) and/or those in contaminated environments such as hospitals (Goldman, 1989; Facon et al. 1993).

NON-PROTEIN NITROGEN

Many non-protein nitrogen components have been identified, including urea, uric acid, ammonia, creatine and creatinine, free amino acids, nucleic acids and nucleotides, polyamines, carnitine, low molecular weight peptide hormones, growth factors, the amino sugars *N*-acetylglucosamine and *N*-acetylneuraminic acid (sialic acid), and the amino alcohols choline and ethanolamine (Carlson, 1985*a*; Atkinson *et al.* 1989). Some of this nitrogen contributes to the pool available for synthesis of non-essential amino acids. For instance urea nitrogen, which accounts for 30–50% of the non-protein nitrogen fraction, is partly hydrolysed to ammonia by intestinal microorganisms, with subsequent intestinal absorption of the released ammonia. From studies with isotopically labelled urea, it has been estimated that about 13–23% of dietary urea is retained and available for amino acid synthesis (Heine *et al.* 1986; Fomon *et al.* 1988).

Other non-protein nitrogen compounds may be involved in the development of the newborn infant. Taurine and nucleotides have been claimed to be conditionally essential substances and are dealt with later. Non-protein nitrogen components with clear trophic characteristics are polyamines (spermine, spermidine, putrescine) and epidermal growth factor. Polyamines are known to be involved in cell proliferation and differentiation in many tissues, including the gastrointestinal tract (Pegg, 1986; Pollack *et al.* 1992). Whereas human milk has been found to contain considerable amounts of putrescine, spermine, and spermidine, standard cows' milk based infant formulas contain no detectable polyamines or only very small amounts (Pollack *et al.* 1992; Romain *et al.* 1992). Additionally, epidermal growth factor may be involved in maturational processes of the newborn (Kidwell & Salomon, 1989). It can be expected that this rapidly emerging research area will give us significant information with regard to the importance of human milk for gut proliferation and maturation.

LIPIDS

LONG CHAIN POLYUNSATURATED FATTY ACIDS

Fats are vital for normal growth and development. In addition to providing energy, fats supply essential fatty acids and are the vehicle for fat-soluble vitamins and hormones in milk. Recently, the essential fatty acid requirements of newborns have received increasing attention. Particular interest has focused on the importance of long chain polyunsaturated fatty acids with 20 and 22 carbon atoms (LCP). These fatty acids are important structural components of cell membrane phospholipids, particularly those of the central nervous system and of retinal photoreceptors, and serve as precursors for the synthesis of eicosanoids (British Nutrition Foundation, 1992).

Docosahexaenoic acid (DHA) and arachidonic acid (AA) constitute a large proportion of the total lipids in brain and retina and their accretion primarily occurs during the last trimester of pregnancy and the first year of life (Clandinin *et al.* 1980*a, b*). Fetal accretion of LCP may result from placental transfer (Kuhn & Crawford, 1986). Postnatally, human milk provides the breast fed infant with preformed AA and DHA. Term human milk has an AA content of about 0·5% and a DHA content of about 0·3% of total fatty acids (Koletzko *et al.* 1992). The mean LCP levels of colostrum and preterm milk are higher (Rönneberg & Skåra, 1992; Foreman-van Drongelen *et al.* 1994).

The available evidence strongly suggests that a dietary supply of LCP is desirable for the preterm infant. Preterm infants fed formulas without LCP develop poor AA and DHA status, suggesting that the preterm infant is unable sufficiently to elongate and desaturate linoleic acid and α-linolenic acid to their long chain derivatives (Carlson *et al.* 1991). The

DHA status of preterm infants is positively related to their visual acuity, whereas their AA status positively correlates with their first-year growth and with scores on development tests (Carlson et al. 1992a, 1993a, b). Feeding formulas supplemented with fish oil to preterm infants improves their DHA status and visual function (Carlson et al. 1991, 1993b; Birch et al. 1992) but their AA status deteriorates and may result in poorer growth (Carlson et al. 1991, 1992b).

Both the ESPGAN Committee on Nutrition (1991) and the British Nutrition Foundation (1992) recommend enrichment of premature formulas with AA and DHA. There are already some premature formulas available which are supplemented with DHA and either AA or its precursor, the δ-6 desaturation product γ-linolenic acid. In term infants, however, supplementation of γ-linolenic acid was not able to prevent the fall in AA, which suggests that it may not be sufficient simply to bypass the first δ-6 desaturation step (Makrides et al. 1993a).

It is to be expected that within the next few years all premature formulas will contain both preformed DHA and AA. In the absence of a clear understanding of the actual LCP requirements of preterm infants, it seems prudent to aim at LCP levels at least approximating to those of preterm human milk, as the most important period for brain AA and DHA accumulation is the third trimester of gestation, and as the preterm infant is exposed to a significantly reduced intake of LCP compared to what it would have obtained by placental transfer had it been born at term.

The degree to which a dietary source of preformed LCP is also essential for term infants is an area of active investigation. As in preterm infants, the levels of AA and DHA in plasma and erythrocyte phospholipids of term, formula fed infants are lower than of breast fed infants (Clark et al. 1992; Makrides et al. 1993a, b). Decreasing the linoleic acid:α-linolenic acid ratio of a term formula to about 4:1 improves the DHA status of the infants, although not to values as in breast fed infants, but it worsens their AA status (Clark et al. 1992). Recently it has been demonstrated that breast fed infants have higher DHA concentrations in their brain cortical phospholipids, and higher AA and DHA concentrations in subcutaneous tissue, compared to infants fed formula (Farquharson et al. 1992, 1993).

The observed differences in LCP levels of brain cortex and subcutaneous tissue between breast fed and formula fed infants may affect physiological function. Term, breast fed infants have been found to have a better visual function (visual evoked potential acuity) at four to five months of age than infants fed a formula devoid of LCP (Birch et al. 1992; Makrides et al. 1993b). Additionally, at 3 years of age, breast fed infants have been found to have a significantly better visual function (stereo acuity and letter matching ability) than infants fed a corn oil formula during the first year of life (Uauy et al. 1992; Birch et al. 1993). The scores on the tests of visual function at 3 years were correlated with the DHA status at 4 months of life. It should be noted, however, that the formula used in this study was deficient in α-linolenic acid. Supplementing infant formula with fish oil and evening primrose oil (containing γ-linolenic acid) improved the DHA status and visual acuity of term infants. AA levels of these infants were reduced below levels of infants fed the non-supplemented formula, which did not, however, negatively affect growth (Makrides et al. 1993a).

The British Nutrition Foundation (1992) recommends that infant formulas should contain AA and DHA in amounts similar to those of human milk, although the Task Force acknowledges that addition of these LCP is of most importance for preterm formulas. Farquharson et al. (1992, 1993) concluded from their studies that a minimum daily requirement of 0·2 g DHA/100 g fatty acids (or 30 mg DHA/d) should be supplied in formulas designed for term infants to prevent the cerebrocortical deficiency of DHA. In their most recent opinion, expressed on 17 September 1993, the Scientific Committee for Food stated not to object to the possibility of adding them in infant formulas provided that the resulting content in n-3 and n-6 LCP is similar to that present in human milk in Europe, and that the eicosapentaenoic acid content does not exceed that of DHA.

It is as yet unknown whether the effect of dietary DHA on neural maturity is long lasting. Longer term effects of lower concentrations of DHA in cerebrocortical phospholipids on neuronal integrity and function need urgent study (Cockburn, 1994). However, the fact that mature human milk contains both AA and DHA in considerable amounts may already serve as a rationale to add AA and DHA to term infant formulas, provided the amounts do not exceed those of human milk.

STRUCTURED LIPIDS

Another topical issue with respect to infant formula fat is the development of structured triacylglycerols. The fatty acids in human milk triacylglycerols have a highly specific positional distribution (Martin *et al.* 1993). Especially the positional distribution of palmitic acid in human milk has received increasing attention. Palmitic acid constitutes about 22% of mature human milk lipids and 70–75% of it is esterified at the sn-2 position (β-position) of the triacylglycerol (Freeman *et al.* 1965; Christie, 1986; Martin *et al.* 1993). The palmitic acid in vegetable oils commonly used in infant formulas and in chicken egg, on the other hand, is predominantly esterified at the sn-1 and sn-3 positions (Freeman *et al.* 1965; Tomarelli *et al.* 1968; Christie, 1986). Pancreatic colipase dependent lipase selectively hydrolyses the fatty acids at the sn-1 and sn-3 positions, yielding free fatty acids and a 2-monoacylglycerol (Bernbäck *et al.* 1990). The 2-monoacylglycerol is a well absorbed form of most fatty acids since it readily forms micelles with bile acids and cannot form insoluble soaps with cations like calcium and magnesium. Therefore, the absorption of palmitic acid is likely to be greater when it is esterified at the sn-2 position than when it is attached predominantly at the sn-1,3 positions (Small, 1991). Indeed, mixtures of coconut oil and palm olein are better absorbed by rats if the proportions of palmitic and stearic acids in the sn-2 position are increased by chemical randomization (Lien *et al.* 1993).

Several studies have focused on the effect of the triacylglycerol configuration on intestinal fat absorption in preterm infants (Brooke 1985; Verkade *et al.* 1989). Carnielli *et al.* (1994) recently evaluated the effects of two formulas for preterm infants, differing only in the isomeric position of palmitic acid, on the absorption of fat and fatty acids and on mineral balance in a crossover study in preterm infants. The palmitic acid content of both formulas was 25–26% of total fatty acids. Although total fat absorption was not significantly different between the two groups, infants who were given the formula with palmitic acid predominantly esterified at the sn-2 position showed a significantly higher palmitic acid absorption compared with the infants that obtained the control formula with palmitic acid mainly attached to the sn-1,3 positions. This effect on palmitic acid strongly correlated with faecal calcium excretion. Absolute calcium retention of infants given the sn-2 palmitate formula was increased by more than 20 mg/kg daily.

Only one study has reported on the influence of the triacylglycerol structure on the absorption of fatty acids in term infants (Filer *et al.* 1969). In this early study, a formula based on natural lard (containing palmitic acid mainly at the sn-2 position) was compared with one based on randomized lard. The absorption of all fatty acids was improved in the infants receiving the formula containing natural lard, the effect being most pronounced for palmitic and stearic acids.

Palmitic acid is not the only fatty acid to show a specific preference for a particular position in human milk triacylglycerols; oleic acid and stearic acid are mainly located at the sn-1 position, whereas linoleic acid is located mainly in the sn-1 and sn-3 positions. AA and DHA are found primarily esterified in the sn-2 ($\sim 50\%$) and sn-3 ($\sim 45\%$) positions of human milk (Martin *et al.* 1993).

Because human milk is the natural source of fat for the newborn, the structure of its triacylglycerols may be used as a reference point for the design of lipid sources for infant formulas. Thus, adaptation of the triacylglycerol structure of infant fat to approximate that of human milk more closely seems a logical step in the further improvement of infant formula. The triacylglycerol structure of infant formula lipids could be modified by the process of 1,3-enzymic interesterification or by chemical randomization (Lien *et al.* 1993; Quinlan & Moore, 1993).

CHOLESTEROL

The widespread occurrence of atherosclerosis in developed countries has increased the emphasis on prevention of this disorder. One question that has been the subject of numerous studies is whether or not the infant's diet influences blood cholesterol or lipoprotein concentrations later in life. It is well documented that infants given human milk have higher plasma total cholesterol and plasma LDL cholesterol concentrations and a higher LDL:HDL cholesterol ratio than formula fed infants (Jooste *et al.* 1991; Kallio *et al.* 1992; Hayes *et al.* 1992). These differences, which gradually diminish at the age of one year, are primarily attributable to the relatively high cholesterol content, 100–150 mg/l, of human milk (Clark & Hundrieser, 1989; Lammi-Keefe *et al.* 1990), rather than to its relatively high level of saturated fatty acids (Carlson *et al.* 1982; Hayes *et al.* 1992). Wong *et al.* (1993) recently observed, using the ^2H_2O method, that the 6-fold greater cholesterol intake of breast fed infants resulted in a 3-fold suppression of cholesterol synthesis compared with formula fed infants. This down-regulation of cholesterol synthesis did not prevent increases in plasma total cholesterol and LDL-cholesterol concentrations in the breast fed group.

Data from animal studies suggest that there may be long term effects of infant diet on cholesterol metabolism. Adult baboons that were breast fed during infancy have lower HDL cholesterol concentrations, higher LDL+VLDL:HDL cholesterol ratios, lower cholesterol production and bile acid excretion rates, more extensive arterial lesions and an increased bile cholesterol saturation index (which may promote gallstone formation) compared with those fed formula (Mott *et al.* 1990, 1991). Additionally, short term exposure to high dietary cholesterol in early life has been found to increase arterial sensitivity to further cholesterol insult in adult rabbits in terms of enhanced atherogenesis, despite normalization of plasma cholesterol in these animals (Subbiah *et al.* 1989).

The question of whether there are also long term effects of breast versus formula feeding on serum cholesterol concentrations in humans has not been solved. Among children of 2·5 years (Ward *et al.* 1980) and of 7–12 years of age (Hodgson *et al.* 1976), those that were breast fed had higher serum cholesterol levels than those fed formula. Other investigators studying children 1–8 years of age observed no differences in serum cholesterol concentrations related to breast *v.* formula feeding (Huttunen *et al.* 1983; Fomon *et al.* 1984; Jooste *et al.* 1991). Measurement of plasma total cholesterol alone, however, may not be sufficient to identify the effects of early postnatal cholesterol ingestion on cholesterol homeostasis later in life (Hamosh, 1988). Fall *et al.* (1992) recently presented data which showed that adult men born during 1911–30 who were either exclusively breast fed during the first year of life or exclusively bottle fed from birth had higher mortality ratios from ischaemic heart disease and higher serum total cholesterol and LDL cholesterol concentrations than men who were either both breast and bottle fed or breast fed but weaned before one year. No information was available on the fat composition of the bottle feeds, but it is to be expected that these formulas contained a high percentage of (cholesterol-rich) milk fat.

Cholesterol is not an essential nutrient; the human fetus is able to synthesize cholesterol endogenously from the 11th week of gestation and the cholesterol needed for brain myelinization is entirely synthesized within the brain (Carr & Simpson, 1981; Edmond et al. 1991). However, as breast feeding is the natural method of feeding an infant, it has been postulated that the cholesterol level of human milk must be considered physiological, and it has been questioned whether infant formulas in this respect are sufficient at the moment (Kallio et al. 1992). The hypothesis that exogenous cholesterol may influence the biochemical composition and function of the small intestinal microvillus membrane could not be confirmed in a study in neonatal pigs (Engelhardt et al. 1991). Van Biervliet et al. (1992) reported that addition of cholesterol to formula affected maturation of the HDL particles and postulated that exogenous cholesterol may promote adequate delivery of cholesterol and AA to the developing brain. Others hypothesize that cholesterol is present in human milk merely because it is needed for the secretion of milk fat, and are not in favour of adding cholesterol to infant formulas as an increased LDL:HDL cholesterol ratio is associated with increased atherogenic risk in adults, and as expansion of the LDL pool may result in a decrease of the LCP-rich HDL pool by down-regulation of hepatic LDL receptors (Hayes et al. 1992).

A clear understanding of the role of cholesterol in human milk will depend on future research. Additional well controlled human studies are warranted to answer the question whether or not addition of cholesterol to infant formulas is desirable.

CARBOHYDRATES

Until now, the carbohydrate contribution of human milk has been attributed for the most part to the disaccharide lactose, generally neglecting the fact that human milk is a rich source of oligosaccharides. The latter are complex sugars composed of D-glucose, D-galactose, L-fucose, N-acetylglucosamine, and N-acetylneuraminic acid (sialic acid). They may be classed either as acidic or neutral according to the presence or absence of sialic acid (Kunz & Rudloff, 1993). Oligosaccharides are synthesized in the mammary gland by the action of several enzymes, which add specific monosaccharides to the core structure. Some of the fucosyloligosaccharides in human milk have structural similarity to blood group determinants of the mother (ABH secretor and Lewis secretor type), as a result of the activity of fucosyltransferases common to the two systems (Viverge et al. 1990).

More than 100 oligosaccharide structures in human milk have been characterized so far (Kunz & Rudloff, 1993). Oligosaccharides represent about 27% of total carbohydrates in colostrum, decreasing to 19% by day 30 until a value of 15–16% is reached by day 60 (Coppa et al. 1991, 1993). The major oligosaccharide in human milk is lacto-N-tetraose (Galβ1-3GlcNAcβ1-3Galβ1-4Glc), followed by monofucosylated lacto-N-fucopentaose I and II. These three carbohydrates add up to approximately 50–70% of the complex carbohydrates (Kunz & Rudloff, 1993).

Sialyllactose is the only complex oligosaccharide present in both human milk and bovine milk (Parkkinen & Finne, 1987; Neeser et al. 1991). Human milk sialyllactose consists primarily of the (α2-6) isomer, whereas bovine sialyllactose is mainly in the (α2-3) form (Parkkinen & Finne, 1987; Kunz & Rudloff, 1993).

The pattern of urinary oligosaccharides in breast fed infants is strongly related to that of the milk they ingest, which might be explained by intestinal absorption of intact oligosaccharides from human milk (Coppa et al. 1990).

It is well recognized that oligosaccharides containing N-acetylglucosamine stimulate the growth of *Bifidobacterium* species. Already in 1954, it was found that such oligosaccharides from human milk stimulated the growth of *Bifidobacterium bifidum* subsp. *pennsylvanicus*, which was originally isolated from the faeces of breast fed infants (Gauhe et al. 1954). However, this strain is exceptional in that it is unable to utilize glucose and requires D-glucosamine derivatives for cell wall synthesis (Veerkamp, 1969). N-acetylglucosamine appears not to promote the *in vitro* growth of *B. infantis*, *B. breve* and *B. longum* (Petschow & Talbott, 1991).

As enteropathogens use the oligosaccharide portion of glycolipids and glycoproteins as targets for attachment of whole bacteria and toxins, human milk oligosaccharides might prevent intestinal attachment of microorganisms by acting as receptor analogues competing with intestinal ligands for binding. Neutral oligosaccharides from human colostrum caused inhibition of adhesion to uroepithelial cells of a strain of *E. coli* isolated from an infant with urinary tract infection (Coppa et al. 1990). Additionally, human milk oligosaccharides inhibited the adherence of *Streptococcus pneumoniae* to human pharyngeal or buccal epithelial cells, the inhibitory activity being in the same concentration range as that of synthetic lacto-N-tetraose and lacto-N-neotetraose (Andersson et al. 1986). In a similar way, fucose containing oligosaccharides from human milk could abolish the binding activity of *Vibrio cholerae* (Holmgren et al. 1983). Sialyl(α2-3)lactose was found to inhibit haemagglutination of *Campylobacter pylori* and of S-fimbriae carrying strains of *E. coli* which may cause meningitis and neonatal sepsis in newborns (Korhonen et al. 1985; Evans et al. 1988), but its concentration in human milk may be too low to exert a significant inhibiting effect (Schroten et al. 1993).

Another possible function of human milk oligosaccharides is to provide the infant with sialic acid. Human milk contains about 1 g/l of oligosaccharide derived sialic acid during the first week of lactation, a value which decreases to about 250 mg/l at 6–8 weeks (Carlson, 1985b). Whey and casein predominant formulas have been reported to contain only 50–70 and 10–30 mg/l of oligosaccharide derived sialic acid respectively (Carlson, 1985b; Neeser et al. 1991). It is known that mammalian species, including man, have the capacity to synthesize sialic acid from simple sugars and phosphoenolpyruvate. However, the relative capacity for synthesis by the neonate has not been studied (Carlson, 1985b). There is evidence from animal studies that exogenous administration of sialic acid can significantly increase its content in synaptosomal regions of the brain, and is associated with desirable early and long term modifications of behaviour (Morgan & Winick, 1980, 1981).

More basic and clinical research is warranted to clarify outstanding questions regarding the possible functions and gastrointestinal metabolism of the various oligosaccharides. Assuming that human milk oligosaccharides indeed contribute to the wellbeing of the baby, it is to be foreseen that eventually infant formulas will be supplemented with oligosaccharides. However, it is as yet too early to decide which of the more than 100 human milk oligosaccharides would be the best candidates for supplementation and in which amounts these complex sugars should be added to infant formulas.

MINERALS AND TRACE ELEMENTS

CALCIUM AND PHOSPHORUS

Compared with cows' milk, human milk is very low in calcium and phosphorus. Values for mature milk range from 200–350 and 110–160 mg/l for calcium and phosphorus respectively (Gross et al. 1980; Anderson, 1992). Although calcium and phosphorus levels of human milk are significantly lower than those of current infant formulas, bone mineralization is similar in breast and formula fed infants (Hillman *et al.* 1988; Mimouni *et al.* 1993).

The low phosphorus content of human milk is held to be advantageous to the infant for several reasons (Manz, 1992). Firstly, owing to the low phosphorus and protein content of human milk, its buffering capacity is poor; this is suggested as one of the factors responsible for the low pH and the characteristic bacterial flora of the intestine of breast fed infants (Bullen & Willis, 1971; Balmer & Wharton, 1989). Secondly, owing to the immature renal handling of phosphates in the newborn, high phosphorus intakes substantially increase serum phosphorus levels (Manz, 1992). Increased serum phosphorus and parathyroid levels, and decreased serum ionized calcium levels have been observed during the first week of age in infants receiving a formula with a high phosphorus content, regardless of its Ca:P ratio (Specker *et al.* 1991). High serum phosphorus levels can have clinical consequences. Occasionally, hypocalcaemic tetany may occur in otherwise healthy term infants receiving a high phosphorus formula (Venkataraman *et al.* 1985). Finally, a high phosphorus intake is a risk factor for the development of metabolic acidosis in high risk term infants (Kalhoff *et al.* 1990).

It is not unlikely that in the future both calcium and phosphorus levels of infant formulas will be further reduced to levels closer to human milk, provided that mineral homeostasis and bone mineralization prove to be adequate. Calcium and phosphorus could be reduced to levels of around 400 and 200 mg/l respectively without negatively affecting bone mineralization of term infants (Vainsel, 1992).

IRON

Human milk has a very low iron content of about 0·5 mg/l, yet term breast fed infants rarely exhaust their iron stores until after four months of age (Saarinen *et al.* 1977; Siimes *et al.* 1979; Calvo *et al.* 1992). This is attributed to the high absorption of iron from breast milk (Saarinen *et al.* 1977; Fomon *et al.* 1993). Iron absorption from infant formulas is significantly lower (Saarinen & Siimes, 1977; Fomon *et al.* 1993), probably at least partly owing to their higher calcium content and the presence of bovine casein and whey proteins (Hurrell *et al.* 1989; Hallberg *et al.* 1992).

The optimal iron level in infant formulas is still a major controversy. The Committee on Nutrition of the American Academy of Pediatrics (1992) recommends that iron fortified formula be used for all formula fed infants during the first year of life. These formulas should contain between 1 and 2 mg of iron per 100 kcal, or between 7 and 13 mg/l (Committee on Nutrition, American Academy of Pediatrics, 1976). Most iron fortified formulas available in the United States contain iron levels at the upper limit of this recommendation. Iron fortified formulas available in Europe generally contain lower amounts of iron, 5–8 mg/l. These lower amounts have been found to be equally effective in supporting normal iron status (Bradley *et al.* 1993; Haschke *et al.* 1993). Even a formula with an iron content as low as 3 mg/l could prevent infants from developing iron deficiency during the first 6 months of life (Haschke *et al.* 1993).

Whether iron fortification of formula is preferable right from birth is uncertain. It can be argued that the body iron content at term birth of most infants is sufficient to support haematopoiesis until about 4 months of age (Dallman, 1986; Aggett *et al.* 1989). Another argument against the initial fortification of formula with iron is that omission of iron gives a faecal flora somewhat closer to breast milk (Mevissen-Verhage *et al.* 1985; Balmer & Wharton, 1991). The impression that low iron formulas are associated with fewer gastrointestinal side effects is not supported by controlled studies (Oski, 1980; Nelson *et al.* 1988).

SELENIUM, MOLYBDENUM AND CHROMIUM

Selenium functions as an integral cofactor for glutathione peroxidase (*EC* 1.11.1.9), which catalyses the destruction of peroxides, is part of the selenoprotein Type I iodothyronine deiodinase, and is a component of several other selenoproteins of which the metabolic functions are not yet understood (Rotruck *et al.* 1973; Arthur *et al.* 1993). At the present time, proper data on the selenium requirement of the infant are lacking; selenium intakes and status of the exclusively breast fed infant should therefore serve as the basis for recommendations on infant feeding. Selenium levels of mature human milk are in the range 12–20 µg/l, whereas unfortified milk based formulas generally contain between 3 and 9 µg/l of intrinsic selenium (Roekens *et al.* 1985; Kumpulainen *et al.* 1987; Dörner *et al.* 1990). The higher selenium intake of the breast fed infant as well as the higher availability of human milk selenium are reflected in higher serum selenium concentrations and a higher selenium retention compared to formula fed infants (Kumpulainen *et al.* 1987; Dörner *et al.* 1990; McGuire *et al.* 1993).

It is to be expected that in the near future all infant formulas will be fortified with selenium to levels found in mature human milk. Supplementation of infant formulas with sodium selenite has been found to be effective in maintaining selenium status comparable to that observed in human milk fed infants (Kumpulainen *et al.* 1987; Litov *et al.* 1989; McGuire *et al.* 1993). American formula manufacturers have already started supplementing their infant formulas with sodium selenite. In Europe, addition of selenium has to await the formalization of the latest 1993 amendment to the EC Directive. Sodium selenite, selenate, selenomethionine, and selenium-enriched yeast, provided that the selenomethionine concentration of these yeasts is well standardized, will then be permitted for use in European infant formulas. The optimal form of selenium supplementation still requires further research. The potentially most promising candidate for organic selenium supplementation could be selenocysteine, the bioactive form of selenium found in glutathione peroxidase and other selenoproteins. Selenomethionine has the drawback that it is metabolized like methionine and therefore is non-specifically incorporated into a large number of proteins (Behne *et al.* 1991). Selenocysteine, on the other hand, is specifically inserted into glutathione peroxidase and other selenoproteins by a selenocysteine specific tRNA which differs from that for cysteine. Selenocysteine can therefore be seen as the 21st amino acid in terms of ribosome mediated protein synthesis (Böck *et al.* 1991).

As the chromium content of infant formulas (5–25 µg/l) is much higher than that of mature human milk (0·2–0·5 µg/l), the addition of chromium to infant formulas is not necessary (Deelstra *et al.* 1988; Foucault *et al.* 1989; Kumpulainen, 1992). The same holds true for molybdenum: levels found in infant formula and mature human milk are 15–200 and 1–3 µg/l respectively (Casey & Neville, 1987; Bougle *et al.* 1988; Foucault *et al.* 1989).

VITAMINS AND CONDITIONALLY ESSENTIAL SUBSTANCES

The vitamin levels of infant formulas are based on the values found in mature human milk, corrected for losses during processing and storage.

β-CAROTENE

Human milk, particularly colostrum, contains considerable amounts of β-carotene. The concentration of β-carotene in human milk decreases from 2·13 mg/l at day 1 to 0·4 mg/l at day 5 of lactation. Owing to the high concentrations of β-carotene in colostrum and early breast milk, the serum level of β-carotene of breast fed infants increases rapidly during this period to normal adult levels, whereas it does not rise in formula fed infants (Ostrea et al. 1986).

It has recently been demonstrated in infants that β-carotene can be converted to retinal by an intestinal mucosal enzyme (Lakshman et al. 1993). Thus, β-carotene may serve as a source of vitamin A in the neonatal period. Additionally, β-carotene may provide the infant's defence against oxygen toxicity by quenching singlet oxygen and free radicals (Ostrea et al. 1986; Krinsky, 1988). Future research should better define the role of dietary β-carotene in protecting the infant against oxygen toxicity, as well as in enhancing immune function (Bendich, 1991).

INOSITOL

Inositol, a component of membrane phospholipids and of compounds involved in signal transduction, is present in mature human milk in an amount of 250–300 mg/l (Bromberger & Hallman, 1986; Pereira et al. 1990). Serum inositol of preterm infants correlates significantly with inositol intake, the concentration being higher in infants receiving human milk than in those receiving unsupplemented formulas, which are low in inositol (Bromberger & Hallman, 1986; Pereira et al. 1990). Recently, it was reported that inositol supplementation to preterm infants with respiratory distress syndrome during the first week of life was associated with increased survival, a lower incidence of bronchopulmonary dysplasia, and a lower incidence of retinopathy of prematurity (Hallman et al. 1992). The authors suggested that inositol may increase surfactant availability through increased synthesis, release or recycling. Varying amounts of inositol have already been added to some formulas. Future research should further evaluate the importance of dietary inositol during the neonatal period, especially among high risk preterm infants.

TAURINE

Taurine is now added to nearly all infant formulas, although the effects of taurine supplementation on cholesterol synthesis, bile acid excretion, fat and vitamin D absorption, and auditory brainstem evoked responses have been shown only in infants born preterm (Tyson et al. 1989; Wasserhess et al. 1993; Zamboni et al. 1993).

NUCLEOTIDES

The importance of dietary nucleotides in infant nutrition has been the subject of active research for the last decade. Human milk is known to contain a significant amount of nucleotides (Gil & Sanchez-Medina, 1982; Janas & Picciano, 1982). Nucleotide supplemented formulas have been available in Spain and Japan for a number of years, and have more recently been introduced in the United States of America.

The clinical evidence for the claimed beneficial effects of dietary nucleotides is yet far from convincing. The results of Gil et al. (1986a) who found that nucleotides enhanced the growth of bifidobacteria in the faecal flora of infants could not be confirmed in a very recent study of Balmer et al. (1994). In this last study, at 2 weeks of age the reverse effect was noted with more nucleotide supplemented infants colonized with E. coli and a reduction in the counts of bifidobacteria. Dietary nucleotides only marginally influenced the essential fatty acid status of infants (Gil et al. 1986b; DeLucchi et al. 1987). Effects of dietary nucleotides on the gastrointestinal system and on hepatic growth and function have only been observed in animal studies (Uauy et al. 1990; Bustamante et al. 1994; Novak et al. 1994). Nucleotide supplementation appeared significantly to increase all plasma lipoprotein concentrations in preterm, but not in term infants (Sanchez-Pozo et al. 1994). A quite fascinating finding that warrants further study has been that dietary nucleotides increased indices of cell mediated immunity in infants without, however, influencing incidence and severity of infections (Carver et al. 1991). Very recently, it was reported that infants living in a contaminated environment, fed a nucleotide supplemented formula, experienced less diarrhoea than controls receiving an unsupplemented formula (Brunser et al. 1994). The only significant difference between the two groups, however, was the number of first episodes of diarrhoea. The total number of episodes of diarrhoea, the total duration of episodes, and the pattern of enteropathogens isolated did not differ between the supplemented infants and the controls. More clinical research needs to be done to determine efficacy, safety, and optimal level of supplementation, and to elucidate what population of infants will derive clear benefit from dietary nucleotides, before routine supplementation of infant formulas with nucleotides can be recommended (Quan & Barness, 1990).

CONCLUDING REMARKS

Through the last decades, infant formulas have been developed that closely approach human milk in nutrient composition. As knowledge has accumulated about the effects and action of different substances in human milk which serve other than nutritional roles, some of them have already been incorporated into infant formulas. Whether all nutritional and metabolic components of human milk confer unequivocal benefit to the infant in terms of growth and development is difficult to determine. For many substances, such as hormones and hormone binding proteins, vitamin binding proteins, growth factors, enzymes, and various non-protein nitrogen components, it will be extremely difficult to demonstrate clear functional advantages of supplementation, while the costs of addition of these substances will generally be substantial. Therefore, attempts to improve the composition of infant formulas should be applauded, but the benefits of compositional modifications should be carefully weighed against the costs. Unlike the rigid USA guidelines, it is to be hoped that the European guidelines for clinical testing of infant formulas, which are currently being drawn up, will provide a reasonable balance between protecting public health against opportunistic supplementations and leaving room for scientific innovations.

Obviously, no formula can supplant mother's milk as the ideal food for healthy term infants. The biological properties of human milk make it uniquely suited to the human infant. In the years to come, researchers will continue to attempt to identify and explain the role of different substances in human milk in the hope of incorporating all the benefits provided by human milk into infant formulas. It will be a major challenge for the industry to concentrate efforts on those substances which have a clear physiological function and thus are worth their price.

REFERENCES

Aggett, P. J., Barclay, S. & Whitley, J. E. (1989). Iron for the suckling. *Acta Paediatrica Scandinavica* Suppl. 361, 96–102.

Anderson, R. R. (1992). Variations in major minerals of human milk during the first 5 months of lactation. *Nutrition Research* **12**, 701–711.

Andersson, B., Porras, O., Hanson, L. Å., Lagergård, T. & Svanborg-Edén, C. (1986). Inhibition of attachment of *Streptococcus pneumoniae* and *Haemophilus influenzae* by human milk and receptor oligosaccharides. *Journal of Infectious Diseases* **153**, 232–237.

Arthur, J. R., Nicol, F. & Beckett, G. J. (1993). Selenium deficiency, thyroid hormone metabolism, and thyroid hormone deiodinases. *American Journal of Clinical Nutrition* **57**, S236–S239.

Atkinson, S. A., Schnurr, C. M., Donovan, S. M. & Lönnerdal, B. (1989). The non-protein nitrogen components in human milk: biochemistry and potential functional role. In *Protein and Non-protein Nitrogen in Human Milk*, pp. 117–133 [S. A. Atkinson and B. Lönnerdal, editors]. Boca Raton, FL: CRC Press.

Balmer, S. E., Hanvey, L. S. & Wharton, B. A. (1994). Diet and faecal flora in the newborn: nucleotides. *Archives of Disease in Childhood* **70**, F137–F140.

Balmer, S. E., Scott, P. H. & Wharton, B. A. (1989). Diet and faecal flora in the newborn: casein and whey proteins. *Archives of Disease in Childhood* **64**, 1678–1684.

Balmer, S. E. & Wharton, B. A. (1989). Diet and faecal flora in the newborn: breast milk and infant formula. *Archives of Disease in Childhood* **64**, 1672–1677.

Balmer, S. E. & Wharton, B. A. (1991). Diet and faecal flora in the newborn: iron. *Archives of Disease in Childhood* **66**, 1390–1394.

Bauchner, H., Leventhal, J. M. & Shapiro, E. D. (1986). Studies of breast-feeding and infections. How good is the evidence? *Journal of the American Medical Association* **256**, 887–892.

Behne, D., Kyriakopoulos, A., Scheid, S. & Gessner, H. (1991). Effects of chemical form and dosage on the incorporation of selenium into tissue proteins in rats. *Journal of Nutrition* **121**, 806–814.

Bellamy, W., Takase, M., Yamauchi, K., Wakabayashi, H., Kawase, K. & Tomita, M. (1992). Identification of the bactericidal domain of lactoferrin. *Biochimica et Biophysica Acta* **1121**, 130–136.

Bendich, A. (1991). β-Carotene and the immune response. *Proceedings of the Nutrition Society* **50**, 263–274.

Bernbäck, S., Bläckberg, L. & Hernell, O. (1990). The complete digestion of human milk triacylglycerol in vitro requires gastric lipase, pancreatic colipase-dependent lipase, and bile salt-stimulated lipase. *Journal of Clinical Investigation* **85**, 1221–1226.

Billeaud, C., Guillet, J. & Sandler, B. (1990). Gastric emptying in infants with or without gastro-oesophageal reflux according to the type of milk. *European Journal of Clinical Nutrition* **44**, 577–583.

Birch, E. E., Birch, D. G., Hoffman, D. R., Hale, L., Everett, M. & Uauy, R. (1993). Breast-feeding and optimal visual development. *Journal of Pediatric Ophthalmology and Strabismus* **30**, 33–38.

Birch, E. E., Birch, D. G., Hoffman, D. R. & Uauy, R. (1992). Dietary essential fatty acid supply and visual acuity development. *Investigative Ophthalmology and Visual Science* **33**, 3242–3253.

Böck, A., Forchhammer, K., Heider, J., Leinfelder, W., Sawers, G., Veprek, B. & Zinoni, F. (1991). Selenocysteine: the 21st amino acid. *Molecular Microbiology* **5**, 515–520.

Boesman-Finkelstein, M. & Finkelstein, R. A. (1991). Bovine lactogenic immunity against pediatric enteropathogens. In *Immunology of Milk and the Neonate*, pp. 361–367 [J. Mestecky, C. Blair and P. L. Ogra, editors]. New York: Plenum Press.

Bougle, D., Bureau, F., Foucault, P., Duhamel, J.-F., Muller, G. & Drosdowsky, M. (1988). Molybdenum content of term and preterm human milk during the first 2 months of lactation. *American Journal of Clinical Nutrition* **48**, 652–654.

Bradley, C. K., Hillman, L., Sherman, A. R., Leedy, D. & Cordano, A. (1993). Evaluation of two iron-fortified, milk-based formulas during infancy. *Pediatrics* **91**, 908–914.

British Nutrition Foundation. (1992). *Unsaturated Fatty Acids. Nutritional and Physiological Significance*. London: Chapman & Hall.

Bromberger, P. & Hallman, M. (1986). Myoinositol in small preterm infants: relationship between intake and serum concentration. *Journal of Pediatric Gastroenterology and Nutrition* **5**, 455–458.

Brooke, O. G. (1985). Absorption of lard by infants. *Human Nutrition: Applied Nutrition* **39A**, 221–223.

Brunser, O., Espinoza, J., Araya, M., Cruchet, S. & Gil, A. (1994). Effect of dietary nucleotide supplementation on diarrhoeal disease in infants. *Acta Paediatrica* **83**, 188–191.

Brunser, O., Espinoza, J., Figueroa, G., Araya, M., Spencer, E., Hilpert, H., Link-Amster, H. & Brussow, H. (1992). Field trial of an infant formula containing anti-rotavirus and anti-*Escherichia coli* milk antibodies from hyperimmunized cows. *Journal of Pediatric Gastroenterology and Nutrition* **15**, 63–72.

Bullen, C. L. & Willis, A. T. (1971). Resistance of the breast-fed infant to gastroenteritis. *British Medical Journal* **iii**, 338–343.

Bustamante, S. A., Sanches, N., Crosier, J., Miranda, D., Colombo, G. & Miller, M. J. S. (1994). Dietary nucleotides: effects on the gastrointestinal system in swine. *Journal of Nutrition* **124**, 149S–156S.

Butte, N. F., Wong, W. W., Ferlic, L., O'Brian Smith, E., Klein, P. D. & Garza, C. (1990). Energy expenditure and deposition of breast-fed and formula-fed infants during early infancy. *Pediatric Research* **28**, 631–640.

Calvo, E. B., Galindo, A. C. & Aspres, N. B. (1992). Iron status in exclusively breast-fed infants. *Pediatrics* **90**, 375–379.

Carlson, S. E. (1985a). Human milk nonprotein nitrogen: occurrence and possible functions. *Advances in Pediatrics* **32**, 43–70.

Carlson, S. E. (1985b). N-Acetylneuraminic acid concentrations in human milk oligosaccharides and glycoproteins during lactation. *American Journal of Clinical Nutrition* **41**, 720–726.

Carlson, S. E., Cooke, R. J., Rhodes, P. G., Peeples, J. M., Werkman, S. H. & Tolley, E. A. (1991). Long-term feeding of formulas high in linolenic acid and marine oil to very low birth weight infants: phospholipid fatty acids. *Pediatric Research* **30**, 404–412.

Carlson, S. E., Cooke, R. J., Werkman, S. H. & Tolley, E. A. (1992b). First year growth of preterm infants fed standard compared to marine oil n-3 supplemented formula. *Lipids* **27**, 901–907.

Carlson, S. E., De Voe, P. W. & Barness, L. A. (1982). Effect of infant diets with different polyunsaturated to saturated fat ratios on circulating high-density lipoproteins. *Journal of Pediatric Gastroenterology and Nutrition* **1**, 303–309.

Carlson, S. E., Werkman, S. H., Peeples, J. M., Cooke, R. J. & Tolley, E. A. (1993a). Arachidonic acid status correlates with first year growth in preterm infants. *Proceedings of the National Academy of Sciences, USA* **90**, 1073–1077.

Carlson, S. E., Werkman, S. H., Peeples, J. M., Cooke, R. J. & Wilson, W. W. (1992a). Plasma phospholipid arachidonic acid and growth and development of preterm infants. In *Recent Advances in Infant Feeding*, pp. 22–27 [B. Koletzko, A. Okken, J. Rey, B. Salle and J. P. Van Biervliet, editors]. Stuttgart: Thieme.

Carlson, S. E., Werkman, S. H., Rhodes, P. G. & Tolley, E. A. (1993b). Visual-acuity development in healthy preterm infants: effect of marine-oil supplementation. *American Journal of Clinical Nutrition* **58**, 35–42.

Carnielli, V. P., Luijendijk, I. H. T., van Goudoever, J. B., Sulkers, E. J., Boerlage, A., Degenhart, H. J. & Sauer, J. J. (1994). Feeding premature newborn infants palmitic acid in amounts and stereo isomeric position similar to human milk: effects on fat and mineral balance. *American Journal of Clinical Nutrition* (submitted).

Carr, B. R. & Simpson, E. R. (1981). Synthesis of cholesterol in the human fetus: 3-hydroxy-3-methylglutaryl coenzyme A reductase activity of liver microsomes. *Journal of Clinical Endocrinology and Metabolism* **53**, 810–812.

Carver, J. D., Pimentel, B., Cox, W. I. & Barness, L. A. (1991). Dietary nucleotide effects upon immune function in infants. *Pediatrics* **88**, 359–363.

Casey, C. E. & Neville, M. C. (1987). Studies in human lactation. 3. Molybdenum and nickel in human milk during the first month of lactation. *American Journal of Clinical Nutrition* **45**, 921–926.

Chierici, R., Sawatzki, G., Tamisari, L., Volpato, S. & Vigi, V. (1992). Supplementation of an adapted formula with bovine lactoferrin. 2. Effects on serum iron, ferritin and zinc levels. *Acta Paediatrica* **81**, 475–479.

Christie, W. W. (1986). The positional distributions of fatty acids in triglycerides. In *Analysis of Oils and Fats*, pp. 313–340 [R. J. Hamilton, J. B. Rossell & D. Reffold, editors]. London: Elsevier.

Chung, T. D. Y. & Raymond, K. N. (1993). Lactoferrin: the role of conformational changes in its iron binding and release. *Journal of the American Chemical Society* **115**, 6765–6768.

Clandinin, M. T., Chappell, J. E., Leong, S., Heim, T., Swyer, P. R. & Chance, G. W. (1980a). Intrauterine fatty acid accretion rates in infant brain: implications for fatty acid requirements. *Early Human Development* **4**, 121–129.

Clandinin, M. T., Chappell, J. E., Leong, S., Heim, T., Swyer, P. R. & Chance, G. W. (1980b). Extrauterine fatty acid accretion rates in infant brain; implications for fatty acid requirements. *Early Human Development* **4**, 131–138.

Clark, K. J., Makrides, M., Neumann, M. A. & Gibson, R. A. (1992). Determination of the optimal ratio of linoleic acid to α-linolenic acid in infant formulas. *Journal of Pediatrics* **120**, S151–S158.

Clark, R. M. & Hundrieser, K. E. (1989). Changes in cholesteryl esters of human milk with total milk lipid. *Journal of Pediatric Gastroenterology and Nutrition* **9**, 347–350.

Cockburn, F. (1994). Neonatal brain and dietary lipids. *Archives of Disease in Childhood* **70**, F1–F2.

Committee on Nutrition, American Academy of Pediatrics. (1976). Commentary on breast-feeding and infant formulas, including proposed standards for formulas. *Pediatrics* **57**, 278–285.

Committee on Nutrition, American Academy of Pediatrics. (1992). The use of whole cow's milk in infancy. *Pediatrics* **89**, 1105–1109.

Coppa, G. V., Gabrielli, O., Giorgi, P., Catassi, C., Montanari, M. P., Varaldo, P. E. & Nichols, B. L. (1990). Preliminary study of breastfeeding and bacterial adhesion to uroepithelial cells. *Lancet* **335**, 569–571.

Coppa, G. V., Gabrielli, O., Pierani, P., Catassi, C., Carlucci, A. & Giorgi, P. L. (1993). Changes in carbohydrate composition in human milk over 4 months of lactation. *Pediatrics* **91**, 637–641.

Coppa, G. V., Gabrielli, O., Pierani, P., Zampini, L., Rottoli, G., Carlucci, A. & Giorgi, P. L. (1991). [Qualitative and quantitative studies of carbohydrates of human colostrum and mature milk.] *Rivista Italiana di Pediatria* **17**, 303–307.

Dallman, P. R. (1986). Iron deficiency in the weanling: a nutritional problem on the way to resolution. *Acta Paediatrica Scandinavica* Suppl. 323, 59–67.

Daniel, H., Vohwinkel, M. & Rehner, G. (1990). Effect of casein and β-casomorphins on gastrointestinal motility in rats. *Journal of Nutrition* **120**, 252–257.

Davidson, G. P., Whyte, P. B. D., Daniels, E., Franklin, K., Nunan, H., McCloud, P. I., Moore, A. G. & Moore, D. J. (1989). Passive immunisation of children with bovine colostrum containing antibodies to human rotavirus. *Lancet* **ii**, 709–712.

Davidson, L.-A. & Lönnerdal, B. (1987). Persistence of human milk proteins in the breastfed infant. *Acta Paediatrica Scandinavica* **76**, 733–740.

Davidsson, L., Kastenmayer, P., Yuen, M., Lönnerdal, B. & Hurrell, R. F. (1994). Influence of lactoferrin on iron absorption from human milk in infants. *Pediatric Research* **35**, 117–124.

Davin, J.-C., Senterre, J. & Mahieu, P. R. (1991). The high lectin-binding capacity of human secretory IgA protects nonspecifically mucosae against environmental antigens. *Biology of the Neonate* **59**, 121–125.

Deelstra, H., Van Schoor, O., Robberecht, H., Clara, R. & Eylenbosch, W. (1988). Daily chromium intake by infants in Belgium. *Acta Paediatrica Scandinavica* **77**, 402–407.

DeLucchi, C., Pita, M. L., Faus, M. J., Molina, J. A., Uauy, R. & Gil, A. (1987). Effects of dietary nucleotides on the fatty acid composition of erythrocyte membrane lipids in term infants. *Journal of Pediatric Gastroenterology and Nutrition* **6**, 568–574.

Department of Health and Social Security (1977). *The Composition of Mature Human Milk* (*Reports on Health and Social Subjects* no. 12). London: HMSO.

Donovan, S. M. & Lönnerdal, B. (1989). Non-protein nitrogen and true protein in infant formulas. *Acta Paediatrica Scandinavica* **78**, 497–504.

Dörner, K., Schneider, K., Sievers, E., Schulz-Lell, G., Oldigs, H.-D. & Schaub, J. (1990). Selenium balances in young infants fed on breast milk and adapted cow's milk formula. *Journal of Trace Elements and Electrolytes in Health and Disease* **4**, 37–40.

Ebina, T., Sato, A., Umezu, K., Ishida, N., Ohyama, S., Oizumi, A., Aikawa, K., Katagiri, S., Katsushima, N., Imai, A., Kitaoka, S., Suzuki, H. & Konno, T. (1985). Prevention of rotavirus infection by oral administration of cow colostrum containing antihumanrotavirus antibody. *Medical Microbiology and Immunology* **174**, 177–185.

Edmond, J., Korsak, R. A., Morrow, J. W., Torok-Both, G. & Catlin, D. H. (1991). Dietary cholesterol and the origin of cholesterol in the brain of developing rats. *Journal of Nutrition* **121**, 1323–1330.

Eigel, W. N., Butler, J E., Ernstrom, C. A., Farrell, H. M., Harwalkar, V. R., Jenness, R. & Whitney, R. McL. (1984). Nomenclature of proteins of cow's milk: fifth revision. *Journal of Dairy Science* **67**, 1599–1631.

Engelhardt, E. L., Sankar, M., Wu-Wang, C. Y., Thomas, M. R., Walker, W. R. & Neu, J. (1991). Effect of cholesterol deprivation on piglet small intestinal and serum lipids. *Journal of Pediatric Gastroenterology and Nutrition* **12**, 494–500.

ESPGAN Committee on Nutrition. (1982). Guidelines on infant nutrition. III. Recommendations for infant feeding. *Acta Paediatrica Scandinavica* Suppl. 302, 1–27.

ESPGAN Committee on Nutrition. (1991). Comment on the content and composition of lipids in infant formulas. *Acta Paediatrica Scandinavica* **80**, 887–896.

Evans, D. G., Evans, D. J., Moulds, J. J. & Graham, D. Y. (1988). *N*-acetylneuraminyllactose-binding fibrillar hemagglutinin of *Campylobacter pylori*: a putative colonization factor antigen. *Infection and Immunity* **56**, 2896–2906.

Facon, M., Skura, B. J. & Nakai, S. (1993). Potential for immunological supplementation of foods. *Food and Agricultural Immunology* **5**, 85–91.

Fairweather-Tait, S. J., Balmer, S. E., Scott, P. H. & Minski, M. J. (1987). Lactoferrin and iron absorption in newborn infants. *Pediatric Research* **22**, 651–654.

Fall, C. H. D., Barker, D. J. P., Osmond, C., Winter, P. D., Clark, P. M. S. & Hales, C. N. (1992). Relation of infant feeding to adult serum cholesterol concentration and death from ischaemic heart disease. *British Medical Journal* **304**, 801–805.

Farquharson, J., Cockburn, F., Patrick, W. A., Jamieson, E. C. & Logan, R. W. (1992). Infant cerebral cortex phospholipid fatty-acid composition and diet. *Lancet* **340**, 810–813.

Farquharson, J., Cockburn, F., Patrick, W. A., Jamieson, E. C. & Logan, R. W. (1993). Effect of diet on infant subcutaneous tissue triglyceride fatty acids. *Archives of Disease in Childhood* **69**, 589–593.

Fazzolari-Nesci, A., Domianello, D., Sotera, V. & Räihä, N. C. R. (1992). Tryptophan fortification of adapted formula increases plasma tryptophan concentrations to levels not different from those found in breast-fed infants. *Journal of Pediatric Gastroenterology and Nutrition* **14**, 456–459.

Filer, L. J., Mattson, F. H. & Fomon, S. J. (1969). Triglyceride configuration and fat absorption by the human infant. *Journal of Nutrition* **99**, 293–298.

Fomon, S. J., Bier, D. M., Matthews, D. E., Rogers, R. R., Edwards, B. B., Ziegler, E. E. & Nelson, S. E. (1988). Bioavailability of dietary urea nitrogen in the breast-fed infant. *Journal of Pediatrics* **113**, 515–517.

Fomon, S. J., Rogers, R. R., Ziegler, E. E., Nelson, S. E. & Thomas, L. N. (1984). Indices of fatness and serum cholesterol at age eight years in relation to feeding and growth during early infancy. *Pediatric Research* **18**, 1233–1238.

Fomon, S. J., Ziegler, E. E. & Nelson, S. E. (1993). Erythrocyte incorporation of ingested ^{58}Fe by 56-day-old breast-fed and formula-fed infants. *Pediatric Research* **33**, 573–576.

Foreman-van Drongelen, M. M. H. P., van Houwelingen, A. C., Kester, A. D. M., de Jong, A. E. P., Blanco, C. E., Hasaart, T. H. M. & Hornstra, G. (1994). Long chain polyene status of preterm infants with regard to the fatty acid composition of their diet: comparison between absolute and relative fatty acid amounts in plasma and red blood cell phospholipids. *British Journal of Nutrition* (in press).

Foucault, P., Bureau, F., Bougle, D., Neuville, D., Duhamel, J. F. & Drosdowsky, M. (1989). [Trace elements in 26 infant formulas.] *Cahiers de Nutrition et de Diététique* **24**, 385–388.

Freeman, C. P., Jack, E. L. & Smith, L. M. (1965). Intramolecular fatty acid distribution in the milk fat triglycerides of several species. *Journal of Dairy Science* **48**, 853–858.

Gauhe, A., György, P., Hoover, J. R. E., Kuhn, R., Rose, C. S., Ruelius, H. W. & Zilliken, F. (1954). Bifidus factor. IV. Preparations obtained from human milk. *Archives of Biochemistry and Biophysics* **48**, 214–224.

Gil, A., Corral, E., Martinez, A. & Molina, J. A. (1986a). Effects of the addition of nucleotides to an adapted milk formula on the microbial pattern of faeces in at term newborn infants. *Journal of Clinical Nutrition and Gastroenterology* **1**, 127–132.

Gil, A., Pita, M., Martinez, A., Molina, J. A. & Sanchez Medina, F. (1986b). Effect of dietary nucleotides on the plasma fatty acids in at-term neonates. *Human Nutrition: Clinical Nutrition* **40C**, 185–195.

Gil, A. & Sanchez-Medina, F. (1982). Acid-soluble nucleotides of human milk at different stages of lactation. *Journal of Dairy Research* **49**, 301–307.

Goldman, A. S. (1989). Immunologic supplementation of cow's milk formulations. *International Dairy Federation Bulletin* no. 244, 38–43.

Goldman, A. S., Garza, C., Nichols, B. L. & Goldblum, R. M. (1982). Immunologic factors in human milk during the first year of lactation. *Journal of Pediatrics* **100**, 563–567.

Goldman, A. S. & Goldblum, R. M. (1989). Immunoglobulins in human milk. In *Protein and Non-protein Nitrogen in Human Milk*, pp. 43–51 [S. A. Atkinson and B. Lönnerdal, editors]. Boca Raton, FL: CRC Press.

Gross, S. J., David, R. J., Bauman, L. & Tomarelli, R. M. (1980). Nutritional composition of milk produced by mothers delivering preterm. *Journal of Pediatrics* 96, 641–644.

Hallberg, L., Rossander-Hulten, L., Brune, M. & Gleerup, A. (1992). Bioavailability in man of iron in human milk and cow's milk in relation to their calcium contents. *Pediatric Research* 31, 524–527.

Hallman, M., Bry, K., Hoppu, K., Lappi, M. & Pohjavuori, M. (1992). Inositol supplementation in premature infants with respiratory distress syndrome. *New England Journal of Medicine* 326, 1233–1239.

Hambraeus, L., Lönnerdal, B., Forsum, E. & Gebre-Medhin, M. (1978). Nitrogen and protein components of human milk. *Acta Paediatrica Scandinavica* 67, 561–565.

Hamosh, M. (1988). Does infant nutrition affect adiposity and cholesterol levels in the adult? *Journal of Pediatric Gastroenterology and Nutrition* 7, 10–16.

Hanning, R. M., Paes, B. & Atkinson, S. A. (1992). Protein metabolism and growth of term infants in response to a reduced-protein 40:60 whey:casein formula with added tryptophan. *American Journal of Clinical Nutrition* 56, 1004–1011.

Hanson, L. Å., Ashraf, R., Zaman, S., Karlberg, J., Lindblad, B. S. & Jalil, F. (1994). Breast feeding is a natural contraceptive and prevents disease and death in infants, linking infant mortality and birth rates. *Acta Paediatrica* 83, 3–6.

Harrison, G. G., Graver, E. J., Vargas, M., Churella, H. R. & Paule, C. L. (1987). Growth and adiposity of term infants fed whey-predominant or casein-predominant formulas or human milk. *Journal of Pediatric Gastroenterology and Nutrition* 6, 739–747.

Harzer, G. & Bindels, J. G. (1985). Changes in human milk immunoglobulin A and lactoferrin during early lactation. In *Composition and Physiological Properties of Human Milk*, pp. 285–293 [J. Schaub, editor]. Amsterdam: Elsevier Science Publishers.

Harzer, G., Haug, M. & Bindels, J. G. (1986). Biochemistry of maternal milk in early lactation. *Human Nutrition: Applied Nutrition* 40A, Suppl. 1, 11–18.

Haschke, F., Vanura, H., Male, C., Owen, G., Pietschnig, B., Schuster, E., Krobath, E. & Huemer, C. (1993). Iron nutrition and growth of breast- and formula-fed infants during the first 9 months of life. *Journal of Pediatric Gastroenterology and Nutrition* 16, 151–156.

Hayes, K. C., Pronczuk, A., Wood, R. A. & Guy, D. G. (1992). Modulation of infant formula fat profile alters the low-density lipoprotein/high-density lipoprotein ratio and plasma fatty acid distribution relative to those with breast-feeding. *Journal of Pediatrics* 120, S109–S116.

Heine, W., Tiess, M. & Wutzke, K. D. (1986). ^{15}N-tracer investigations of the physiological availability of urea nitrogen in mother's milk. *Acta Paediatrica Scandinavica* 75, 439–443.

Heinig, M. J., Nommsen, L. A., Peerson, J. M., Lönnerdal, B. & Dewey, K. G. (1993). Energy and protein intakes of breast-fed and formula-fed infants during the first year of life and their association with growth velocity: the DARLING study. *American Journal of Clinical Nutrition* 58, 152–161.

Hillman, L. S., Chow, W., Salmons, S. S., Weaver, E., Erickson, M. & Hansen, J. (1988). Vitamin D metabolism, mineral homeostasis, and bone mineralization in term infants fed human milk, cow milk-based formula, or soy-based formula. *Journal of Pediatrics* 112, 864–874.

Hodgson, P. A., Ellefson, R. D., Elveback, L. R., Harris, L. E., Nelson, R. A. & Weidman, W. H. (1976). Comparison of serum cholesterol in children fed high, moderate, or low cholesterol milk diets during neonatal period. *Metabolism* 25, 739–746.

Holmgren, J., Svennerholm, A.-M. & Lindblad, M. (1983). Receptor-like glycocompounds in human milk that inhibit classical and El Tor *Vibrio cholerae* cell adherence (hemagglutination). *Infection and Immunity* 39, 147–154.

Howie, P. W., Forsyth, J. S., Ogston, S. A., Clark, A. & Florey, C. du V. (1990). Protective effect of breast feeding against infection. *British Medical Journal* 300, 11–16.

Hurrell, R. F., Lynch, S. R., Trinidad, T. P., Dassenko, S. A. & Cook, J. D. (1989). Iron absorption in humans as infuenced by bovine milk proteins. *American Journal of Clinical Nutrition* 49, 546–552.

Huttunen, J. K., Saarinen, U. M., Kostiainen, E. & Siimes, M. A. (1983). Fat composition of the infant diet does not influence subsequent serum lipid levels in man. *Atherosclerosis* 46, 87–94.

Iyer, S. & Lönnerdal, B. (1993). Lactoferrin, lactoferrin receptors and iron metabolism. *European Journal of Clinical Nutrition* 47, 232–241.

Janas, L. M. & Picciano, M. F. (1982). The nucleotide profile of human milk. *Pediatric Research* 16, 659–662.

Janas, L. M., Picciano, M. F. & Hatch, T. F. (1985). Indices of protein metabolism in term infants fed human milk, whey-predominant formula, or cow's milk formula. *Pediatrics* 75, 775–784.

Janas, L. M., Picciano, M. F. & Hatch, T. F. (1987). Indices of protein metabolism in term infants fed either human milk or formulas with reduced protein concentration and various whey/casein ratios. *Journal of Pediatrics* 110, 838–848.

Järvenpää, A. L., Räihä, N. C. R., Rassin, D. K. & Gaull, G. E. (1982a). Milk protein quantity and quality in the term infant. I. Metabolic responses and effects on growth. *Pediatrics* 70, 214–220.

Järvenpää, A. L., Räihä, N. C. R., Rassin, D. K. & Gaull, G. E. (1982b). Milk protein quantity and quality in the term infant. II. Effects on acidic and neutral amino acids. *Pediatrics* 70, 221–230.

Jason, J. M., Nieburg, P. & Marks, J. S. (1984). Mortality and infectious disease associated with infant-feeding practices in developing countries. *Pediatrics* **74**, 702–727.

Jooste, P. L., Rossouw, L. J., Steenkamp, H. J., Rossouw, J. E., Swanepoel, A. S. P. & Charlton, D. O. (1991). Effect of breast feeding on the plasma cholesterol and growth of infants. *Journal of Pediatric Gastroenterology and Nutrition* **13**, 139–142.

Kalhoff, H., Manz, F., Diekmann, L. & Stock, G. J. (1990). Suboptimal mineral composition of cow's milk formulas: a risk factor for the development of late metabolic acidosis. *Acta Paediatrica Scandinavica* **79**, 743–749.

Kallio, M. J. T., Salmenperä, L., Siimes, M. A., Perheentupa, J. & Miettinen, T. A. (1992). Exclusive breast-feeding and weaning: effect on serum cholesterol and lipoprotein concentrations in infants during the first year of life. *Pediatrics* **89**, 663–666.

Kidwell, W. R. & Salomon, D. S. (1989). Growth factors in human milk: sources and potential physiological roles. In *Protein and Non-protein Nitrogen in Human Milk*, pp. 117–133 [S. A. Atkinson and B. Lönnerdal, editors]. Boca Raton, FL: CRC Press.

Koletzko, B., Thiel, I. & Abiodun, P. O. (1992). The fatty acid composition of human milk in Europe and Africa. *Journal of Pediatrics* **120**, S62–S70.

Korhonen, T. K., Valtonen, M. V., Parkkinen, J., Väisänen-Rhen, V., Finne, J., Ørskov, F., Ørskov, I., Svenson, S. B. & Mäkelä, P. H. (1985). Serotypes, hemolysin production, and receptor recognition of *Escherichia coli* strains associated with neonatal sepsis and meningitis. *Infection and Immunity* **48**, 486–491.

Krimpenfort, P. (1993). The production of human lactoferrin in the milk of transgenic animals. *Cancer Detection and Prevention* **17**, 301–305.

Krinsky, N. I. (1988). The evidence for the role of carotenoids in preventive health. *Clinical Nutrition* **7**, 107–112.

Kuhn, D. C. & Crawford, M. (1986). Placental essential fatty acid transport and prostaglandin synthesis. *Progress in Lipid Research* **25**, 345–353.

Kumpulainen, J. T. (1992). Chromium content of foods and diets. *Biological Trace Element Research* **32**, 9–18.

Kumpulainen, J., Salmenperä, L., Siimes, M. A., Koivistoinen, P., Lehto, J. & Perheentupa, J. (1987). Formula feeding results in lower selenium status than breast-feeding or selenium supplemented formula feeding: a longitudinal study. *American Journal of Clinical Nutrition* **45**, 49–53.

Kunz, C. & Lönnerdal, B. (1990). Casein and casein subunits in preterm milk, colostrum, and mature human milk. *Journal of Pediatric Gastroenterology and Nutrition* **10**, 454–461.

Kunz, C. & Lönnerdal, B. (1992). Re-evaluation of the whey protein/casein ratio of human milk. *Acta Paediatrica* **81**, 107–112.

Kunz, C. & Rudloff, S. (1993). Biological functions of oligosaccharides in human milk. *Acta Paediatrica* **82**, 903–912.

Lakshman, M. R., Johnson, L. H., Okoh, C., Attlesey, M., Mychkovsky, I. & Bhagavan, H. N. (1993). Conversion of *all trans* β-carotene to retinal by an enzyme from the intestinal mucosa of human neonates. *Journal of Nutritional Biochemistry* **4**, 659–663.

Lammi-Keefe, C. J., Ferris, A. M. & Jensen, R. G. (1990). Changes in human milk at 06.00, 10.00, 18.00 and 22.00 h. *Journal of Pediatric Gastroenterology and Nutrition* **11**, 83–88.

Li-Chan, E. & Nakai, S. (1989). Enzymic dephosphorylation of bovine casein to improve acid clotting properties and digestibility for infant formula. *Journal of Dairy Research* **56**, 381–390.

Lien, E. L., Yuhas, R. J., Boyle, F. G. & Tomarelli, R. M. (1993). Corandomization of fats improves absorption in rats. *Journal of Nutrition* **123**, 1859–1867.

Litov, R. E., Sickles, V. S., Chan, G. M., Hargett, I. R. & Cordano, A. (1989). Selenium status in term infants fed human milk or infant formula with or without added selenium. *Nutrition Research* **9**, 585–596.

Lönnerdal, B. & Chen, C.-L. (1990). Effects of formula protein level and ratio on infant growth, plasma amino acids and serum trace elements. 1. Cow's milk formula. *Acta Paediatrica Scandinavica* **79**, 257–265.

Lucas, A., Davies, P. S. W. & Phil, M. (1990). Physiologic energy content of human milk. In *Breastfeeding, Nutrition, Infection and Infant Growth in Developed and Emerging Countries*, pp. 337–357 [S. A. Atkinson, L. Å. Hanson and R. K. Chandra, editors]. Newfoundland: ARTS Biomedical Publishers and Distributors.

McGuire, M. K., Burgert, S. L., Milner, J. A., Glass, L., Kummer, R., Deering, R., Boucek, R. & Picciano, M. F. (1993). Selenium status of infants is influenced by supplementation of formula or maternal diets. *American Journal of Clinical Nutrition* **58**, 643–648.

Makrides, M., Neumann, M. A., Simmer, K. & Gibson, R. A. (1993*a*). Dietary supply of polyunsaturated fats and neural function of infants. *XII National Conference of the Dieticians Association of Australia*. Abstract.

Makrides, M., Simmer, K., Goggin, M. & Gibson, R. A. (1993*b*). Erythrocyte docosahexaenoic acid correlates with the visual response of healthy, term infants. *Pediatric Research* **33**, 425–427.

Manz, F. (1992). Why is the phosphorus content of human milk exceptionally low? *Monatsschrift für Kinderheilkunde* **140**, Suppl. 1, S35–S39.

Martin, J.-C., Bougnoux, P., Antoine, J.-M., Lanson, M. & Couet, C. (1993). Triacylglycerol structure of human colostrum and mature milk. *Lipids* **28**, 637–643.

Mevissen-Verhage, E. A. E., Marcelis, J. H., Harmsen-Van Amerongen, W. C. M., de Vos, N. M., Berkel, J. & Verhoef, J. (1985). Effect of iron on neonatal gut flora during the first week of life. *European Journal of Clinical Microbiology* **4**, 14–19.

Migliore-Samour, D. & Jollès, P. (1988). Casein, a prohormone with an immunomodulating role for the newborn? *Experientia* **44**, 188–193.

Mimouni, F., Campaigne, B., Neylan, M. & Tsang, R. C. (1993). Bone mineralization in the first year of life in infants fed human milk, cow-milk formula, or soy-based formula. *Journal of Pediatrics* **122**, 348–354.

Morgan, B. L. G. & Winick, M. (1980). Effects of administration of N-acetylneuraminic acid (NANA) on brain NANA content and behavior. *Journal of Nutrition* **110**, 416–424.

Morgan, B. L. G. & Winick, M. (1981). The subcellular localization of administered N-acetylneuraminic acid in the brains of well-fed and protein restricted rats. *British Journal of Nutrition* **46**, 231–238.

Mott, G. E., Jackson, E. M. & McMahan, C. A. (1991). Bile composition of adult baboons is influenced by breast versus formula feeding. *Journal of Pediatric Gastroenterology and Nutrition* **12**, 121–126.

Mott, G. E., Jackson, E. M., McMahan, C. A. & McGill, H. C. (1990). Cholesterol metabolism in adult baboons is influenced by infant diet. *Journal of Nutrition* **120**, 243–251.

Nakai, S. & Li-Chan, E. (1987). Effect of clotting in stomachs of infants on protein digestibility of milk. *Food Microstructure* **6**, 161–170.

Neeser, J.-R., Golliard, M. & Del Vedovo, S. (1991). Quantitative determination of complex carbohydrates in bovine milk and in milk-based infant formulas. *Journal of Dairy Science* **74**, 2860–2871.

Nelson, S. E., Ziegler, E. E., Copeland, A. M., Edwards, B. B. & Fomon, S. J. (1988). Lack of adverse reactions to iron-fortified formula. *Pediatrics* **81**, 360–364.

Nichols, B. L., McKee, K. S., Henry, J. F. & Putman, M. (1987). Human lactoferrin stimulates thymidine incorporation into DNA of rat crypt cells. *Pediatric Research* **21**, 563–567.

Novak, D. A., Carver, J. D. & Barness, L. A. (1994). Dietary nucleotides affect hepatic growth and composition in the weanling mouse. *Journal of Parenteral and Enteral Nutrition* **18**, 62–66.

Oski, F. A. (1980). Iron-fortified formulas and gastrointestinal symptoms in infants: a controlled study. *Pediatrics* **66**, 168–170.

Ostrea, E. M., Balun, J. E., Winkler, R. & Porter, T. (1986). Influence of breast-feeding on the restoration of the low serum concentration of vitamin E and β-carotene in the newborn infant. *American Journal of Obstetrics and Gynecology* **154**, 1014–1017.

Parkkinen, J. & Finne, J. (1987). Isolation of sialyl oligosaccharides and sialyl oligosaccharide phosphates from bovine colostrum and human urine. *Methods in Enzymology* **138**, 289–300.

Pegg, A. E. (1986). Recent advances in the biochemistry of polyamines in eukaryotes. *Biochemical Journal* **234**, 249–262.

Pereira, G. R., Baker, L., Egler, J., Corcoran, L. & Chiavacci, R. (1990). Serum myoinositol concentrations in premature infants fed human milk, formula for infants, and parenteral nutrition. *American Journal of Clinical Nutrition* **51**, 589–593.

Petschow, B. W. & Talbott, R. D. (1991). Response of bifidobacterium species to growth promotors in human and cow milk. *Pediatric Research* **29**, 208–213.

Picone, T. A., Benson, J. D., Moro, G., Minoli, I., Fulconis, F., Rassin, D. K. & Räihä, N. C. R. (1989). Growth, serum biochemistries, and amino acids of term infants fed formulas with amino acid and protein concentrations similar to human milk. *Journal of Pediatric Gastroenterology and Nutrition* **9**, 351–360.

Pollack, P. F., Koldovský, O. & Nishioka, K. (1992). Polyamines in human and rat milk and in infant formulas. *American Journal of Clinical Nutrition* **56**, 371–375.

Prentice, A. M., Lucas, A., Vasquez-Velasquez, L., Davies, P. S. W. & Whitehead, R. G. (1988). Are current dietary guidelines for young children a prescription for overfeeding? *Lancet* **ii**, 1066–1069.

Quan, R. & Barness, L. A. (1990). Do infants need nucleotide supplemented formula for optimal nutrition? *Journal of Pediatric Gastroenterology and Nutrition* **11**, 429–434.

Quinlan, P. & Moore, S. (1993). Modification of triglycerides by lipases: process technology and its application to the production of nutritionally improved fats. *Inform* **4**, 580–585.

Räihä, N. C. R. (1985). Nutritional proteins in milk and the protein requirement of normal infants. *Pediatrics* **75**, S136–S141.

Räihä, N., Minoli, I. & Moro, G. (1986a). Milk protein intake in the term infant. I. Metabolic responses and effects on growth. *Acta Paediatrica Scandinavica* **75**, 881–886.

Räihä, N., Minoli, I., Moro, G. & Bremer, H. J. (1986b). Milk protein intake in the term infant. II. Effects on plasma amino acid concentrations. *Acta Paediatrica Scandinavica* **75**, 887–892.

Roekens, E., Robberecht, H., Van Caillie-Bertrand, M., Deelstra, H. & Clara, R. (1985). Daily intake of selenium by bottle-fed infants in Belgium. *European Journal of Pediatrics* **144**, 45–48.

Romain, N., Dandrifosse, G., Jeusette, F. & Forget, P. (1992). Polyamine concentration in rat milk and food, human milk, and infant formulas. *Pediatric Research* **32**, 58–63.

Rönneberg, R. & Skåra, B. (1992). Essential fatty acids in human colostrum. *Acta Paediatrica* **81**, 779–783.

Rotruck, J. T., Pope, A. L., Ganther, H. E., Swanson, A. B., Hafeman, D. G. & Hoekstra, W. G. (1973). Selenium: biochemical role as a component of glutathione peroxidase. *Science* **179**, 588–590.

Rubin, D. H., Leventhal, J. M., Krasilnikoff, P. A., Kuo, H. S., Jekel, J. F., Weile, B., Levee, A., Kurzon, M. & Berget, A. (1990). Relationship between infant feeding and infectious illness: a prospective study of infants during the first year of life. *Pediatrics* **85**, 464–471.

Saarinen, U. M. & Siimes, M. A. (1977). Iron absorption from infant milk formula and the optimal level of iron supplementation. *Acta Paediatrica Scandinavica* **66**, 719–722.

Saarinen, U. M., Siimes, M. A. & Dallman, P. R. (1977). Iron absorption in infants: high bioavailability of breast milk iron as indicated by the extrinsic tag method of iron absorption and by the concentration of serum ferritin. *Journal of Pediatrics* **91**, 36–39.

Saito, H., Miyakawa, H., Tamura, Y., Shimamura, S. & Tomita, M. (1991). Potent bactericidal activity of bovine lactoferrin hydrolysate produced by heat treatment at acidic pH. *Journal of Dairy Science* **74**, 3724–3730.

Sanchez-Pozo, A., Morillas, J., Molto, L., Robles, R. & Gil, A. (1994). Dietary nucleotides influence lipoprotein metabolism in newborn infants. *Pediatric Research* **35**, 112–116.

Schroten, H., Plogmann, R., Hanisch, F. G., Hacker, J., Nobis-Bosch, R. & Wahn, V. (1993). Inhibition of adhesion of S-fimbriated *E. coli* to buccal epithelial cells by human skim milk is predominantly mediated by mucins and depends on the period of lactation. *Acta Paediatrica* **82**, 6–11.

Schulz-Lell, G., Dörner, K., Oldigs, H. D., Sievers, E. & Schaub, J. (1991). Iron availability from an infant formula supplemented with bovine lactoferrin. *Acta Paediatrica Scandinavica* **80**, 155–158.

Shield, J., Meville, C., Novelli, V., Anderson, G., Scheimberg, I., Gibb, D. & Milla, P. (1993). Bovine colostrum immunoglobulin concentrate for cryptosporidiosis in AIDS. *Archives of Disease in Childhood* **69**, 451–453.

Siimes, M. A., Vuori, E. & Kuitunen, P. (1979). Breast milk iron: a declining concentration during the course of lactation. *Acta Paediatrica Scandinavica* **68**, 29–31.

Small, D. M. (1991). The effects of glyceride structure on absorption and metabolism. *Annual Review of Nutrition* **11**, 413–434.

Specker, B. L., Tsang, R. C., Ho, M. L., Landi, T. M. & Gratton, T. L. (1991). Low serum calcium and high parathyroid hormone levels in neonates fed 'humanized' cow's milk-based formula. *American Journal of Diseases of Children* **145**, 941–945.

Steinberg, L. A., O'Connell, N. C., Hatch, T. F., Picciano, M. F. & Birch, L. L. (1992). Tryptophan intake influences infants' sleep latency. *Journal of Nutrition* **122**, 1781–1791.

Subbiah, M. T. R., Sprinkle, J. D., Rymaszewski, Z. & Yunker, R. L. (1989). Short-term exposure to high dietary cholesterol in early life: arterial changes and response after normalization of plasma cholesterol. *American Journal of Clinical Nutrition* **50**, 68–72.

Tomarelli, R. M., Meyer, B. J., Weaber, J. R. & Bernhart, F. W. (1968). Effect of positional distribution on the absorption of the fatty acids of human milk and infant formulas. *Journal of Nutrition* **95**, 583–590.

Turner, R. B. & Kelsey, D. K. (1991). Passive immunization for prevention of rotavirus infection. *Pediatric Research* **29**, 187A.

Tyson, J. E., Lasky, R., Flood, D., Mize, C., Picone, T. & Paule, C. L. (1989). Randomized trial of taurine supplementation for infants $\leq$ 1,300-gram birth weight: effect on auditory brainstem-evoked responses. *Pediatrics* **83**, 406–415.

Uauy, R., Birch, E., Birch, D. & Peirano, P. (1992). Visual and brain function measurements in studies of n-3 fatty acid requirements of infants. *Journal of Pediatrics* **120**, S168–S180.

Uauy, R., Stringel, G., Thomas, R. & Quan, R. (1990). Effect of dietary nucleosides on growth and maturation of the developing gut in the rat. *Journal of Pediatric Gastroenterology and Nutrition* **10**, 497–503.

Vainsel, M. (1992). Evaluation of a low-phosphate cow's milk diet on growth and bone mineralization of full-term infants. *Monatsschrift für Kinderheilkunde* **140**, Suppl. 1, S45–S50.

Van Biervliet, J.-P., Vinaimont, N., Vercaemst, R. & Rosseneu, M. (1992). Serum cholesterol, cholesteryl ester, and high-density lipoprotein development in newborn infants: response to formulas supplemented with cholesterol and γ-linolenic acid. *Journal of Pediatrics* **120** (4), S101–S108.

Veerkamp, J. H. (1969). Uptake and metabolism of derivatives of 2-deoxy-2-amino-D-glucose in *Bifidobacterium bifidum* var. *pennsylvanicus*. *Archives of Biochemistry and Biophysics* **129**, 248–256.

Venkataraman, P. S., Tsang, R. C., Greer, F. R., Noguchi, A., Laskarzewski, P. & Steichen, J. J. (1985). Late infantile tetany and secondary hyperparathyroidism in infants fed humanized cow milk formula. *American Journal of Diseases of Children* **139**, 664–668.

Verkade, H. J., van Asselt, W. A., Vonk, R. J., Bijleveld, C. M. A., Fernandes, J., de Jong, H., Fidler, V. & Okken, A. (1989). Fat absorption in premature infants: the effect of lard and antibiotics. *European Journal of Pediatrics* **149**, 126–129.

Viverge, D., Grimmonprez, L., Cassanas, G., Bardet, L. & Solere, M. (1990). Discriminant carbohydrate components of human milk according to donor secretor types. *Journal of Pediatric Gastroenterology and Nutrition* **11**, 365–370.

Ward, S. D., Melin, J. R., Lloyd, F. P., Norton, J. A. & Christian, J. C. (1980). Determinants of plasma cholesterol in children – a family study. *American Journal of Clinical Nutrition* **33**, 63–70.

Wasserhess, P., Becker, M. & Staab, D. (1983). Effect of taurine on synthesis of neutral and acidic sterols and fat absorption in preterm and full-term infants. *American Journal of Clinical Nutrition* **58**, 349–353.

Wong, W. W., Hachey, D. L., Insull, W., Opekun, A. R. & Klein, P. D. (1993). Effect of dietary cholesterol on cholesterol synthesis in breast-fed and formula-fed infants. *Journal of Lipid Research* **34**, 1403–1411.

Wright, A. L., Holberg, C. J., Martinez, F. D., Morgan, W. J., Taussig, L. M. & Group Health Medical Associates. (1989). Breast feeding and lower respiratory tract illness in the first year of life. *British Medical Journal* **299**, 946–949.

Zamboni, G., Piemonte, G., Bolner, A., Antoniazzi, F., Dall'Agnola, A., Messner, H., Gambaro, G. & Tato, L. (1993). Influence of dietary taurine on vitamin D absorption. *Acta Paediatrica* **82**, 811–815.

Printed in Great Britain

FETO-MATERNAL INTERACTION OF ANTIBODY AND ANTIGEN TRANSFER, IMMUNITY AND ALLERGY DEVELOPMENT

JULIE A. LOVEGROVE AND JANE B. MORGAN

Nutritional Metabolism Research Group, School of Biological Sciences, University of Surrey, Guildford, Surrey GU2 5XH

CONTENTS

INTRODUCTION	25
DEFINITIONS	26
INCIDENCE	26
FETO-MATERNAL RELATIONSHIP	26
PLACENTAL ANTIBODY TRANSFER	27
PLACENTAL ANTIGEN TRANSFER	29
HUMAN MILK ANTIBODY TRANSFER	29
HUMAN MILK ANTIGEN TRANSFER	31
DEVELOPMENT OF TOLERANCE IN THE NEONATE	32
HUMAN MILK AS A PROTECTION AGAINST ALLERGY	33
MATERNAL ELIMINATION DIETS	34
MATERNAL ELIMINATION DIETS DURING PREGNANCY	36
MATERNAL ELIMINATION DIETS DURING LACTATION	36
MATERNAL ELIMINATION DIETS AND NUTRITIONAL STATUS	37
CONCLUSION	38
REFERENCES	38

INTRODUCTION

Although the human infant is immunologically complete at birth, the levels of immunoglobulins present are only a small fraction of those in adulthood. Through placental transfer, immunoglobulin G (IgG) passes from the mother to her fetus. The newborn infant receives secretory immunoglobulin A (sIgA) and, to a much lesser extent, IgG from colostrum and mature human milk. This phenomenon of passive immunity is vital for the infant's defence system against pathogenic microorganisms and potential allergens. In some infants, often those born into a family with a history of allergy, this harmonious state of affairs can be disrupted. In this case the infant may already be sensitized to allergens transferred *in utero* from the mother *via* the placenta. Furthermore the breast fed infant may suffer hypersensitivity to human milk born allergens derived from foods the mother has eaten during lactation.

In this review we examine the part played by feto–maternal interactions in the immunological development of the infant. In particular the development of tolerance is explored and the possible mechanisms involved when an infant becomes hypersensitive to food proteins or other allergens. The usefulness of maternal elimination diets, as a prophylactic measure in the prevention of the development of adverse reactions to foods including allergy in infants, is assessed.

DEFINITIONS

Over the past decade a better understanding of the classification of adverse reactions to environmental substances has led to clearer definitions. Ferguson (1992) has described the terms for the various syndromes and diseases associated with reactions to foods. One of the problems associated with defining the word 'allergy' is that it can be used to describe a different condition in different countries (Brostoff & Gamlin, 1990). The accepted definition of a food allergy in the United Kingdom is that it is a form of food intolerance that has an immunological basis, and that there is evidence of an immunological reaction mediated by an antibody or T-lymphocytes or both. Mechanisms involved in allergic reactions can be further subdivided into four types (Bleumink, 1983), although in any particular disease state more than one mechanism may operate. There are five different immunoglobulin (Ig) classes: IgA, IgD, IgE, IgG and IgM. Some of the immunoglobulin classes have a further subclass division, e.g. IgG1–IgG4. Atopy is associated with high circulating levels of IgE in response to common environmental allergens and individuals with atopic disease may or may not develop one or more of the following: asthma, hay fever, eczema (Savin, 1993). The atopic phenomenon is known to be familial. Aas (1989) contends that symptoms associated with an allergic reaction depend on the degree of reactivity of the involved tissue receptors and of the effector cells.

Current knowledge of mechanisms related to allergic reactions is poor, and several mechanisms may interact. Although it is beyond the scope of this article to detail our understanding of all possible mechanisms, it is of interest to note that the risk of developing an immediate or delayed allergic reaction is higher in infants and children compared with adults. This implies that developmental factors operative during gestation and after birth are likely to play an important, though as yet poorly understood, role in the development of allergic disease (Strobel, 1988).

INCIDENCE

A comprehensive review of immunological hypersensitivity and other untoward effects related to the chemical composition of foods has highlighted the difficulty of assessing the incidence of these conditions and the lack of a definitive clinical test to diagnose allergy (Wood, 1986). There are, however, studies that have attempted to quantify the incidence of many of these conditions. It has been estimated that 10% of the childhood population have an atopic constitution (Wood, 1986), and that the prevalence of atopic diseases is increasing in North West Europe (Croner, 1992). There is evidence to suggest that the incidence of atopic eczema is rising in Britain (Ferguson & Watret, 1988). The incidence of allergic reactions to food in the paediatric population is variously estimated to be in the range 0·5–6·0% (Chandra & Prasad, 1991).

FETO-MATERNAL RELATIONSHIP

The ability to mount an antibody response and an immune T-cell mediated response, as shown by allograft rejection, is reasonably well developed by birth. Antibody levels, except that of IgG, are low in the absence of an intra-uterine infection. The transfer of antibodies, antigens and cells from the mother to the fetus and infant *via* the placenta and breast milk respectively is of great significance in the immunological development of the child.

PLACENTAL ANTIBODY TRANSFER

The mother and fetus possess a unique relationship *via* the placenta. In addition to the essential function of supplying the fetus with nutrients and oxygen, and removing any waste, the placenta affords a degree of protection to the fetus by the transfer of antibodies and specific cells and preventing the transfer of toxins and potentially harmful substances. The placenta is also involved with metabolism and endocrine secretion (Hurley, 1980).

The only known immunoglobulin class that is transferred across the placenta is IgG. The fragment crystallizable portion of IgG plays a major role in the transport process (Gitlin *et al*. 1964). The transfer of IgG is believed to be achieved *via* an active transport mechanism (Harlow & Lane, 1988) which involves a receptor mediated transcytosis of IgG across the syncytiotrophoblast, and a transcellular pathway through the endothelium (Leach *et al*. 1990). There is evidence that this transfer begins at approximately 20 weeks of gestation (Leach *et al*. 1990).

The majority of the IgG transfer occurs in the third trimester. This presents a potential problem for the prematurely born infant as the time available for antibody transfer is reduced compared with that for the full term infant. Premature infants do not obtain the full benefit of maternal IgG antibodies to help protect them against potentially life threatening infections or the development of possible allergies (Papadatos *et al*. 1970). Babies born with intra-uterine growth retardation also have low levels of IgG. The growth retardation is due to impaired placental transfer of nutrients to the fetus and there is also impaired transport of IgG (Papadatos *et al*. 1970). The level of IgG found in infants who are small-for-gestational-age is significantly less than that of those who are of the same gestational age but of normal weight. However the levels of IgG in the small-for-gestational-age infant are still greater than those of infants of the same body weight who are born prematurely.

Studies of IgG levels in premature infants and aborted fetuses show that fetuses of less than 20 weeks' gestational age have IgG levels under 1 g/l (Hobbs & Davis, 1967). At 16 weeks, IgG1 may be detected in fetal serum, followed by IgG2 and IgG3; IgG4 is detected by 22 weeks (Chandra, 1976). Fetuses or premature infants of less than 32 weeks gestational age usually have an IgG level of less than 4 g/l. The levels of IgG antibodies at full term are usually above 14 g/l (Chandra, 1976). After 32 weeks of gestation, total IgG concentrations in the mother and fetus are similar, and at birth IgG levels in cord blood of a full term infant tend to be higher than the corresponding paired maternal samples (Kohler & Farr, 1966; Carlsson *et al*. 1976; Iikura *et al*. 1989; Lovegrove, 1991; Lovegrove *et al*. 1994).

Fig. 1 shows results from a study of 20 normal mothers and their offspring of cows' milk β-lactoglobulin IgG antibody levels in maternal and cord serum samples (Lovegrove, 1991). The concentration of cows' milk β-lactoglobulin IgG antibodies in cord serum was higher compared with the concentration in the maternal serum. The same relationship was found if the levels of total IgG were compared. This implies that IgG antibodies are actively transported from mother to fetus across a concentration gradient. Maternal and cord serum antibody levels were found to be significantly correlated.

The significance of this antibody transfer is not clearly understood. It affords a degree of passive immunity for the first 3 months of the infant's life. Gill *et al*. (1983) demonstrated that immunization of the mother at 5–9 months of pregnancy with tetanus toxoid produced

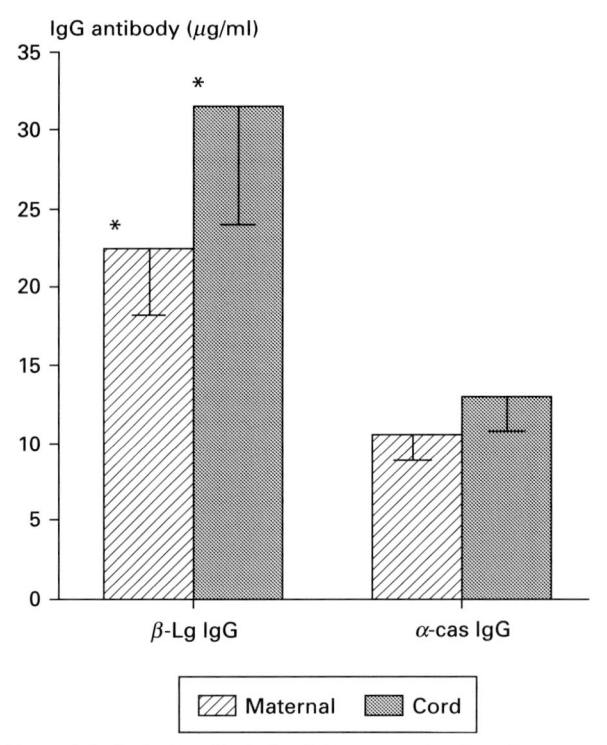

Fig. 1. Cows' milk β-lactoglobulin IgG antibody levels in maternal and cord serum samples. Cord serum β-lactoglobulin IgG levels were significantly higher than maternal serum β-lactoglobulin IgG levels ($P < 0.05$). Modified from Lovegrove (1991). β-Lg, β-lactoglobulin; IgG, immunoglobulin G; α-cas, α-casein.

active immunization of the fetus. The importance of maternally derived IgG antibodies as a protection against neonatal infection is supported by the finding that there is a correlation between low levels of type-specific serum antibodies and the risk of sepsis with type III group B *Streptococcus* (Baker & Kasper, 1976; Baker *et al.* 1981; Oxelius *et al.* 1983).

The significance of the transfer of IgG antibodies *via* the placenta and the subsequent development of allergy is less clear. Casimir *et al.* (1985) reported that a high level of IgG antibodies to β-lactoglobulin in cord blood protected against the development of cows' milk protein allergy in the infant. Another study, however, showed an association between increased cord blood total IgG antibody levels and later development of atopic disease (Iikura *et al.* 1989). In contrast, Høst *et al.* (1992) found no association between cord blood total IgG and IgG subclass antibody levels against β-lactoglobulin and bovine whey and the risk of cows' milk protein allergy development in the infant. The discrepancies in these studies may be explained by different definitions of the term allergy, diagnostic criteria and the specificity of the IgG quantified. Therefore the association of maternally derived IgG antibodies in relation to the development of allergies is far from certain.

Maternal antibodies, in addition to providing passive immunity for a limited period, may be able to enhance infant antibody production when present in small amounts (Levi *et al.* 1969; Dawe *et al.* 1971), thus playing a role in advancing immune competence in the infant. In addition to maternally derived IgG antibodies, the passage of intact maternal lymphocytes *via* the placenta to the fetus has also been implicated (Mohr, 1972). Although lymphocytes traverse the placenta in small numbers and are viable for a short time only, they might play a role in passing information to the fetal cells or recruiting fetal cells during an immune reaction.

PLACENTAL ANTIGEN TRANSFER

During gestation the developing fetal immune system has the protection of the maternal environment. The placenta provides a physical barrier, selectively preventing the passage of potentially damaging agents from reaching the fetus. However, clinical evidence is accumulating which suggests that antigens may pass across the placenta in a form which could elicit an immune response in the developing neonate. This was initially suggested when reports of shared neonatal and maternal infections were observed at birth (Gill, 1973).

Pathogenic antigens are not the only ones which are involved in shared sensitization in the mother and infant. Antigens which are non-pathogenic are also involved (Leiken & Oppenheim, 1971). Common allergens such as grass pollen, ragweed, moulds and dust have been implicated in shared sensitization (Kaufman, 1971; Hashem, 1972). Indirect evidence for placental antigen transfer was presented by Ratner (1922) and by Van Asperen *et al.* (1983) implicating intra-uterine sensitization in infants in whom the first contact with a specific food led to allergic symptoms.

Evidence for the placental transfer of antigens including foods (egg and bovine proteins) in rabbits was documented as early as 1902 by Ascoli. Stronger evidence for this placental antigen transfer was provided when maternally derived food proteins, β-lactoglobulin, gliadin and hens' egg ovalbumin were quantified in cord blood samples of atopic and non-atopic mothers (Lovegrove *et al.* 1991; Morris, 1991). Transfer of potentially harmful antigens across the placenta seems somewhat surprising as the placenta is believed to be a protective barrier for the developing fetus. However, the transfer of food derived antigens across other physiological barriers, for example the intact gut mucosa, has been well documented in healthy adults (Paganelli & Levinsky, 1980; Husby *et al.* 1986; Lovegrove *et al.* 1993a), in individuals suffering from eczema (Jackson *et al.* 1981) and in individuals with cows' milk protein allergy (Jackson *et al.* 1981; Heyman *et al.* 1988; Husby *et al.* 1990; Schrander *et al.* 1990).

The significance of the transfer of antigens from mother to fetus appears in some cases to be associated with the development of tolerance in the fetus (Firer *et al.* 1987). Tolerance is defined as the inability of an individual to produce a specific antibody in response to an ingested antigen and is the most important protective, homeostatic function of the gut associated lymphoid tissues. Its induction is probably the most important factor preventing development of food allergy in animals and humans (Ferguson & Watret, 1988). However, the transfer of antigens may in some cases be associated with unwanted sensitization, which could have important clinical implications regarding possible fetal sensitization and subsequent allergy development (discussed below).

HUMAN MILK ANTIBODY TRANSFER

Discussion of the physiology of gut maturation with respect to permeability to macronutrients, e.g. proteins, is beyond the scope of this review. The timing of the reduction of permeability to proteins has been determined with some degree of certainty in experimental animals (e.g. mice and rats) and also domestic animals (e.g. pig) but not as yet in humans (Mehrishi, 1976). Until the reduction in permeability to proteins occurs, the

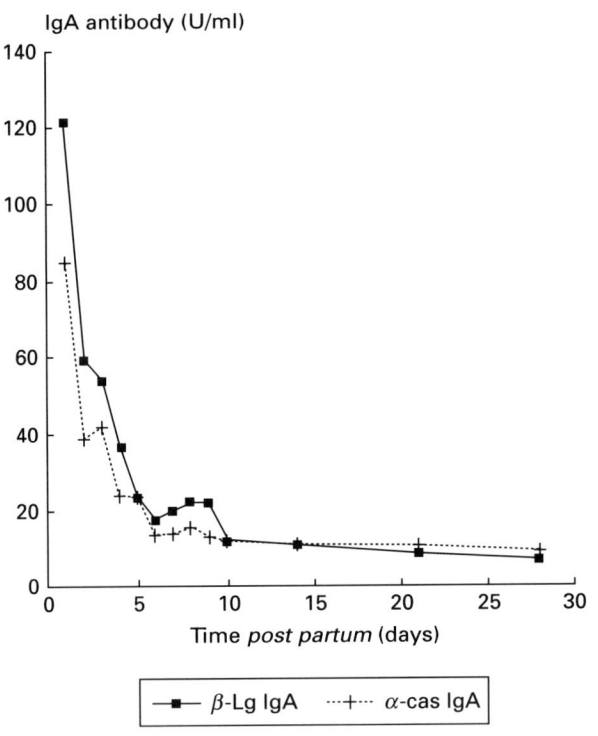

Fig. 2. Cows' milk β-lactoglobulin and α-casein specific IgA breast milk antibody levels for a 28 day period postnatally. Significant reduction in β-lactoglobulin and α-casein specific IgA levels from day 1 to 5 ($P < 0.001$). Modified from Lovegrove, 1991. β-Lg, β-lactoglobulin; IgA, immunoglobulin A; α-cas, α-casein.

infant is immunologically immature, and the transport of maternal immunoglobulins *via* breast milk makes a very important contribution to an infant's immunocompetence.

The unique properties of human milk have received much attention over the decades; it is believed to be the most appropriate food for a normal healthy infant. In addition to the provision of energy and nutrients human milk also contains components such as lactoferrin, lysozyme and antibodies, the levels of which depend on a number of factors, of which maternal nutritional status is the most important (Hennart *et al.* 1991). The predominant antibodies in breast milk are sIgA antibodies with lower levels of IgG and IgM (Hanson & Winberg, 1972; Ahlstedt *et al.* 1975; Jatsyk *et al.* 1985). The breast milk sIgA in the gut of the suckled infant appears to be involved with blocking adhesion of potential pathogens to mucosal epithelial cells and complexing with potential pathogens to facilitate their clearance (Slade & Schwartz, 1987). It has recently been reported that breast milk sIgA antibodies benefit the infant by activating infant monocytes *via* receptors (Padeh *et al.* 1991), in addition to stimulating gastrointestinal humoral immunogenic development (Koutras & Vigorita, 1989). These factors illustrate some of the benefits of breast milk over formula milk to the immunologically immature infant.

The transferred sIgA antibodies act against a wide variety of microorganisms and their products, as well as harmless substances such as food proteins (Machtinger & Moss, 1986; Ladjeva *et al.* 1989). Fig. 2 shows the change in the concentration of specific IgA antibodies against cows' milk β-lactoglobulin and α-casein in breast milk from twenty normal healthy women (Lovegrove, 1991). The initial mean value for β-lactoglobulin specific IgA

(121 U/ml) fell, so that by day 5 *post partum* the mean value was 24 U/ml and by day 28 the mean value was 7 U/ml. This represents a typical profile over time for milk specific sIgA antibodies in breast milk (Cruz & Arévalo, 1985). Earlier, work by Carlsson *et al.* (1976) reported a similar profile for sIgA against other antigens such as *Escherichia coli*. However Cruz & Arévalo (1985) did not confirm this; the profile they reported showed no regular pattern with time in the concentration of sIgA against *E. coli*.

An increased risk of developing an adverse reaction to foods in infants has been reported to be associated with a lack of sIgA antibodies specific to food antigens in maternal breast milk, although the mechanism of this action is not clear (Hanson *et al.* 1977; Machtinger & Moss, 1986; Renz *et al.* 1990). It has been postulated, however, that to quantify maternal breast milk sIgA antibody levels may act as an important screening technique for the identification of high risk infants. Infants born to women who present with low levels of breast milk sIgA may have a higher risk of allergy development than those infants born to women who present with average levels of breast milk sIgA (Machtinger & Moss, 1986; Lovegrove, 1991). This is an interesting concept that requires further research, and may prove to be an easy, non-invasive screening technique for infants potentially at risk for the development of adverse reactions to food or other allergy symptoms.

The levels of sIgA antibodies in mature breast milk are approximately 20 times those of IgG (Jatsyk *et al.* 1985). Recent research shows that the latter are principally responsible for surface phagocytosis in the infant, a function not associated with sIgA. This action of the IgG antibodies illustrates their invaluable benefit to the infant (Avery & Gordon, 1991).

In addition to preventing the development of allergy, antibodies transferred from the mother to the infant are believed to contribute significantly to the prevention of infantile infection (Carlsson *et al.* 1976; Hanson *et al.* 1977; Cruz & Arévalo, 1985; Machtinger & Moss, 1986).

Human colostrum and mature milk also contain a variety of cells. Most of these cells are neutrophilic granulocytes and macrophages. Their numbers decrease sharply after the fourth day and more gradually thereafter over several months (Ogra & Ogra, 1978). These cells include short lived and long lived cells, antibody forming cells, primed (memory) cells, and unprimed cells. It seems likely that some of these cells exert a protective role in the gastrointestinal tract of the newborn infant, and that others penetrate the circulation of the recipient newborn infant.

HUMAN MILK ANTIGEN TRANSFER

The presence of antigens in breast milk, both in individuals who have a genetic predisposition to allergy development and those who do not, is undisputed. More than 75 years ago Talbot (1918) reported that eczema in a 3-week infant was due to chocolate eaten by the mother who is breast feeding. Jakobsson & Lindberg (1978) showed that infantile colic could often be relieved by removing cows' milk from the mother's diet. Lifschitz *et al.* (1988) showed that removal of cows' milk from the mother's diet resolved all allergic symptoms previously encountered by her infant. More recently the direct quantification of food antigens in breast milk has been performed by a number of groups. The presence of antigens in breast milk has been reported thus: ovalbumin (Kilshaw & Cant, 1984; Morris, 1991); β-lactoglobulin (Kilshaw & Cant, 1984; Jakobsson *et al.* 1985; Axelsson *et al.* 1986; Machtinger & Moss, 1986; Cavagni *et al.* 1988; Høst *et al.* 1988; Lovegrove, 1991); and gliadin (Troncone *et al.* 1987; Morris, 1991). Using sensitive immunoassays, levels of antigens have been reported to be in the range 0·2–800 ng/ml. The levels vary greatly, and are influenced by the specific food protein assayed, the type of assay used, inherent individual variation and, to a much lesser extent, the dietary intake of the individual.

The total amount of antigen in milk varies according to the antigen measured. So, for example, Harmatz *et al.* (1986) and Telemo *et al.* (1986) reported, in rodent models, that significantly less bovine serum albumin and bovine γ globulin were present in the breast milk compared with ovalbumin and β-lactoglobulin. These differences may be explained by preferential clearance of antigens into maternal tissue other than mammary glands, or by the existence of specific transport mechanisms for certain proteins in the mammary gland (Harmatz *et al.* 1986; Telemo *et al.* 1986).

The presence of antigens, including food antigens, in breast milk is not restricted to atopic individuals, and is therefore a natural occurrence. A possible explanation for the existence of antigens in breast milk could be the development of oral tolerance in the suckled infant (see below). Exclusive breast feeding does not prevent some infants from allergy development, suggesting a possible disadvantage of this antigen transfer (Hattevig *et al.* 1989). It is probable that antigens transferred to the infant *via* breast milk may have the ability to sensitize the high risk infant, resulting in subsequent allergy development (Jakobsson & Lindberg, 1978; Cant *et al.* 1985; Machtinger & Moss, 1986; Lifschitz *et al.* 1988). It has been speculated that since the antibody titre to numerous dietary substances decreases with age and reaches a stable low level at about 40–50 years for most individuals, the eventual development of oral tolerance could be the result of continued antigen absorption (Rothberg & Farr, 1965; André *et al.* 1975).

DEVELOPMENT OF TOLERANCE IN THE NEONATE

Typically, oral (intestinal) exposure to foods or other foreign protein antigens either induces systemic immunological hyporesponsiveness (oral tolerance) or sensitizes the immune system, possibly by gut associated lymphoid tissue. The acquisition of tolerance to foods is a multifactorial process and is coregulated by genetic influences.

Cramer *et al.* (1974) suggested several mechanisms by which the maternal environment may influence the immunological development of the offspring, none of which, the authors claimed, need be mutually exclusive: first, the passage of maternal antibody *via* the placenta with alteration of fetal immune response by specific antibody; secondly, the passage of immunological information products, such as transfer factor, from mother *via* the placenta; thirdly, active migration of sensitized maternal lymphocytes to the fetus, with proliferation and participation in the immune response or recruitment of fetal lymphocytes for specific response; lastly, the transplacental passage of antigen to the fetus and direct sensitization of its lymphocytes. In addition to the above mechanisms of maternal influence, maternal anti-idiotype IgG antibody transfer has been implicated as being the most likely stimulus in the fetus for the production of antibodies to allergens before birth (Hanson *et al.* 1989; Hahn-Zoric *et al.* 1993).

For ethical reasons many studies investigating the immunological mechanisms of oral tolerance development have been performed in animals. Using rodent animal models, different groups have studied the role of the liver in modulating the immune response to dietary antigens (Thomas *et al.* 1976), have implicated suppressor T-cells in Peyer's patches (Ngan & Kind, 1978), and have investigated circulating complexes of dietary antigen and antibody (André *et al.* 1975). Pathirana *et al.* (1981) reported the importance of perinatal exposure to dietary proteins in the induction of tolerance to food antigens in rabbits. More recently investigators have attempted to evaluate, in humans, the role and integrity of the gastrointestinal mucosa in the host response after oral antigen exposure (Walker-Smith, 1992; Businco *et al.* 1992). To extrapolate evidence derived from animal experiments to man is questionable, but the placental transfer of maternally derived antigens and their role in immune tolerance induction seems a plausible theory. Strobel (1992) (Table 1) has

Table 1. *Factors which influence the induction and maintenance of systemic immunologic hyporesponsiveness (oral tolerance). Modified from Strobel (1992)*

- Genetic background
- Nature of antigen (soluble, particulate, replicating)
- Dose of antigen
- Frequency of administration
- Age (maturity v. immaturity) at first antigen exposure
- Immunologic status (virus infection, gastroenteropathy)
- Previous dietary exposure of the mother
- Antigen transmission *via* breast feeding

suggested factors which could be important in determining whether the antigen will induce stimulation or tolerance.

Although acquisition of tolerance is usual, it is known that some infants become sensitized to certain antigens. Miller *et al.* (1973) demonstrated for the first time the capability of the fetus to synthesize IgE as early as 11 weeks in lung and liver, and at 21 weeks in the spleen. Elevated cord serum total IgE antibody levels have been implicated as a risk factor for the development of allergy or intolerance to food or other substances (Høst *et al.* 1992). Cord serum total IgE levels are now routinely used as a screening tool in studies investigating 'at risk' infants (Arshad *et al.* 1992). Reagins (antibodies responsible for anaphylactic reactions), presumably IgE, are known not to cross the placenta (Kuhns, 1965), so IgE present in cord serum is probably derived from the fetal circulation as long as blood mixing has not occurred at birth. Synthesis of IgE in the fetus may be due to a number of factors none of which has been elucidated.

The mechanisms involved in the development of tolerance or sensitization in the neonate have been reviewed by Strobel (1992). Results from studies in rodents suggested that perinatal antigen exposure was more likely to prime the immune system rather than to induce tolerance, and that continuous feeding beyond the critical neonatal period may lead to induction of tolerance (Strobel, 1992).

Evidence seems to indicate that antigen exposure in the fetus *via* the placenta, or in the infant *via* breast milk or formula milk can sensitize those who have a predisposition to the development of allergy. It therefore seems prudent to protect the high risk infant from exposure to foreign antigens by the elimination of antigen exposure.

HUMAN MILK AS A PROTECTION AGAINST ALLERGY

There is still controversy concerning the protective effect of breast milk as opposed to cows' milk formula, against the development of atopic diseases. Since the hallmark study of Grulee & Sanford (1936), which demonstrated that breast feeding compared to cows' milk formula ingestion reduced the development of eczema 7-fold in a group of 20000 infants, many dietary prevention studies have attempted to confirm their findings but they have produced conflicting conclusions.

Two extensive publications reviewed the role of breast feeding in protecting the infant from allergy development (Zeiger, 1987; Kramer, 1988). Zeiger discussed the large number of confounding variables and exposed weaknesses in the numerous studies reviewed. These included: timing and control of solid food introduction; lack of blind assessments of allergy development; failure to document compliance; lack of immunological documentation; differential environmental control measures in groups; high dropout rates; small sample size; brief duration of breast feeding. Due to these confounding variables, comparison of the studies was, in many cases, inappropriate. However Zeiger concluded that breast feeding seemed to have a protective effect, compared with formula feeding, against infant allergy development.

Kramer (1988) evaluated 36 original studies published between 1936 and 1986 and sought to find whether those studies which confirmed a protective effect of breast feeding in the development of allergy were better conducted than those which reported no protective effect. Biological and methodological aspects of cows' milk exposure, outcome (allergic conditions) and statistical analysis were assessed. The author concluded that most studies had weaknesses in various areas, preventing direct comparison one with another, and went on to state that "the inconsistent findings, even among the better studies, prevents any firm inferences, although it seems likely that the results from the larger high-quality studies could be compatible with a large protective effect of breast feeding".

Feeding an infant human milk does seem to have some protective effect on infant allergy development, although even exclusive breast feeding does not eliminate the risk of allergy development (Jakobsson & Lindberg, 1978; Chandra et al. 1986). A possible explanation for the lack of conclusive evidence for the benefit of breast milk, compared with infant formula feeding, in the risk of allergic disease development is that in non-randomized trials atopic mothers are more inclined to breast feed their infants than are non-atopic mothers. Additionally, the possible influence of maternal dietary protein transfer to the infant *via* the placenta or breast milk has not, in the majority of studies, been considered.

So, although it appears that breast feeding may reduce the incidence of allergy during early childhood (Chandra et al. 1989; Morris, 1991), this protection seems to be lost at about 9 years of age (Pöysä et al. 1989, 1991). This theory was supported by Åberg et al. (1989), who reported a significantly lower cumulative incidence of allergic disease during the first three years of life in breast fed children with a double-parental history of allergy, compared with children having a history of allergy in one or in neither parent. However, at the ages of 7, 10 and 14 years, diet in infancy had no influence on the incidence of atopy, irrespective of parental status. These studies suggest that the effect of breast feeding is to delay the onset of allergic disease, rather than to decrease the overall risk of becoming allergic during childhood, but more strictly controlled studies are necessary to confirm this conclusion.

MATERNAL ELIMINATION DIETS

Before introducing the subject of maternal elimination diets it seems appropriate to highlight the problem of how to define the allergic condition from which an individual is suffering, in these types of studies. In most publications mothers are described as being 'atopic', although there is no evidence presented that the mothers have inherited the allergic condition. Strictly speaking the words 'atopy' and 'atopic' should be used only to refer to an inherited condition of adverse reactions involving the immune system (Savin, 1993).

There are reports of studies which have observed sensitization of infants by transferred food proteins (Van Asperen et al. 1983; Cant et al. 1985; Lifschitz et al. 1988). It would therefore seem that elimination of the suspect protein or proteins from the mother's diet during pregnancy, lactation, or both could reduce the risk of the infant developing an adverse reaction to foods.

The value of maternal dietary elimination in reducing or preventing the risk of allergy development in the infant during pregnancy and/or lactation has been studied. Owing to the restrictive nature of any dietary elimination regimen, the timing of dietary intervention must be carefully defined for maximum benefit to the infant and minimum inconvenience to the mother. In 1986, Chandra and colleagues performed a study on 121 women who already had an offspring with an allergic condition. These women were randomly allocated into an antigen avoidance group or a control group. Milk, dairy products, egg, fish, beef and peanuts were eliminated from the diet throughout pregnancy and lactation in the food avoidance group. Infants born to mothers in this group had a reduced occurrence, and milder form, of atopic eczema compared with those infants in the control group. The benefit of a reduced exposure of infants to allergenic proteins during gestation and lactation was also observed by Zeiger et al. (1989), who reported a more ambitious programme. Mothers with a history of allergy followed a diet excluding milk, dairy products, eggs and peanuts during the last trimester of pregnancy and lactation. A casein hydrolysate formula was given to the infant at weaning, with the avoidance of solid food until the infant was 6 months of age. Despite this protocol, the accumulated incidence of atopic disease up to 36 months of age in the offspring treated prophylactically was: food allergy development, 14%; eczema, 12%; infectious asthma, 12%; allergic rhinitis, 7%. The authors concluded that the two most obvious factors responsible for the above findings were a genetic predisposition to allergy development and parental non-compliance.

Lovegrove (1991) and Lovegrove et al. (1993b, 1994) studied a group of 38 mothers from 36 weeks gestation to 18 months *post partum*. They investigated the effect of a maternal milk free and milk product free diet (cows', sheep and goat) during late pregnancy and lactation on the incidence of eczema in the offspring. Women who presented with clinically diagnosed allergic conditions, defined for the purposes of this review as atopic, were randomly allocated to a prophylactic group, on a milk and milk product free diet (with calcium supplementation and a whey hydrolysate formula consumption advised, $n = 12$), or an unrestricted diet group ($n = 14$). These women were compared with a third group of non-atopic women on an unrestricted diet ($n = 12$). At 18 months of age, the incidence of eczema in the offspring of the atopic women on the unrestricted diet was significantly higher than in those of the atopic group prophylactically treated with the restricted diet ($P < 0.005$); the eczema incidence in the infants in the prophylactic group was similar to that of the non-atopic women on an unrestricted diet. No significant differences were observed for β-lactoglobulin or α-casein IgG levels in cord serum or infant serum between the three groups of infants. However, there was a trend for infants born to atopic women on the unrestricted diet to have elevated serum cows' milk specific IgG antibody levels compared with serum levels from infants born to non-atopic women or to atopic women on a restricted diet.

Contrary to the findings of the studies described above, another group has reported that maternal dietary restriction during pregnancy and lactation had a limited effect on the atopic outcome of the infants (Lilja et al. 1989, 1991). This group investigated the immunological consequences and atopic outcome of a range of maternal dietary regimens on 163 infants born to atopic women. The diets ranged from high quantities of cows' milk (1 litre per day) and egg (1 per day) to low quantities of cows' milk and egg (although the diet was not completely free from these foods) during late pregnancy and lactation. This 'high' and 'low' antigen intake was in contrast to a total exclusion diet reported in most other studies. Results of the study showed that maternal intake of cows' milk and eggs did not affect total IgE antibody levels in the infants up to 18 months of age (Lilja et al. 1991) and dietary manipulation did not affect the development of atopic symptoms in the infants at 18 months *post partum* (Lilja et al. 1989). It is possible that the less than rigorous nature of the dietary regimen contributed to these inconclusive results.

The studies described above have observed varying degrees of benefit to the high risk infant in terms of the development of adverse reactions to foods resulting from maternal dietary avoidance during pregnancy and lactation. Although the quantity of milk and eggs had limited effect on the development of allergy, in order to evaluate the importance of intervention in either pregnancy or lactation, studies have been conducted to investigate each period in isolation.

MATERNAL ELIMINATION DIET DURING PREGNANCY

Fälth-Magnusson & Kjellman (1987) and Fälth-Magnusson *et al.* (1987, 1988) conducted a study on 212 atopic women in Sweden. The women were randomly allocated to a diet group taking no cows' milk or eggs from week 28 until delivery, with extra calcium and casein hydrolysate formula consumption advised, or a control group following an unaltered diet. No significant differences were observed in specific cord blood IgE levels against cows' milk and eggs in either of the groups, although a significant decrease in maternal serum total IgG levels was observed in the group that complied with the restricted diet. The mothers' ingestion of cows' milk and egg during lactation did affect the immunological response of the infant to these antigens as the IgG levels were lower in the diet restricted infants compared with the control group. In addition the authors reported that the specific IgG levels to gliadin were lower in the diet restricted group than in the control group even though this food protein had not been restricted in the mothers' diets. However no protection was subsequently observed against the development of clinical signs and symptoms associated with allergy in infants up to 18 months of age.

Lilja *et al.* (1988), in an early publication of their longitudinal study, described above, investigated the effect of the consumption of various quantities of egg and milk protein in a maternal diet during the last trimester of pregnancy on the outcome of allergy in their offspring. Women ($n = 165$) with respiratory allergies and their infants were observed. Although the maternal serum total IgG antibody levels were reduced, due to the elimination of these foods from the maternal diet, the cord blood total IgE and total IgG antibody levels did not change significantly. They also observed no significant differences in the distribution of atopic disease among the infants in relation to maternal diet during pregnancy.

MATERNAL ELIMINATION DIETS DURING LACTATION

According to Jakobsson & Lindberg (1978), the removal of cows' milk from the mother's diet while she is breast feeding will 'cure' colic in her infant. Cant *et al.* (1986) reported that maternal dietery exclusion of cows' milk and eggs benefited some breast fed babies with eczema.

In an ambitious study undertaken by Chandra *et al.* (1989), mothers who were breast feeding their infants ($n = 97$) were randomly allocated to a restricted diet group avoiding milk, dairy products, eggs, fish, peanuts and soya ($n = 48$), or an unrestricted diet group ($n = 49$). Mothers who bottle fed their infants ($n = 124$) were directed to use cows' milk formula or soya formula or a casein hydrolysate. The incidence of signs and symptoms associated with an allergic condition was greatly affected by the infant formula used, with the incidence of eczema at 70%, 63% and 21% to cows' milk formula, soya formula and casein hydrolysate respectively. The authors also observed that eczema was less common and milder in babies who were breast fed and whose mothers were following the elimination diet.

Hattevig and colleagues (1989, 1990) investigated the effect of a strict maternal elimination of milk, eggs, and fish during the first 3 months of lactation on 155 babies. The cumulative prevalence of allergic manifestations up to 6 months of age was higher (28%) in the control babies compared to the elimination diet babies (11%; $P < 0.05$). The elimination diet also improved the relationship between mild and severe atopic dermatitis. This difference was only transitory, for after 6 months the difference between the two groups had disappeared.

In a prenatal, randomized, control study of 120 infants with a family history of allergy and high cord blood IgE levels the prophylactic group avoided the consumption of milk, egg, fish and nuts during lactation (Arshad et al. 1992). As the presence of house dust mite in an infant's room is known to influence infant allergy development (Sporik et al. 1990), Arshad and colleagues also instructed the mothers in the prophylactic group to reduce the levels of house dust mite by vigilant house cleaning. The result of this combined regimen was to reduce the incidence of allergy by 50% within the first 12 months of the infant's life although long term follow-up was not reported.

Thus it appears that maternal dietary elimination during lactation, rather than in pregnancy, may afford some protection to the high risk infant. In the majority of studies in which elimination diets are used by mothers nursing high risk infants a significant decrease in the incidence and severity of atopic eczema and other allergic manifestations that occur during the first year of life are reported. Whether these diets will afford long term protection from the development of atopic disease in high risk individuals is unclear and requires further long term studies.

MATERNAL ELIMINATION DIETS AND NUTRITIONAL STATUS

Prophylactic elimination diets prescribed for mothers in pregnancy or lactation are not usually as rigorously enforced as those prescribed for adults with a suspected food allergy. Although no one specific food is required in a diet, individual foods (e.g. cows' milk) are useful sources of essential nutrients (e.g. calcium and riboflavin). Care must be taken to ensure that a nutritionally complete diet is consumed when any type of elimination diet is prescribed. Medical and/or dietetic supervision is essential.

In the dietary intervention study conducted by Lovegrove and colleagues (1994) in which atopic mothers excluded milk and milk products in the latter stages of pregnancy and during lactation, their energy and nutrient intakes were assessed by a 7-day weighed food intake and compared with the intakes of mothers following an unrestricted diet. The mean energy and nutrient intakes in both groups were similar with the exception of the lower mean daily intake of calcium in the diet-restricted group ($P < 0.001$, Table 2). Indeed, none of the subjects in the diet-restricted group received, from a dietary source, a mean daily intake of calcium that reached the reference nutrient intake of calcium (700 mg/d; Department of Health, 1991). However, all women in this group were supplied with a calcium supplement (1 g/d derived from $CaCO_3$). One factor responsible for the high polyunsaturated to saturated (p:s) fat ratio in the diet restricted group was the omission of milk and milk products from their diet.

It is therefore important to monitor energy and nutrient intakes during any dietary restriction, and to give supplements where appropriate. Supplementation with calcium is usually carried out in studies where milk and milk products are eliminated from the diet (Fälth-Magnusson et al. 1987; Lilja et al. 1991; Lovegrove et al. 1994).

Table 2. *Daily energy and nutrient intake of atopic women in restricted diet group (diet) and women in unrestricted diet group (control) at approximately 36 weeks gestation calculated from seven day weighed food inventory. (Modified from Lovegrove, 1991)*

Nutrient	Control group ($n = 12$)		Diet group ($n = 14$)	
Energy (MJ)	10·0 (2·1)	[7·0–14·3]	8·9 (3·2)	[3·8–15·7]
Protein (g)	84 (19)	[50–122]	77 (29)	[34–151]
Carbohydrate (g)	300 (98)	[201–579]	285 (105)	[122–459]
Fat (g)	103 (22)	[67–151]	86 (40)	[33–172]
P:S ratio	0·20 (0·10)*	[0·12–0·40]	0·50 (0·20)*	[0·19–1·00]
Calcium (mg)	1261 (411)*	[860–2297]	434 (135)*	[236–667]

Values are given as mean (SD) [range].
P:S, polyunsaturated:saturated fat.
* Significant difference between diet and control groups ($P < 0.001$).

CONCLUSION

Parents who are known to be atopic should be made aware that their offspring may develop allergic symptoms. They may wish to minimize this risk by adopting certain strategies to decrease the chance of sensitizing the infant. One strategy that has been studied, and which appears to be successful, is for the breastfeeding mother to avoid foods which contain allergenic proteins. The benefit to the high risk infant of maternal dietary intervention during pregnancy is less evident. Some workers advise the avoidance of allergenic foods in the last two or three months of pregnancy to decrease the chances of sensitizing the infant *in utero*. There are other factors which also influence the development of conditions associated with adverse reaction to food and other environmental substances in infancy. These include the timing and types of weaning foods introduced into the infant's diet, inhaled airborne allergens and maternal smoking habits. As no single mechanism is sufficient to explain the many forms of reactivity which are present in this condition, it would appear that a simple means of preventing the development of allergy in infancy does not at present exist.

The authors are grateful for the very helpful comments of Professor J. W. T. Dickerson, Dr S. M. Hampton, Dr M. Murphy and Mrs S. Smith.

We gratefully acknowledge support for the studies undertaken by the authors, reported in this review, from Cow and Gate Nutricia, Ltd, Trowbridge, Wilts.

REFERENCES

Aas, K. (1989). Chemistry of food allergens. In *Food Intolerance in Infancy* (Carnation Nutrition Education Series 1), pp. 9–21 [R. N. Hamburger, editor]. New York: Raven Press.
Åberg, N., Engström, I. & Lindberg, U. (1989). Allergic diseases in Swedish school children. *Acta Paediatrica Scandinavica* **78**, 246–252.
Ahlstedt, S., Carlsson, B., Hanson, L. Å. & Goldblum, R. M. (1975). Antibody production by human colostral cells. 1. Immunoglobulin class, specificity and quantity. *Scandinavian Journal of Immunology* **4**, 535–539.
André, C., Heremans, J. F., Vaerman, J. P. & Cambiaso, C. L. (1975). A mechanism for the induction of immunological tolerance by antigen feeding: antigen-antibody complexes. *Journal of Experimental Medicine* **142**, 1509–1519.
Arshad, S. H., Matthews, S., Gant, C. & Hide, D. W. (1992). Effect of allergen avoidance on development of allergic disorders in infancy. *Lancet* **339**, 1493–1497.
Ascoli, A. (1902). [Do proteins pass the placental barrier?] *Zeitschrift für Physiologische Chemie* **36**, 498.
Avery, V. M. & Gordon, D. L. (1991). Antibacterial properties of breast milk: requirements for surface phagocytosis and chemiluminescence. *European Journal of Clinical Microbiology and Infectious Diseases* **10**, 1034–1039.

Axelsson, I., Jakobsson, I., Lindberg, T. & Benediktsson, B. (1986). Bovine β-lactoglobulin in the human milk: a longitudinal study during the whole lactation period. *Acta Paediatrica Scandinavica* **75**, 702–707.

Baker, C. J., Edwards, M. S. & Kasper, D. L. (1981). Role of antibody to native type III polysaccharide of group B streptococcus in infant infection. *Pediatrics* **68**, 544–549.

Baker, C. J. & Kasper, D. L. (1976). Correlation between maternal antibody deficiency with susceptibility to neonatal group B streptococcal infection. *New England Journal of Medicine* **294**, 753–756.

Bleumink, E. (1983). Immunological aspects of food allergy. *Proceedings of the Nutrition Society* **42**, 219–231.

Brostoff, J. & Gamlin, L. (1990). Another man's poison. *Food Allergy and Intolerance*, pp. 1–21. London: Bloomsbury Publishing Ltd.

Businco, L., Bruno, G., Giampietro, P. G. & Cantani, A. (1992). Allergenicity and nutritional adequacy of soy protein formulas. *Journal of Pediatrics* **121** (5), S21–S28.

Cant, A. J., Bailes, J. A. & Marsden, R. A. (1985). Cow's milk, soy milk, and goat's milk in a mother's diet causing eczema and diarrhoea in her breast fed infant. *Acta Paediatrica Scandinavica* **74**, 467–468.

Cant, A. J., Bailes, J. A., Marsden, R. A. & Hewitt, D. (1986). Effect of maternal dietary exclusion on breast fed infants with eczema: two controlled studies. *British Medical Journal* **293**, 231–233.

Carlsson, B., Gothefors, L., Ahlstedt, S., Hanson, L. Å. & Winberg, J. (1976). Studies of *Escherichia coli* antigen specific antibodies in human milk, maternal serum and cord blood. *Acta Paediatrica Scandinavica* **65**, 216–224.

Casimir, G., Gossart, B., Vis, H. L. & Duchateau, J. (1985). Antibody against betalactoglobulin (IgG) and cow's milk allergy. *Journal of Allergy and Clinical Immunology* **75**, 206.

Cavagni, G., Paganelli, R., Caffarelli, C., D'Offizi, G. P., Bertolini, P., Aiuti, F. & Giovannelli, G. (1988). Passage of food antigens into circulation of breast-fed infants with atopic dermatitis. *Annals of Allergy* **61**, 361–365.

Chandra, R. K. (1976). Levels of IgG subclasses, IgA, IgM, and tetanus antitoxin in paired maternal and fetal sera: findings in healthy pregnancy and placental insufficiency. In *Maternofoetal Transmission of Immunoglobulins*, pp. 77–90 [W. A. Hemmings, editor]. Cambridge: Cambridge University Press.

Chandra, R. K. & Prasad, C. (1991). Food allergy: diagnosis and strategies for prevention. *Beiträge zur Infusionstherapie* **27**, 142–151.

Chandra, R. K., Puri, S. & Hamed, A. (1989). Influence of maternal diet during lactation and use of formula feeds on development of atopic eczema in high risk infants. *British Medical Journal* **299**, 228–230.

Chandra, R. K., Puri, S., Suraiya, C. & Cheema, P. S. (1986). Influence of maternal food antigen avoidance during pregnancy and lactation on incidence of atopic eczema infants. *Clinical Allergy* **16**, 563–569.

Cramer, D. V., Kunz, H. W. & Gill, T. J. (1974). Immunologic sensitization prior to birth. *American Journal of Obstetrics and Gynecology* **120**, 431–439.

Croner, S. (1992). Prediction and detection of allergy development: influence of genetic and environmental factors. *Journal of Pediatrics* **121** (5), S58–S63.

Cruz, J. R. & Arévalo, C. (1985). Fluctuations of specific IgA antibodies in human milk. *Acta Paediatrica Scandinavica* **74**, 897–903.

Dawe, D. L., Myers, W. L. & Segre, D. (1971). Enhancement of the immune response in rabbits by administration of IgG. *Immunology* **18**, 897–907.

Department of Health (1991). Dietary reference values for food energy and nutrients for the United Kingdom. *Report on Health and Social Subjects* no. 41. London: HMSO.

Fälth-Magnusson, K. & Kjellman, N.-I. M. (1987). Development of atopic disease in babies whose mothers were receiving exclusion diet during pregnancy – a randomized study. *Journal of Allergy and Clinical Immunology* **80**, 868–875.

Fälth-Magnusson, K., Kjellman, N.-I. M. & Magnusson, K.-E. (1988). Antibodies IgG, IgA, and IgM to food antigens during the first 18 months of life in relation to feeding and development of atopic disease. *Journal of Allergy and Clinical Immunology* **81**, 743–749.

Fälth-Magnusson, K., Öman, H. & Kjellman, N.-I. M. (1987). Maternal abstention from cow milk and egg in allergy risk pregnancies. Effect on antibody production in the mother and the newborn. *Allergy* **42**, 64–73.

Ferguson, A. (1992). Definitions and diagnosis of food intolerance and food allergy: consensus and controversy. *Journal of Pediatrics* **121** (5), S7–S11.

Ferguson, A. & Watret, K. C. (1988). Cows' milk intolerance. *Nutrition Research Reviews* **1**, 1–23.

Firer, M. A., Hoskings, C. S. & Hill, D. J. (1987). Humoral immune response to cow's milk in children with cow's milk allergy. Relationship to the time of clinical response to cow's milk challenge. *International Archives of Allergy and Applied Immunology* **84**, 173–177.

Gill, T. J. (1973). Maternal/fetal interactions and the immune response. *Lancet* **i**, 133–135.

Gill, T. J., Repetti, C. F., Metlay, L. A., Rabin, B. S., Taylor, F. H., Thompson, D. S. & Cortese, A. L. (1983). Transplacental immunization of the human fetus to tetanus by immunization of the mother. *Journal of Clinical Investigation* **72**, 987–996.

Gitlin, D., Kumate, J., Urrusti, J. & Morales, C. (1964). The selectivity of the human placenta in the transfer of plasma proteins from mother to fetus. *Journal of Clinical Investigation* **43**, 1938–1951.

Grulee, C. G. & Sanford, H. N. (1936). The influence of breast and artificial feeding on infantile eczema. *Journal of Pediatrics* **9**, 223–225.

Hahn-Zoric, M., Carlsson, B., Jeansson, S., Ekre, H. P., Osterhaus, A. D. M. E., Robertson, D. & Hanson, L. Å. (1993). Anti-idiotypic antibodies to poliovirus antibodies in commercial immunoglobulin preparations, human serum and milk. *Pediatric Research* **33**, 475–480.

Hanson, L. Å., Ahlstedt, S., Carlsson, B. & Fällström, S. P. (1977). Secretory IgA antibodies against cow's milk proteins in human milk and their possible effect in mixed feeding. *International Archives of Allergy and Applied Immunology* **54**, 457–462.

Hanson, L. Å., Carlsson, B., Ekre, H. P., Hahn-Zoric, M., Osterhaus, A. D. M. E. & Robertson, D. (1989). Immunoregulation mother-fetus/newborn, a role for anti-idiotypic antibodies. *Acta Paediatrica Scandinavica* Suppl. 351, 38–41.

Hanson, L. Å. & Winberg, J. (1972). Breast milk and defence against infection in the newborn. *Archives of Disease in Childhood* **47**, 845–848.

Harlow, E. D. & Lane, D. (1988). Antibodies. In *A Laboratory Manual*, pp. 1–139 [E. D. Harlow and D. Lane, editors]. Dorking, UK: Cold Spring Harbor Laboratory.

Harmatz, P. R., Hanson, D. G., Walsh, M. K., Kleinman, R. E., Bloch, K. J. & Walker, W. A. (1986). Transfer of protein antigens into milk after intravenous injection into lactating mice. *American Journal of Physiology* **251**, E227–233.

Hashem, N. (1972). Is maternal lymphocyte sensitisation passed to the child? *Lancet* **i**, 40–41.

Hattevig, G., Kjellman, B., Sigurs, N., Bjorkstén, B. & Kjellman, N.-I. M. (1989). Effect of maternal avoidance of eggs, cow's milk and fish during lactation upon allergic manifestations in infants. *Clinical and Expeirmental Allergy* **19**, 27–32.

Hattevig, G., Kjellman, B., Sigurs, N., Grodzinsky, E., Hed, J. & Bjorkstén, B. (1990). The effect of maternal avoidance of eggs, cow's milk, and fish during lactation on the development of IgE, IgG, and IgA antibodies in infants. *Journal of Allergy and Clinical Immunology* **85**, 108–115.

Hennart, P. F., Brasseur, D. J., Delogne-Desnoeck, J. B., Dramaix, M. M. & Robyn, C. E. (1991). Lysozyme, lactoferrin, and secretory immunoglobulin A content in breast milk: influence of duration of lactation, nutritional status, prolactin status, and parity of mother. *American Journal of Clinical Nutrition* **53**, 32–39.

Heyman, M., Grasset, E., Ducroc, R. & Desjeux, J.-F. (1988). Antigen absorption by the jejunal epithelium of children with cow's milk allergy. *Pediatric Research* **24**, 197–202.

Hobbs, J. R. & Davis, J. A. (1967). Serum γG-globulin levels and gestational age in premature babies. *Lancet* **i**, 757–759.

Høst, A., Husby, S., Gjesing, B., Larsen, J. N. & Lowenstein, H. (1992). Prospective estimation of IgA, IgG subclasses and IgE antibodies to dietary proteins in infants with cow milk allergy. Levels of antibodies to whole milk protein, BLG and ovalbumin in relation to repeated milk challenge and clinical course of cow milk allergy. *Allergy* **47**, 218–229.

Høst, A., Husby, S. & Østerballe, O. (1988). A prospective study of cow's milk allergy in exclusively breast-fed infants. Incidence, pathogenetic role of early inadvertent exposure to cow's milk formula and characterization of bovine milk protein in human milk. *Acta Paediatrica Scandinavica* **77**, 663–670.

Hurley, L. S. (1980). The placenta and placental transfer. *Developmental Nutrition* pp. 49–64. Englewood Cliffs, NJ: Prentice-Hall Inc.

Husby, S., Høst, A., Teisner, B. & Svehag, S.-E. (1990). Infants and children with cow milk allergy/intolerance. Investigation of the uptake of cow milk protein and activation of the complement system. *Allergy* **45**, 547–551.

Husby, S., Jensenius, J. C. & Svehag, S.-E. (1986). Passage of undegraded dietary antigen into the blood of healthy adults. Further characterisation of the kinetics of uptake and the size distribution of the antigen. *Scandinavian Journal of Immunology* **24**, 447–455.

Iikura, Y., Akimoto, K., Odajima, Y., Akazawa, A. & Nagakura, T. (1989). How to prevent allergic disease. 1. Study of specific IgE, IgG, and IgG4 antibodies in serum of pregnant mothers, cord blood, and infants. *International Archives of Allergy and Applied Immunology* **88**, 250–252.

Jackson, P. G., Lessof, M. H., Baker, R. W. R., Ferrett, J. & MacDonald, D. M. (1981). Intestinal permeability in patients with eczema and food allergy. *Lancet* **i**, 1285–1286.

Jakobsson, I. & Lindberg, T. (1978). Cow's milk as a cause of infantile colic in breast-fed infants. *Lancet* **ii**, 437–440.

Jakobsson, I., Lindberg, T., Benediktsson, B. & Hansson, B.-G. (1985). Dietary bovine β-lactoglobulin is transferred to human milk. *Acta Paediatrica Scandinavica* **74**, 342–345.

Jatsyk, G. V., Kuvaeva, I. B. & Gribakin, S. G. (1985). Immunological protection of the neonatal gastrointestinal tract: the importance of breast feeding. *Acta Paediatrica Scandinavica* **74**, 246–249.

Kaufman, H. (1971). Allergy in the new-born: skin test reactions confirmed by the Prausnitz-Kustner test at birth. *Clinical Allergy* **1**, 363–367.

Kilshaw, P. J. & Cant, A. J. (1984). The passage of maternal dietary proteins into human breast milk. *International Archives of Allergy and Applied Immunology* **75**, 8–15.

Kohler, P. F. & Farr, R. S. (1966). Elevation of cord over maternal IgG immunoglobulin: evidence for an active placental IgG transport. *Nature* **210**, 1070–1071.

Koutras, A. K. & Vigorita, V. J. (1989). Fecal secretory immunoglobulin A in breast milk versus formula feeding in early infancy. *Journal of Pediatric Gastroenterology and Nutrition* **9**, 58–66.

Kramer, M. S. (1988). Does breast feeding help protect against atopic disease? Biology, methodology, and a golden jubilee of controversy. *Journal of Pediatrics* **112**, 181–190.

Kuhns, W. J. (1965). Studies of immediate wheal reactions and of reaginic antibodies in pregnancy and in the newborn infant. *Proceedings of the Society for Experimental Biology and Medicine* **118**, 377–380.

Ladjeva, I., Peterman, J. H. & Mestecky, J. (1989). IgA subclasses of human colostral antibodies specific for microbial and food antigens. *Clinical and Experimental Immunology* **78**, 85–90.

Leach, L., Eaton, B. M., Firth, J. A. & Contractor, S. F. (1990). Uptake and intracellular routing of peroxidase-conjugated immunoglobulin-G by the perfused human placenta. *Cell and Tissue Research* **261**, 383–388.

Leikin, S. & Oppenheim, J. J. (1971). Prenatal sensitisation. *Lancet* **ii**, 876–877.

Levi, M. I., Kravitzov, F. E., Levova, T. M. & Fomenko, G. A. (1969). The ability of maternal antibody to increase the immune response in infants. *Immunology* **16**, 145–148.

Lifschitz, C. H., Hawkins, H. K., Guerra, C. & Byrd, N. (1988). Anaphylactic shock due to cow's milk protein hypersensitivity in a breast-fed infant. *Journal of Pediatric Gastroenterology and Nutrition* **7**, 141–144.

Lilja, G., Dannaeus, A., Fälth-Magnusson, K., Graff-Lonnevig, V., Johansson, S. G. O., Kjellman, N.-I. M. & Öman, H. (1988). Immune response of the atopic woman and foetus: effect of high- and low-dose food allergen intake during late pregnancy. *Clinical Allergy* **18**, 131–142.

Lilja, G., Dannaeus, A., Foucard, T., Graff-Lonnevig, V., Johansson, S. G. O. & Öman, H. (1989). Effects of maternal diet during late pregnancy and lactation on the development of atopic diseases in infants up to 18 months of age – in vivo results. *Clinical and Experimental Allergy* **19**, 473–479.

Lilja, G., Dannaeus, A., Foucard, T., Graff-Lonnevig, V., Johansson, S. G. O. & Öman, H. (1991). Effects of maternal diet during late pregnancy and lactation on the development of IgE and egg- and milk-specific IgE and IgG antibodies in infants. *Clinical and Experimental Allergy* **21**, 195–202.

Lovegrove, J. A. (1991). *Studies of Milk Antibodies and Antigens in Human Adults and Infants*. PhD thesis, University of Surrey.

Lovegrove, J. A., Hampton, S. M. & Morgan, J. B. (1993*b*). Does a maternal milk-free diet prevent allergy in the 'at risk' infant? *Proceedings of the Nutrition Society* **52**, 217A.

Lovegrove, J. A., Hampton, S. M. & Morgan, J. B. (1994). The immunological and long term atopic outcome of infants born to women following a milk-free diet during pregnancy and lactation: a pilot study. *British Journal of Nutrition* **72**, 223–238.

Lovegrove, J. A., Hampton, S. M., Morgan, J. B. & Marks, V. (1991). Effect of a milk-free diet on fetal cord blood milk antibody levels. A study of normal and atopic mothers. *Proceedings of the Nutrition Society* **50**, 9A.

Lovegrove, J. A., Osman, D. L., Morgan, J. B. & Hampton, S. M. (1993*a*). Transfer of cows' milk β-lactoglobulin to human serum after a milk load: a pilot study. *Gut* **34**, 203–207.

Machtinger, S. & Moss, R. (1986). Cow's milk allergy in breast-fed infants: the role of allergen and maternal secretory IgA antibody. *Journal of Allergy and Clinical Immunology* **77**, 341–347.

Mehrishi, J. N. (1976). Cell surface membrane interactions in reproductive physiology. In *Maternofoetal Transmission of Immunoglobulins*, pp. 25–35 [W. A. Hemmings, editor]. Cambridge: Cambridge University Press.

Miller, D. L., Hirvonen, T. & Gitlin, D. (1973). Synthesis of IgE by the human conceptus. *Journal of Allergy and Clinical Immunology* **52**, 182–187.

Mohr, J. A. (1972). Lymphocyte sensitisation passed to the child from the mother. *Lancet* **i**, 688.

Morris, E. R. (1991). *A Study of Food Antibodies in Infants and Adults*. PhD Thesis, University of Surrey.

Ngan, J. & Kind, L. S. (1978). Suppressor T cells for IgE and IgG in Peyer's patches of mice made tolerant by the oral administration of ovalbumin. *Journal of Immunology* **120**, 861–865.

Ogra, S. S. & Ogra, P. L. (1978). Immunologic aspects of human colostrum and milk. II Characterics of lymphocyte reactivity and distribution of E-rosette forming cells at different times after the onset of lactation. *Journal of Pediatrics* **92**, 550–555.

Oxelius, V.-A., Lindén, V., Christensen, K. K. & Christensen, P. (1983). Deficiency of IgG subclasses in mothers of infants with group B streptococcal septicemia. *International Archives of Allergy and Applied Immunology* **72**, 249–252.

Padeh, S., Jaffe, C. L. & Passwell, J. H. (1991). Activation of human monocytes via their sIgA receptors. *Immunology* **72**, 188–193.

Paganelli, R. & Levinsky, R. J. (1980). Solid phase radioimmunoassay for detection of circulating food protein antigens in human serum. *Journal of Immunological Methods* **37**, 333–341.

Papadatos, C., Papaevangelou, G. J., Alexiou, D. & Mendris, J. (1970). Serum immunoglobulin G levels in small-for-dates new-born babies. *Archives of Disease in Childhood* **45**, 570–572.

Pathirana, C., Goulding, N. J., Gibney, M. J., Pitts, J. M., Gallagher, P. J. & Taylor, T. G. (1981). Immune tolerance produced by pre- and postnatal exposure to dietary antigens. *International Archives of Allergy and Applied Immunology* **66**, 114–118.

Pöysä, L., Korppi, M., Remes, K. & Juntunen-Backman, K. (1991). A nine-year follow-up study. I. Clinical manifestations. *Allergy Proceedings* **12**, 107–111.

Pöysä, L., Remes, K., Korppi, M. & Juntunen-Backman, K. (1989). Atopy in children with and without a family history of atopy. I. Clinical manifestations, with special reference to diet in infancy. *Acta Paediatrica Scandinavica* **78**, 896–901.

Ratner, B. (1922). Certain aspects of eczema and asthma in infancy and childhood from the standpoint of allergy. *Medical Clinics of North America* **6**, 815–830.

Renz, H., Vestner, R., Petzoldt, S., Brehler, C., Prinz, H. & Rieger, C. H. L. (1990). Elevated concentrations of salivary secretory immunoglobulin A anti-cow's milk protein in newborns at risk of allergy. *International Archives of Allergy and Applied Immunology* **92**, 247–253.

Rothberg, R. M. & Farr, R. S. (1965). Anti-bovine serum albumin and anti-alpha lactalbumin in the serum of children and adults. *Pediatrics* **35**, 571–588.

Savin, J. A. (1993). Atopy and its inheritance. *British Medical Journal* **307**, 1019–1020.

Schrander, J. J. P., Unsalan-Hooyen, R. W. M., Forget, P. P. & Jansen, J. (1990). [51Cr]EDTA intestinal permeability in children with cow's milk intolerance. *Journal of Pediatric Gastroenterology and Nutrition* **10**, 189–192.

Slade, H. B. & Schwartz, S. A. (1987). Mucosal immunity: the immunology of breast milk. *Journal of Allergy and Clinical Immunology* **80**, 348–358.

Sporik, R., Holgate, S. T., Platts-Mills, T. A. E. & Cogswell, J. J. (1990). Exposure to house dust mite allergen (*Der p* 1) and the development of asthma in childhood: a prospective study. *New England Journal of Medicine* **323**, 502–507.

Strobel, S. (1988). Developmental aspects of food allergy. In *Food Allergy* (Nestlé Nutrition Workshop series 17), pp. 99–117 [D. Reinhardt and E. Schmidt, editors]. New York: Raven Press.

Strobel, S. (1992). Dietary manipulation and induction of tolerance. *Journal of Pediatrics* **121** (5), S74–S79.

Talbot, F. B. (1918). Eczema in childhood. *Medical Clinics of North America* **1**, 985–996.

Telemo, E., Weström, B., Dahl, G. & Karlsson, B. (1986). Transfer of orally or intravenously administered proteins to the milk of the lactating rat. *Journal of Pediatric Gastroenterology and Nutrition* **5**, 305–309.

Thomas, H. C., Ryan, C. J., Benjamin, I. S., Blumgart, L. H. & MacSween, R. N. M. (1976). The immune response in cirrhotic rats. The induction of tolerance to orally administered protein antigens. *Gastroenterology* **71**, 114–117.

Troncone, R., Scarcells, A., Donatiello, A., Cannataro, P., Tarabuso, A. & Auricchio, S. (1987). Passage of gliadin into human breast milk. *Acta Paediatrica Scandinavica* **76**, 453–456.

Van Asperen, P. P., Kemp, A. S. & Mellis, C. M. (1983). Immediate food hypersensitivity reactions on the first known exposure to food. *Archives of Disease in Childhood* **58**, 253–256.

Walker-Smith, J. A. (1992). Cow milk sensitive enteropathy: predisposing factors and treatment. *Journal of Pediatrics* **121** (5), S111–S115.

Wood, C. B. S. (1986). How common is food allergy? *Acta Paediatrica Scandinavica* Suppl. 323, 76–83.

Zeiger, R. S. (1987). Challenges in the prevention of allergic disease in infancy. *Clinical Reviews in Allergy* **5**, 349–373.

Zeiger, R. S., Heller, S., Mellon, M. H., Forsythe, A. B., O'Connor, R. D., Hamburger, R. N. & Schatz, M. (1989). Effect of combined maternal and infant food allergen avoidance on development of atopy in early infancy: a randomized study. *Journal of Allergy and Clinical Immunology* **84**, 72–89.

Printed in Great Britain

PHYSICAL ACTIVITY, NOT DIET, SHOULD BE THE FOCUS OF MEASURES FOR THE PRIMARY PREVENTION OF CARDIOVASCULAR DISEASE

ALEXANDER L. MACNAIR

20 Wimpole Street, London W1

CONTENTS

INTRODUCTION	43
KEY RISK FACTORS	44
BLOOD LIPIDS	44
Do elevated blood lipids cause atherosclerosis?	44
What causes plasma cholesterol to rise?	45
CARBOHYDRATE METABOLISM AND INSULIN RESISTANCE	48
Does diabetes cause atherosclerosis?	48
What causes insulin resistance?	49
BLOOD PRESSURE	51
Does hypertension cause atherosclerosis?	51
What causes hypertension?	51
ENERGY BALANCE AND ADIPOSITY	54
Does excess adiposity cause atherosclerosis?	55
What causes failure of energy balance?	55
CONCLUSIONS	58
REFERENCES	60

INTRODUCTION

Severe atherosclerosis is found at *post mortem* in the majority of cardiovascular deaths. Although this does not imply cause and effect, atherosclerosis is nevertheless an essential contributor to most of the disability and death associated with ischaemic heart disease (IHD) and its prevention must be a cornerstone of true primary prevention of IHD.

Progressive degeneration of the lining of the great vessels appears to begin in the first few years of life with the formation of fatty streaks beneath the endothelial lining, first of the aorta and later of the main arterial trunks. It continues to extend and to encroach upon the vascular lumen in susceptible subjects, often for half a century or more before occlusion of the vessel is sufficient finally to cause ischaemic injury to critical tissues downstream. Modest regression of lesions appears to be possible (Ornish *et al.* 1990) but atherosclerosis is very largely irreversible once fibrous plaque formation has become established (Davies, 1992). In consequence, primary prevention is increasingly being directed towards healthy subjects in their first two or three decades. Its objective is to encourage the making of relevant changes in behaviour that will continue for life on the basis that whatever initiates the atherosclerotic process may also sustain it. There is thus a need to identify the agents which operate first in the atherogenic process in the hope that their control may inhibit the emergence of the later stages.

The classic risk factors for atherosclerosis are chronic degenerative diseases in their own right and convincing evidence for any of them playing an individual causal role in atherogenesis has very largely eluded researchers in this complex area. The environmental causes of atherosclerosis itself, or of the dyslipidaemia, hypertension, disordered carbohydrate metabolism and obesity which may contribute to it, are by no means clearly understood, but in childhood at least there is little else to consider other than diet and the level of physical activity. Both are difficult to characterize in any single subject but there are easily measured surrogate physiological variables which show strong statistical associations with later cardiovascular morbidity and mortality. These are, in fact, the well established metabolic variables which relate to the uptake, transport, storage and utilization of fuel and oxygen to meet the energy requirements of the organism.

The objective of this paper is to consider each of the key risk factors for evidence of a causal role in atherosclerosis and at the same time to identify the cause or causes of the increase in the risk factor itself. It may be possible thus to determine not only the causes but also the order in which they become effective and to pinpoint the earliest triggering mechanism for atherosclerosis upon which true primary prevention might most fruitfully be focused.

THE KEY RISK FACTORS

BLOOD LIPIDS

Do elevated blood lipids cause atherosclerosis?

How good is the evidence that the classic changes in lipid and lipoprotein concentrations associated with a high risk of IHD (elevated plasma total cholesterol, low density lipoprotein (LDL) and triacylglycerol (TG) together with diminished high density lipoprotein (HDL)), actually cause atherosclerosis? Steinberg (1983) argues that elevated plasma concentrations of LDL are alone sufficient to cause atherosclerosis. Patients with homozygous familial hypercholesterolaemia, who have high plasma concentrations of LDL and no other identified risk factors, may suffer from severe atherosclerosis and die from IHD at a very early age. Familial hypercholesterolaemia is, however, an inherited metabolic disease characterized by LDL receptor dysfunction and plasma cholesterol concentrations which may approach six times normal, leaving deposits of cholesterol in connective tissue throughout the body. It does not necessarily represent a model for the dyslipidaemia associated with idiopathic atherosclerosis in which LDL elevation is much more modest and lowered HDL an important if not essential concomitant. The absence of normal LDL receptor function in these unfortunate individuals results in the prolonged sojourn in the circulation of large numbers of LDL particles. Henriksen *et al.* (1981) demonstrated that LDL may be converted by endothelial cells to a modified form involving the peroxidation of lipids in the LDL particle. Modified LDL is both cytotoxic (Hessler *et al.* 1979) and unrecognizable by the LDL receptor. Nevertheless, it is avidly taken up by the scavenger receptors of a high capacity alternative pathway deployed by monocyte derived macrophages, by Kupffer's cells in the liver and by vascular endothelial cells (Goldstein *et al.* 1979). It is not known how free radical attack is mounted by the endothelial cells (although they are the main source of very highly reactive nitric oxide) but lipid peroxidation may be an explanation for the atherogenicity of familial severe and prolonged hypercholesterolaemia. Much of the LDL in the circulation may be abnormal, quite apart from the possibility that the endothelial cells themselves, with absent or abnormal LDL receptors, may differ functionally; neither should it be overlooked that the vasculature as a whole is exposed to the abnormal lipid levels and yet most vessels remain free of atherosclerosis.

LDL is a normal plasma constituent, and where there is normal receptor function it is not easy to understand how a small elevation in LDL can contribute significantly to atherogenesis. Not only are antioxidants widely available both in LDL particles and in the plasma, but HDL also inhibits the cytotoxicity of oxidized LDL (Hessler et al. 1979). It may be that the duration of exposure of LDL particles to the endothelium, as much as their absolute concentration, permits greater oxidative modification and enhances their atherogenic potential. In normal healthy subjects, plasma half-lives of the lower density lipoprotein particles vary according to their fractional catabolic rates, which are as much dependent on the demand, principally by skeletal and cardiac muscle, for fatty acids for energy production, as upon the rate of recycling of LDL receptors and uptake by the liver.

Most widely accepted as the fundamental basis of atherosclerosis is the endothelial injury hypothesis (Ross, 1986). Ross and Glomset's original hypothesis (Ross & Glomset, 1976) was predicated upon hypercholesterolaemic injury to the endothelium. Their primate model (Faggiotto et al. 1984; Faggiotto & Ross, 1984) exhibited grossly exaggerated levels of plasma cholesterol (six times normal) achieved by giving a high fat, high cholesterol diet to animals with a genetic predisposition (the animals were selected for experiment by their hypercholesterolaemic response to the diet). Furthermore, they were taken from a breeding colony and confined in single cages for the duration of the study. The animals were weighed monthly but changes in their body weights were not reported. Kramsch et al. (1981), employing a similar protocol, studied a closely related strain of macaque. As well as atherogenic diet and control groups, they added a third which was given the high fat, high cholesterol diet and animals in this group were also exercised on a treadmill, three times per week, eventually for an hour. Monkeys in the exercise group lost about 12% body weight (0·6 kg from a starting weight of about 5 kg) while the sedentary groups gained about as much. Most monkeys on the lipid-rich diet, whether exercising or not showed a 6-fold increase in total cholesterol. However, the exercise group alone showed an increase in HDL cholesterol with less of a rise in LDL, VLDL and plasma TG than their inactive fellows. Examination by electrocardiogram, by coronary angiography and at *post mortem*, showed striking protection from atherosclerosis in the exercising animals as compared with their sedentary counterparts. These observations reinforce the notion that hypercholesterolaemia *per se* may not be atherogenic, and it is worth considering that Ross's model, with confinement of the animals throughout the study, depends more on physical inactivity for the development of atherosclerosis than on the elevated blood lipids.

What causes plasma cholesterol to rise?

In classic human atherosclerosis, the more modest derangement in blood lipid levels includes a disturbed ratio of high to low density lipoproteins and often an excess of TG. Since human diets in affluent societies provide more or less equivalent proportions of energy from fat and carbohydrate, while muscles employ a fuel mix which approximates to the ratio in the diet (Flatt, 1987), it is possible that the atherosclerosis seen in man with its origins in childhood, where clustering of metabolic risk factors is common (Webber et al. 1979; Berenson et al. 1989), includes inactivity in its pathogenesis and that hypercholesterolaemia is but one facet of a metabolic derangement which involves both fat and carbohydrate metabolism.

Compared with all other mammals, man is hypercholesterolaemic from soon after birth (Brown & Goldstein, 1986). At birth, the human blood concentration of LDL is similar to that of other mammals in which, by contrast with man, there is little change throughout life. This initial concentration is adequate to provide for the physiological functions of LDL in relation to LDL receptor populations and their affinities (Reichl et al. 1978). In the human, however, immediately after birth in all races studied, the cholesterol concentration rises rapidly, doubling in the first six month of life and redoubling by two to three years when it approaches that of the young adult (Berenson et al. 1979; Sporik et al. 1991). The fasting TG concentration also rises rapidly, increasing by some two-and-a-half-fold by six months (Berenson et al. 1979).

This early rise in blood lipids may have its roots in the evolution of modern man. Brain development, particularly growth of the forebrain, is very recent and has required a considerable increase in the capacity of the cranium. Although gestation may have shortened from perhaps 50 to 40 weeks, the female pelvis has had to become significantly modified to permit the passage of the larger head during parturition, as is shown in the fossil record (Pilbeam, 1984). A consequence of these changes for the human infant is that maturation of the central nervous system is far from complete at birth when the brain is still only a quarter of its adult size. By the end of the first year, the greater part of the postnatal increase in brain weight has taken place and it is about 75% of its adult size. Most of the increase is the result of nerve axon myelination. Myelin is largely lipid of which about 40% is cholesterol. The exceptional cholesterol requirement of the neonatal central nervous system may account for the high blood cholesterol concentration during the first few months of life.

Body fat accumulation also is very rapid in the newborn human; at 1–2 months, when blood lipids are rising fast, fat deposition is about 14 g/d. At 5–6 months the rate falls to 1 g/d but by then body fat has doubled from 13% of body weight at birth to 25% (Fomon et al. 1982). The absence of myelin renders the long motor tracts within the central nervous system inoperative until myelination is complete. As a result, the human infant's voluntary musculature is, except for simple spinal reflexes, largely uncoordinated at birth. The less coordinated early activities of the child are at most lightly loaded and, although apparently vigorous, actually require a low level of energy expenditure. In contrast, purposive locomotion in bipeds lifts the total body weight at each step and demands a high energy expenditure. Thus the human infant, unlike its lower primate cousins, is relatively sedentary with low energy expenditure for up to a year. A possible explanation for the high concentration of TG in the circulation in the first few months is the obligatory low level of energy expenditure. Taken together, these consequences of the evolution of a large brain and its late maturation may account for the exceptionally high concentration of blood fats and the unexpected but universal finding in humans of fatty streaks in the aortae of virtually all two to three year olds (Strong & McGill, 1969).

Whether or not there is an evolutionary explanation for the high blood levels of LDL cholesterol and TG in all young children, by the time they reach adult life the blood lipid pattern differs significantly between those from rural economies and those living in urban communities. An additional environmental explanation is clearly required, and it is traditional, when discussing the causes of elevation of plasma cholesterol in adults, to cite Kinsell et al. (1952), Ahrens et al. (1954), Keys et al. (1957) and Hegsted et al. (1965), for they and others showed, under metabolic ward conditions, that the proportion or the saturation of fatty acids in the diet has an effect upon the concentration of cholesterol in the blood. Experiments in free living subjects, however, cast doubt on the relevance of these early findings, showing a capacity in most individuals to compensate for dietary change by altering absorption, synthesis or elimination of cholesterol (McNamara et al. 1987). Furthermore, cross-sectional studies in large numbers of free living subjects have consistently failed to show an important relationship between any dietary component and plasma cholesterol (Morris et al. 1963; Kahn et al. 1969; Gordon, 1970; Nichols et al. 1976;

Frank *et al.* 1978). Agreement about which fatty acids in the diet have effects upon the blood cholesterol level has still to be reached (Sundram *et al.* 1994), although Dietschy *et al.* (1993) have proposed that individual dietary fatty acids have specifically defined effects on LDL receptor activity which directly determines the plasma concentration of LDL.

Considered functionally, both fats and sugars in the blood are fuels *en route* ultimately to the muscles for oxidation. Exercise training increases the capacity of skeletal muscle tissue to take up oxygen and to release energy from these metabolic fuels. It also increases the ratio of fat to carbohydrate oxidized by muscle (Romijn *et al.* 1993). A small but meticulous study, comparing trained locomotor muscle groups against the contralateral untrained muscle groups in individuals exercising one leg over an 8 week period, measured the lipids in the blood entering and leaving trained and untrained muscles, both at rest and during exercise (Kiens & Lithell, 1989). At rest in the trained leg, there was a markedly higher arteriovenous difference in VLDL TG. There was also a difference, albeit smaller, in the production of LDL at rest. There was a higher production of HDL during exercise in the trained leg.

There are more conventional studies showing the effects of exercise on plasma lipids (Wood *et al.* 1983; Rauramaa *et al.* 1984) but they tend to be confounded by changes in body composition. However, Weintraub *et al.* (1989), in a carefully controlled study, kept weight constant during a treadmill training programme and altered the total energy but not the proportions of the diet to match the increased expenditure. Although weight remained constant, fitness increased as measured by maximal oxygen uptake. Fasting TG concentrations fell significantly and there was a tendency for VLDL and LDL to fall while HDL rose but the latter changes were not significant in this small group of 6 subjects. Exercise significantly increased lipoprotein lipase activity while postprandial total lipoprotein levels were reduced by 32%.

The textbook presentation of plasma lipid transport (Brown & Goldstein, 1991) is as a system for both fatty acids and cholesterol. However, examination of the structure of an LDL particle (Brown & Goldstein, 1984) shows that the cholesterol is functionally involved both in the formation with fatty acids of hydrophobic esters and in the hydrophilic limiting membrane in which bipolar molecules of free cholesterol and phospholipid share in the make-up of the detergent monolayer. The exogenous pathway for lipid transport is manifestly a TG transport system while the endogenous pathway (of which LDL is a part) is remarkably parallel. Chylomicrons are synthesized in the intestine from TG of dietary origin while VLDL are synthesized in the liver from endogenous TG; both incorporate apoproteins E, CII and B. Apo CII activates lipoprotein lipase in adipose and muscle tissue capillaries, releasing TG for uptake by these tissues. Chylomicron remnants bearing apo E and apo B complete the exogenous pathway and are rapidly taken out of circulation by the liver. In the endogenous pathway VLDL particles, although smaller than chylomicrons, have a very similar structure and fate. Apo CII sees to the release of TG in adipose and muscle tissue and the VLDL remnant, now renamed intermediate density lipoprotein, is also taken up by the liver, albeit much more slowly than chylomicron remnants. A large proportion continues to circulate in the blood and progressively loses its remaining TG and apoproteins (except B) to become LDL. Thus although it is generally stated that LDL is a transporter of cholesterol, it is in effect a remnant of endogenous TG transport. This question of the function of lipoprotein particles is of key importance since the accumulation of LDL in the circulation may be at least as much the consequence of disparities in supply and demand for TG as failure of down-regulated or absent LDL receptors to clear it. Chylomicrons and their remnants are cleared from the plasma with a half-life of 4–5 min; the half-life of VLDL particles is 1–3 h. LDL particles, on the other hand, have a prolonged stay in the circulation with a fractional catabolic rate of about 45% of the plasma pool in

24 h. In a study of LDL balance in the whole organism in several species, Dietschy and his colleagues (1993) concluded that, contrary to earlier assumptions, LDL does not function in a major way to deliver cholesterol from the liver to extrahepatic organs, and the amount of cholesterol synthesis taking place in extrahepatic tissues is 3–5 times greater than the amount being delivered to these organs by LDL. Thus far from being a cholesterol transport particle, LDL may simply be the ultimate degradation product of the TG transport system. Some cholesterol is of course utilized outside the liver in steroid hormone and vitamin D synthesis but the absolute amounts are small.

The studies of Weintraub et al. (1989) and Kiens & Lithell (1989) indicate that changes in blood cholesterol concentrations reflect the turnover of TG and lipoprotein lipase activity which are in this context an expression both of the level of muscle training and of the level of physical activity. Lipoprotein lipase, although activated by apo CII, is feedback inhibited by local accumulation of the fatty acid products of hydrolysis of TG as muscle uptake becomes satiated (Quinn et al. 1982), a point quickly reached in muscle tissue continuously at rest and in particular if it is also untrained. Such inhibition of hydrolysis of lipoprotein TG will not only raise the concentration of lower density lipoprotein particles in the circulation but will also prolong the duration of their stay, increasing their exposure to reactive oxygen species released from the endothelium and elsewhere. If cholesterol is present in lipoprotein particles as a functional element rather than as a passenger, then cholesterol concentrations and the associated cardiovascular risk may depend on fatty acid turnover, which is a function of the level of physical activity and the state of training of the skeletal muscle. The fact that some plasma cholesterol is associated with increased risk and some with risk reduction further supports the view that cholesterol is itself not causally associated with IHD.

There are as many studies pointing to exercise and physical activity as a key determinant of the plasma lipoprotein profile as there are metabolic ward studies of dietary fats in this role. A meta-analysis involving 95 such studies (Tran & Weltman, 1985) showed that there are, indeed, significant relationships between physical activity and changes in blood lipids and lipoproteins. Decreases in total cholesterol, LDL, TG and the total cholesterol:HDL-C ratio were correlated significantly with exercise training especially when accompanied by weight loss, but the most consistent results were seen in studies in which there was no weight change. Longer exercise times were also significantly associated with greater changes in lipid and lipoprotein levels. In children also, a low level of physical activity is associated with a high total plasma cholesterol concentration (Wong et al. 1992).

In the light of these data it seems possible that the cause of the classic changes in concentration and metabolism of plasma lipoprotein particles associated with increased risk of atherosclerotic cardiovascular disease may be the failure to maintain some personal or individual minimum threshold level of training that is probably genetically determined.

CARBOHYDRATE METABOLISM AND INSULIN RESISTANCE

Does diabetes cause atherosclerosis?

If, as seems increasingly possible, the fundamental origin of atherosclerosis is in a failure of metabolic fuel regulation, it is logical having considered the lipids to review the role of carbohydrate metabolism and in particular non-insulin dependent diabetes (NIDDM) in this context since in affluent societies, where a choice of foods is freely available, energy is derived about equally from dietary fats and carbohydrates. In childhood and adolescence, when atherosclerosis is believed to have its origin, NIDDM is not common. It is a classic disorder of middle age in which poor control of blood glucose concentration is a consequence both of insulin resistance and of relative failure of pancreatic production of insulin (DeFronzo, 1988) but the origins of NIDDM may, nevertheless, be in childhood. There are few studies examining glucose tolerance or insulin resistance in children but Freedman *et al.* (1987), as part of their monitoring of the classic risk factors in children in the Bogalusa Heart Study, showed that a hyperinsulinaemic response one hour after an oral glucose load is associated strongly with the amount of centrally located adipose tissue. In adults there is an established relationship between central obesity, glucose intolerance, hypertriacylglycerolaemia and hypertension. Kaplan (1989) described this cluster of risk factors as 'The Deadly Quartet' and considered that central obesity was the cause of the disturbances in lipid and carbohydrate metabolism and a contributor to the hypertension. Reaven (1988), however, described as 'syndrome X' a group of metabolic disturbances comprising glucose intolerance, hyperinsulinaemia, hypertriacylglycerolaemia with a rise in VLDL and a decline in HDL together with hypertension. Citing very largely his own work, he argued that the metabolic changes and the hypertension could all be seen as secondary to insulin resistance. The original description of syndrome X did not include obesity, and it is interesting in this context that patients with insulinoma may manifest marked visceral fat deposition, apparently under the influence of the hyperinsulinaemia (Inadera *et al.* 1993). It is possible that hyperinsulinaemia leads to deposition of visceral fat rather than the more usually accepted reverse sequence. Indeed, it may be that visceral fat is the more potent risk factor than peripheral fat precisely because, unlike the gluteofemoral adiposity in women which has an oestrogenic basis, it is a consequence of insulin resistance and elevated insulin levels.

Over 25 years ago Stout (1968, 1991) and Vallance-Owen (Stout & Vallance-Owen, 1969) argued that glucose and insulin play a major role in the genesis of atherosclerosis. In the light of current evidence, Jarrett (1988) considered the question 'Is insulin atherogenic?' He concluded that the case was not proven and, in his view, the hypothesis requires more experimental and observational evidence before it can be supported. In a later review, Jarrett (1992) also provided a well argued case against elevated insulin levels as a cause of hypertension in syndrome X.

There is little evidence either that hyperglycaemia, rather than hyperinsulinaemia, is atherogenic. Persistently raised blood glucose has adverse effects on the microvasculature, especially in the retina (Kohner, 1993) and renal glomerulus (Lorenzi, 1992), but there is no good evidence that the vessels classically involved in atherosclerotic disease are damaged by hyperglycaemia. Thus although syndrome X and Kaplan's quartet comprise the main risk factors for IHD and both include impaired glucose tolerance, the epidemiological evidence linking NIDDM to IHD is not sufficient for a causal relationship to be seriously considered. Nevertheless, it is clear that the combined disorders of lipid and carbohydrate metabolism, central obesity and hypertension are together linked with atherosclerosis, and an understanding of the way these disorders arise may help in elucidating the origins of disease in the blood vessels.

What causes insulin resistance?

Insulin resistance or reduced insulin-mediated glucose uptake is defined as a subnormal response to a given concentration of insulin and measured usually by the euglycaemic, hyperinsulinaemic glucose clamp developed by DeFronzo *et al.* (1979). Although insulin has integrated actions on carbohydrate, protein and lipid metabolism, it exerts a dominant effect on glucose homeostasis and the term 'insulin resistance' typically refers only to the latter (Moller & Flier, 1991). Reaven (1988) makes it clear, however, that insulin resistance includes a reduced ability to regulate the plasma free fatty acid concentration. Indeed he shows that failure to regulate non-esterified fatty acids is in part responsible for fasting hyperglycaemia in NIDDM since elevated blood levels of non-esterified fatty acids stimulate hepatic gluconeogenesis.

Shulman et al. (1990) obtained ^{13}C nuclear magnetic resonance spectra of human muscle glycogen *in vivo*. By this means they were able to show that muscle glycogen synthesis is the principal non-oxidative pathway for glucose disposal. It was about 60% lower in their insulin resistant subjects than in normals and total body glucose uptake was likewise depressed. Shulman and his colleagues concluded that impaired glycogen synthesis in skeletal muscle appears to be the dominant intracellular metabolic basis of insulin resistance. Björntorp (1987) analysed the sequence of events leading to NIDDM and observed that the earliest evident metabolic anomaly in the development of NIDDM is increased insulin resistance in skeletal muscle. Young et al. (1988), using radiolabelled glucose, determined that, in insulin resistant subjects without diabetes, glycogen accumulation in muscle during infusions of glucose and insulin is diminished in comparison with normal subjects. Eriksson et al. (1989) compared NIDDM subjects with their first degree relatives and a group of normal controls. They found a high prevalence of insulin resistance, even in relatives with normal oral glucose tolerance. In perfused rat hindquarters, prolonged exposure of skeletal muscle to high plasma levels of insulin and glucose causes muscle glycogen to increase and then plateau while glycogen synthesis, rapid at first, attenuates to zero after about 5 h (Richter et al. 1988a). The mechanism of this acute insulin resistance involves both membrane transport of glucose and glycogen synthesis, apparently in response to high intracellular glucose and glycogen concentrations. *In vitro* studies with cultured rat adipocytes provide evidence that a high level of insulin alone may induce insulin resistance (Garvey et al. 1986), while Richter et al. (1988b) have shown that high plasma levels of glucose alone lead to marked inhibition of insulin action on muscle glucose transport. In effect, any check on the flux through the system taking glucose from the blood, polymerizing it to glycogen and oxidizing the product to release energy may cause a local excess of substrate, product or regulator and down-regulation of insulin receptors as a result. Man has little capability to dispose of excess levels of glucose by conversion to fatty acids in adipose tissue (Björntorp & Sjöström, 1978), something that happens freely in the rat. Unlike liver glycogen, that in muscle is unavailable for release back into the circulation as glucose and once synthesized in muscle it can only be mobilized by utilization for energy release within the cell. When muscle glycogen stores are replete, there is feedback inhibition of further glycogen synthesis. If blood glucose is elevated following a meal, insulin resistance occurs when the rising blood insulin level fails to move glucose out of the circulation. Once the relatively small glycogen storage capacity of the liver is surpassed, any excess glucose from the diet is shunted through into the peripheral circulation where, in spite of a raised insulin concentration, it produces prolonged hyperglycaemia before the tissues take it up finally by mass action.

Insulin resistance can be seen to be the normal physiological response to relative carbohydrate overfeeding. It stems from failure to utilize available stores of glycogen in muscular work, the consequence either of a low level of energy expenditure in physical activity or of dietary intake of carbohydrate in excess of that needed to make good normal carbohydrate turnover. These observations suggest that insulin resistance is the consequence of inactivity coupled with a relative excess of dietary energy in the form of carbohydrate while the obesity, that may be seen in association with syndrome X and is a part of the deadly quartet, may simply be the consequence of an excess of dietary energy intake in the form of fat. The usual location of this obesity in the abdominal area suggests an association with insulin resistance and hyperinsulinaemia. Although there is a dearth of good studies in children, there is no reason to suppose that insulin resistance does not occur in childhood or that its fundamental cause is other than a low level of physical activity.

The enforced idleness of the human infant may contribute to the high level of fat deposition in the first year. One may speculate that if that inactivity is voluntarily prolonged it may adversely affect a critical period of musculoskeletal development which perhaps culminates in the growth spurt of puberty. Barker (1992) has proposed that the pathogenesis of cardiovascular disease is programmed during the period of rapid growth in early life; however, his data are retrospective and confined to anthropometric measures routinely recorded during the first year. They are not accompanied by metabolic risk factor data recorded concurrently. Young athletes undergoing vigorous training are taller and slimmer than sedentary controls (Ekblom, 1971). Structural and metabolic differences in trained as opposed to untrained muscle in young adults are profound (Ingjer & Brodal, 1978; Kiens & Lithell, 1989) but we do not know whether the level of activity or training plays an important role in the early development of skeletal muscle. There are no data from studies comparing either carbohydrate metabolism or muscular development in active as opposed to inactive children. Such studies would be uniquely valuable in elucidating the relevance of early nutrition and exercise in achieving optimum muscular development in childhood.

BLOOD PRESSURE

Does hypertension cause atherosclerosis?

In the pantheon of risk factors for IHD, blood pressure vies with plasma cholesterol for first place (Martin *et al.* 1986). The key question is whether there is evidence that it is a cause of atherosclerosis and therefore of fundamental importance for prevention. In fact, this is one area in the complex question of the aetiology of atherosclerosis where the evidence is rather clearcut. There is no doubt that atherosclerosis is confined to blood vessels where the pressure is highest. Typically it starts in the aorta, especially where branches and flow dividers cause turbulence (Texon, 1980). Later it involves the main branches at the points where they leave the aorta. It affects, in particular, the coronary arteries in which flow ceases during ventricular systole, giving an exaggerated and possibly more damaging pulsatile component to the effects of pressure. Atherosclerosis does not appear to involve lower pressure vessels on the arterial side of the circulation or any of the veins except when a vein is used as a bypass graft and becomes subjected to arterial pressure.

Schwartz & Benditt (1977) carried out careful observational studies of the effects of blood pressure on the endothelial cells of the aorta in the rat and showed that the earliest sign of endothelial injury in hypertension is a ten-fold increase in cell replication. Texon (1986) showed in man that the hydraulic forces acting at bends and branches tend to lift the endothelial layer from its basement membrane. The loss of laminar flow induces local negative pressure which causes cell junctions to gape or, by actually removing cells, exposes the underlying collagen. Platelets and white blood cells, including monocytes, flock to this lesion, described first as an inflammatory arteritis by Virchow (1858) and later by Ross & Glomset (1976) as the beginning of atherosclerosis. There seems to be little doubt that even quite modest increases in blood pressure induce loss of normal function in endothelial cells and an increase in their normally slow turnover, which may be the first step in the long journey to the development of the classic atherosclerotic fibrous plaque and ultimately vascular occlusion.

What causes hypertension?

If the relationship between hypertension and atherosclerosis seems relatively clear, the cause of the hypertension itself remains in dispute. The second report of the Task Force on

Blood Pressure Control in Children (1987) shows that in urban societies systolic blood pressure rises rapidly in the first six months of life. Both diastolic and systolic pressures increase steadily thereafter to reach adult levels at the end of the second decade when linear growth ceases. Blood pressure is closely tied to height (World Health Organization, 1985a).

The rise in blood pressure in childhood appears to be a universal phenomenon (Grobbee et al. 1990). Apart from an inherited predisposition, the cause of hypertension, particularly that which appears early, is not established. Ambard & Beaujard (1904), in a treatise on the causes of arterial hypertension, described several cases in whom chloride retention was associated with elevated blood pressure and where salt deprivation led to a decline. Allen (1925) advocated salt restriction for the treatment of established hypertension, as did Kempner (1948). The point, however, is not whether salt deprivation will reduce elevated blood pressure so much as whether excess dietary salt is a cause of essential hypertension in the young. Here, the animal studies of Dahl et al. (1968) are compelling. Dahl and his colleagues demonstrated that rats given saline to drink may become hypertensive. By selective inbreeding they produced strains of rats which, given salt in their diet, became rapidly, permanently and severely hypertensive, or conversely were totally resistant to the hypertensive effects of salt. Dahl & Love (1954) concluded that the same interaction between genetic predisposition and environmental conditions applies in man, and present understanding of the role of the kidney in maintaining blood volume and ionic concentration along with blood pressure control suggests that they were right. Both salt sensitive and salt resistant subjects can be found but the nature of the inherited defect remains unknown. In populations in whom salt intake is high, the proportion with hypertension is high in comparison with populations with a low salt intake (Dahl, 1972). Although in large groups of subjects there appears to be a dose response effect (Law et al. 1991), the individual blood pressure responses to dietary salt adjustment are extremely varied and heavily reinforced by multiple feed-back systems. Miller et al. (1987) undertook a twelve week intervention study in 82 individuals with normal blood pressure. Baseline sodium intake was 157 mmol/d and this was reduced to 68 mmol during the restriction phase of the study. For the group as a whole, mean arterial pressure fell from 83 mmHg to 81 mm. Thus moderately severe sodium restriction produced a small reduction in mean arterial pressure for the group as a whole, as might be expected. However, when the individual responses are examined, they present a normal distribution about the mean and extend from a 15 mm reduction to a 17 mm increase in mean arterial pressure.

In reviews of the mechanisms involved in the genesis of chronic idiopathic hypertension, Guyton (1989) and de Wardener (1990) showed that it does not occur without abnormal function of the renal machinery for the maintenance of extracellular fluid volume and cation concentrations. In communities with a high salt intake, those with diminished renal capacity to excrete sodium and water respond to higher levels of salt intake by a greater rise in pressure than those with normal renal function. This phenomenon may explain why blood pressure rises with age in high salt consuming societies and may also be the basis of salt sensitivity, either genetic (e.g. a family trait towards a small renal tissue mass) or congenital such as inhibited growth and development *in utero* of genetically normal kidneys (Barker et al. 1990). Although there is no doubt that the blood volume and the basal blood pressure are affected physiologically and directly by habitual sodium intake, in those with normal renal function, the renin–angiotensin–aldosterone system is so effective that huge changes in salt intake are reflected by a millimetre or two change up or down in the blood pressure. More important, at a given salt intake the setpoint does not drift significantly, so that an increase in pressure with age in healthy normal subjects is to be expected only if their salt intake also rises. Thus, in subjects with normal renal function, a high salt intake is unlikely to be the cause of hypertension.

It is well accepted that high blood pressure in childhood tends to track, frequently remaining elevated in adolescence (Voors *et al.* 1979) and on into adult life. Suggestions for the key determinant of blood pressure in childhood include familial predisposition (Van Hooft *et al.* 1988), increased relative weight (Stamler *et al.* 1978), lack of physical fitness (Fraser *et al.* 1983; Hofman *et al.* 1987), sodium and potassium intake (Geleijnse *et al.* 1990) and insulin resistance (Slater, 1991). Hypertension in childhood is associated with elevated lipids and arterial fatty streaks (Newman *et al.* 1986) as well as with obesity (Havlik *et al.* 1983) and hyperinsulinaemia (Freedman *et al.* 1987). Hofman *et al.* (1987) showed that children with declining physical fitness have above average rises in blood pressure, confirming the results of the earlier study by Fraser *et al.* (1983). Intervention to reduce weight in the overweight (Krotkiewski *et al.* 1979) or to restore fitness to the unfit (Somers *et al.* 1991) is, of course, desirable in its own right and can be an effective non-pharmacological means of reducing blood pressure in adolescents (Hagberg *et al.* 1984) or young adults (Jennings *et al.* 1986). There is at present little evidence from prospective long term childhood studies that avoiding fatness and maintaining fitness will prevent a rise in blood pressure above the normal for age and height, even if we knew what range of levels of adiposity and physical performance were normal or optimal. Furthermore, there is little real understanding at any age of the mechanisms by which elevation in one or more of the several risk factors for hypertension translates into an actual rise in blood pressure.

Dahl and his colleagues (Dahl *et al.* 1958) suggested that the hypertension of obesity is salt dependent and not simply a consequence of increased blood volume. However, others have questioned this thesis (Messerli, 1982; Frohlich *et al.* 1983), proposing increased cardiac output and blood volume as the principal determinants of blood pressure elevation in the obese. Dustan (1985) considered that various endocrine factors may be responsible for the relationship between obesity and hypertension. In her view none was a wholly satisfactory explanation but the data pointing to insulin in this role were the least unconvincing (Miller & Bogdonoff, 1954; DeFronzo *et al.* 1975). DeFronzo and his colleagues studied the effects of an insulin infusion on sodium excretion during a euglycaemic clamp. There was a highly significant reduction in sodium excretion by the kidney as insulin levels rose, without changes in the filtered loads of sodium or glucose or in the glomerular filtration rate, renal blood flow or plasma aldosterone.

Modan and his associates examined a representative group of Jewish adults in a study of glucose intolerance, obesity and hypertension, the Israel GOH Study (Modan *et al.* 1987) and demonstrated that hyperinsulinaemia is independently associated with hypertension, glucose intolerance and a disturbed lipoprotein profile. Although in this cross-sectional study the causal chain of events could not be determined, Modan opined that obesity leads to hyperinsulinaemia which leads to the other conditions. Kaplan (1989), as we have noted, also proposed that obesity leads to glucose intolerance and hyperinsulinaemia which in turn precipitate hypertension and dyslipidaemia, but Reaven (1988), in a review of his own extensive studies on the relationship between insulin resistance and other diseases, noted that increases both in plasma catecholamines and in renal tubular reabsorption of sodium and water can be seen in response to hyperinsulinaemia and that either may lead to an increase in blood pressure. In support of the concept that insulin resistance leads on to hypertension, Beatty *et al.* (1993) showed that the young normotensive offspring of hypertensive parents were already insulin resistant by comparison with matched offspring of normotensive parents, even though they showed no signs of hypertension.

Julius *et al.* (1991) offered a hypothesis for the link between insulin resistance and hypertension. They proposed that the pathophysiological fault is in skeletal muscle and is 'vascular rarefaction', involving loss of vessels from the muscle capillary bed. Pressure induced restriction of the microcirculation limits nutritient flow and thereby impairs glucose uptake in skeletal muscle. However, they were not entirely clear which was cause and which effect. They suggested that hypertension causes microvascular damage with narrowing or loss of nutritive vessels (those which supply oxygen and fuel rather than non-nutritive or 'shunt' vessels) and this, by limiting blood flow through the muscle, led to reduced glucose uptake and insulin resistance. Baron *et al.* (1990 & 1991) put the boot on the other foot with the measurement of glucose uptake and blood flow in the leg muscles in lean and obese men following an oral glucose load. Glucose uptake in muscle is part insulin receptor mediated and in part driven by mass action, the result of the difference in concentration of glucose in plasma and within the skeletal muscle cell. The obese subjects had impaired glucose tolerance. They also had hyperinsulinaemia both while fasting and after the glucose load. However, the arteriovenous differences in glucose concentration both while fasting and during disposal of the glucose load were the same in both lean and obese. Blood flow through the leg increased markedly in the lean subjects during the glucose load but was unchanged in the obese subjects. Thus the impaired glucose tolerance and apparent insulin resistance were partly, at least, the effects of failure of blood flow in skeletal muscle to increase rather than failure of transport of glucose into the myocyte. This mechanism, involving what seems at first sight to be a direct effect of insulin on blood vessel tone, may also be involved in the processes by which hyperinsulinaemia and hypertension appear to be connected.

Ingjer & Brodal (1978) and Kiens & Lithell (1989) showed how training increases the size of the muscle capillary bed and thus loss of fitness or 'detraining' may independently lead to reduced muscle capillary density, insulin resistance (Rosenthal *et al.* 1983) and an increase in blood pressure (Somers *et al.* 1991). It is not clear whether there is a direct or causal relationship, but if so it is logical to suppose that the low nutrient capillary density and reduced blood flow in untrained muscle is part of the cause both of the insulin resistance and of the enhanced resistance to flow which may be expressed as hypertension. Conversely, the increased capillary density in trained muscle may reduce the peripheral resistance to blood flow and contribute to the decline in resting blood pressure and heart rate associated with training.

We have in these observations a plausible hypothesis to explain the onset of a hypertension in children without hereditary or congenital predisposition. It appears likely that insufficient muscular activity to turn over glycogen stores with consequent insulin resistance and hyperinsulinaemia leads over time to a rise in resting blood pressure. Hyperinsulinaemia may in addition increase renal salt and water retention, while the increase in peripheral resistance that accompanies declining physical fitness may be associated with increases in both renin/angiotensin and sympathetic nervous system activity. Taken together with the relationship between physical activity or training and the utilization of lipids and carbohydrates by skeletal muscle, one can see how a level of activity and energy expenditure below some individual or personal threshold may lead to accumulation of metabolic fuels in blood and tissues along with insulin resistance and hypertension and their consequences.

ENERGY BALANCE AND ADIPOSITY

An excess of body fat is evidence of failure of energy balance. It is of limited value debating whether it is due to consuming too much energy or expending too little. Of considerable importance, however, are the answers to the questions, does excess adiposity cause atherosclerotic heart disease for which it is an established risk factor, and if so what causes the breakdown in energy balance which the majority of people manage exquisitely accurately without conscious effort for most of their lives?

Does excess adiposity cause atherosclerosis?

Obesity is a relative latecomer to the risk factors for IHD. Hubert (1984), reviewing data recorded during 26 years' follow-up of the original Framingham cohort, proposed that the degree of adiposity is a significant and independent long term predictor of disease. The risk of cardiovascular disease is most pronounced in those younger than 50 who are most overweight and is independent of the influence of age, blood pressure, serum cholesterol, tobacco use, glucose intolerance and electrocardiographic evidence of left ventricular hypertrophy. No allowance was made in Hubert's analysis for the level of physical activity or energy expenditure. She reviewed also 10 major studies of the relationship between adiposity and IHD risk, concluding that their results, like her own, were consistent with the notion that adiposity plays a fundamental role in the causation of disease, but she made no attempt to speculate on the mechanism.

Lapidus *et al.* (1984), reviewing the data from their studies of adipose tissue distribution and the risk of IHD in both men and women, considered that the strength and independence of the association between excess abdominal fat and IHD spoke in favour of a causal relationship. They, too, offered no explanation other than a suggestion that a sex hormone imbalance may determine the site of the excess adipose tissue.

Dustan (1985) reviewed the association between hypertension and obesity. She pointed out that although there is a common view that obesity causes hypertension, it is quite possible that both are separate manifestations of some other underlying disorder. Physiological changes associated with excess adiposity include an increase in blood volume and cardiac output, hyperinsulinaemia, insulin resistance and impaired glucose tolerance, increased serum TG and cholesterol, increases both in cortisol secretion and in its metabolic clearance and also an increase in sympathetic nervous system activity. Weight reduction may restore most or all of these derangements to normal. However, weight change is itself the result of a fundamental shift in balance between energy intake and energy expenditure. Altering intake of certain nutrients or of energy, as well as changing energy expenditure by a variety of means, may have effects in their own right on any or all of these variables. Dustan was unable to explain the relationship between body weight and blood pressure but seemed to favour, with reservations, an endocrine factor (insulin or aldosterone) as its basis.

If we return to the concept that body fat is an indicator of energy imbalance, our analysis should consider energy balance *per se*. The evidence reviewed so far supports the notion that persistently reduced physical activity and energy expenditure may, if below a given individual's basic fitness threshold, lead to a variety of adverse effects on blood lipids, insulin resistance with its consequences and hypertension. The relationship between obesity and the other key factors may simply be that they are all secondary to physical inactivity. This suggestion, hinted at by Dustan (1985) in regard to hypertension, is made more plausible by Hubert's finding that excess adiposity is often the first risk factor to emerge and may be independent of all the others.

What causes failure of energy balance?

What information do we have pointing to the causes of energy balance failure and excess adiposity? Since both energy intake and energy expenditure (other than basal) are aspects of behaviour, the fundamental cause of their failure to balance, albeit considerably affected by genetic influences, is most probably also behavioural. The evidence from twin studies (Stunkard *et al.* 1986; Sørensen *et al.* 1989; Bouchard *et al.* 1990) underlines the influence of genetic factors but does not identify them. They are frequently assumed to be metabolic but there is no reason to suppose that they may not also involve behavioural characteristics such as an inclination to an active lifestyle or a preoccupation with food. Ravussin *et al.* (1988) showed that a low rate of energy expenditure at rest tends to cluster in families and to predict an increased risk of weight gain in adults. Roberts *et al.* (1988) demonstrated that a low level of energy expenditure is an important factor in excessive weight gain in the first year of life and is, at least in part, an inherited characteristic. It is clear that a low resting energy expenditure is significantly determined by genetic predisposition. What is not clear is why some people avoid physical activity and others find it an essential requirement for their wellbeing.

There has been a wide variety of animal models of obesity but the simplicity of Ingle's (1949) obese rat model is appealing. He confined normal laboratory rats in small individual cages and fed *ad lib.* a palatable, largely carbohydrate diet. On this regimen the animals achieved 200% of ideal weight. With either a palatable diet or restriction of activity alone, weight gain was minimal. In a more complex study in rats, using graded treadmill exercise, Mayer *et al.* (1954) showed that energy expenditure and intake are closely correlated at all levels above a threshold of minimal activity. Like Ingle, he found that inactive animals given a palatable diet lose their precise control of food intake and gain weight. In a later study in workers in a jute mill near Calcutta, Mayer and his colleagues (1956) plotted the energy costs of individuals working at the mill, and commuting on foot or by bicycle, against their food energy intake. Those who were the most active consumed more but were no heavier than those whose efforts were more modest. Here too, energy intake and expenditure correlated at all levels of activity above a 'sedentary threshold'. Below this level, inactive employees (clerks and stall holders) tended to eat more than they required to replace the energy they expended, and were overweight.

On the basis of these and other studies in rats and in man, Mayer (1953) proposed a glucostatic mechanism for regulating food intake. Both protein and fat reserves can sustain the energy needs of the body for months without food intake whereas carbohydrate stores, mostly in the form of muscle glycogen, provide less than a day's energy requirement. Mayer postulated that appetite regulation depended on the extent to which glycogen stores are depleted by the day-to-day muscular work, since food seeking behaviour in humans (and in rats) is initiated several times each day. He showed that hunger was associated with the arteriovenous difference in glucose concentration, a reflection of insulin mediated glucose uptake. Mayer did not consider insulin resistance in this context, but he did attempt to elucidate the apparent anomaly of the hunger experienced by untreated diabetics which had initially been a stumbling block to acceptance of a theory of appetite regulation by blood sugar. Subjects with high blood sugar but decreased utilization (because of insulin deficiency) report feelings of hunger, presumably because insulin dependent glucose uptake into the 'glucostat' sensor cells is low and is perceived as a low availability of blood sugar. One may speculate that in obese subjects the same situation applies and hunger may be experienced because of insulin resistance and reduced glucose uptake in spite of the elevated blood glucose levels that are frequently seen in overweight subjects.

Mayer proposed also that fat and amino acid utilization were in some way regulated by this system, but it has become apparent that fat needs some other form of regulation while the small contribution to total energy intake from protein may be largely unregulated except insofar as it provides a substrate for gluconeogenesis. Flatt (1987) argues that at stable body weight the dietary ratio of fat to carbohydrate must approximate the ratio metabolized to provide energy. In consequence, changes in body composition may occur when there is a mismatch between the ratio of fuels oxidized and the nutrient proportions of the diet. Flatt's model, like Mayer's, requires that food intake is geared primarily to the maintenance of adequate glycogen stores. However, a mechanism by which fat stores are controlled is required not only because fat may provide as much of the daily energy intake as does carbohydrate, but more importantly because for most people body weight is precisely controlled over long periods and change in weight, when it occurs, is predominantly in the adipose tissue mass.

In summary, in a normal healthy subject food intake is probably controlled physiologically by the ebb and flow of glycogen, the bulk of which is stored in and used by skeletal muscle with a smaller reserve in the liver from which the fasting blood sugar level is maintained. Total expended energy is derived in approximately the proportions that fats and carbohydrates contribute energy to the diet (after inclusion of the contribution from protein). Carbohydrate utilization has priority and fat becomes the buffer between the daily intake and expenditure of energy. Fat stored in the portal adipose tissue mass, rich in insulin receptors (Björntorp, 1990), is especially easily mobilized if there is an energy shortfall, and this tissue is also a convenient short term storage site for fatty acids temporarily in excess of requirement. A normally nourished subject maintains fat stores in the region of 20% of body weight (about 15% for men, 25% for women) and daily fluctuations are thus a fraction of 1% of total body fat. However, in subjects who persistently consume an excess of energy it is stored as fat and body weight increases progressively. Should a normal 70 kg man gain 30 kg, his fat store will show a 400% increase while body weight goes up by only 40%. Such a proportional increase in adipose tissue causes an elevation in the blood level of non-esterified fatty acids. Even though in an overweight subject the rate of non-esterified fatty acid release per unit of adipose tissue mass is inhibited by an elevated blood insulin, the large overall increase in the fat mass still produces an absolute elevation in blood non-esterified fatty acids (Flatt, 1987). A feature of the glucose fatty acid cycle (Randle *et al.* 1963) is the restriction imposed on glucose metabolism in muscle by the enhanced release of fatty acids from adipose tissue. Thus there is a clearcut mechanism, which may require a change in body composition, by which an excessive intake of dietary fat will in the long term increase the ratio of fat to carbohydrate metabolized to release energy. A steady state will be reached in which body weight and composition will be maintained until the habitual dietary ratios change or the basic level of activity is altered.

Both the ratio of metabolic fuels metabolized and the total expenditure of energy depend also on the habitual level of activity of the subject. The level of physical training determines the demand for glucose and triacylglycerols and influences the lipoprotein levels in the plasma (Kiens & Lithell, 1989). Endurance training increases the size and number of mitochondria as well as the bulk of the contractile fibres within the cell. It also increases the proportion of type I (slow twitch) fibres which are dependent on the oxidation of fatty acids for most of their energy requirement. At the same time, capillary density in skeletal muscle increases markedly with training (Ingjer & Brodal, 1978). Lipoprotein lipase activity also increases in parallel since this enzyme is deployed on the luminal surface of capillaries in both muscle and adipose tissue. Physical training increases insulin sensitivity by increasing the number of receptors on skeletal muscle cells (Soman *et al.* 1979). According to the provisions of the glucose–fatty acid cycle, the enhanced uptake of glucose inhibits the release of non-esterified fatty acids. Thus training significantly modifies the ratio of metabolic fuels oxidized as well as increasing both basal and total energy flux through the organism.

We can begin to see that Mayer's 'sedentary threshold' may hold the key to avoidance of energy imbalance and an increase in adiposity. Subjects who fail to maintain a minimum level of training lose muscle contractile tissue along with some of the mitochondria which provide ATP from glucose and fatty acids. In addition, nutritive capillaries in muscle disappear along with their inherent lipoprotein lipase activity. Insulin receptors become less responsive and fewer in number. As a result of failure to utilize stored glycogen by muscular work, the normal glucostatic regulation system no longer provides hunger and satiety cues to eat and to stop eating. The ready availability of a highly palatable diet in a culture in which food has important social and hedonic functions may mean that appetite becomes controlled by intellectual triggers which lack both the sensor driven feedback and the precision of normal glucose–insulin regulated appetite. In particular, satiety signals may be absent or overridden and adipose tissue may thus begin to accumulate. In addition, insulin resistance is likely to emerge and encourage a 'false hunger'. Research in this area is very limited and yet it seems that the proximate cause of excess adiposity is a sedentary lifestyle in a congenitally susceptible subject with access *ad lib.* to a palatable diet. Prevention of weight gain very probably requires a level of physical activity or training above an individual's 'sedentary threshold' which may be determined by genetic predisposition.

CONCLUSIONS

The primary prevention of most cardiovascular disease requires that the initial cause of atherosclerosis be identified in order that chronic degenerative changes in the blood vessels can be prevented from developing. Several elements in the vascular occlusive process, once established, are not reversible in any important sense and may be self-sustaining if not actually self-amplifying.

There is insufficient evidence to confirm or deny a causal role for disorders of fat and carbohydrate metabolism or, indeed, for obesity in the initiation of the chronic degenerative disease process in atherosclerosis. However, the predilection of atherosclerotic lesions for sites in blood vessels under the greatest pressure, in those with intermittent flow and at points of greatest hydrodynamic stress, singles out hypertension as the principal aetiological agent in the initiation and maintenance of atherosclerosis.

Primary hypertension appears to have a multiplicity of causes but the time course in both human and experimental hypertension suggests that whatever initiates the early pressure elevation, it is structural change, medial hypertrophy and vascular amplifier effects (Korner, 1994) which contribute increasingly and very largely irreversibly to the persistent elevation in blood pressure. The changes in the vessel wall may be reinforced by the simultaneous development of left ventricular hypertrophy. Thus we are left with the objective of pinpointing the cause of the initial elevation in blood pressure before structural changes in both heart and blood vessels supervene.

The key elements of both the deadly quartet and syndrome X can be seen to follow from habitual inactivity. *Hypertension* is central to both and appears to follow muscle wasting. On the one hand, reduced capillary density in large muscle masses and diminished vasodilator responsiveness in the absence of demand for oxygen increase peripheral vascular resistance. On the other, insulin resistance reduces muscle blood flow, adding to peripheral resistance, while hyperinsulinaemia encourages sodium retention and may stimulate smooth muscle proliferation in the resistance vessels (Stout, 1991).

Glucose intolerance stems from the relatively tiny capacity of the carbohydrate stores, while the pathway for conversion of excess glucose to fatty acids is rudimentary in non-obese humans. If muscle glycogen stores are not regularly depleted by physical activity, prolonged hyperglycaemia and insulin resistance are inevitable consequences.

Hypertriacylglycerolaemia is a direct result of diminished lipoprotein lipase activity which follows reduced capillary density in wasting skeletal muscle. It is also the result of visceral obesity which readily provides the liver with an increased supply of fatty acids as substrate for VLDL synthesis and secretion. Kaplans' quartet includes *upper body obesity* which may be a consequence of positive energy balance combined with hyperinsulinaemia. Reaven's syndrome 'X' identifies *decreased HDL-cholesterol* as a related element. Studies of endurance training show that either the associated weight loss or increased fitness or both raise HDL (Wood *et al.* 1988), while monitoring muscle metabolism during physical training shows that HDL is produced in trained muscle tissue in parallel with the rise in lipoprotein lipase activity consequent upon the increase in capillary density (Kiens & Lithell, 1989). These interconnected consequences of loss of training and muscle wasting (with or without energy imbalance) may explain why risk factors tend to cluster in individuals in the second and third decades when the high energy requirements for growth decline and ultimately cease. Social and economic pressures in affluent societies frequently operate at this time with the change from full time education to economic independence with often marriage, children and home making to limit opportunities for physical exercise sufficient to maintain a minimum level of fitness.

Quite separately, pathological inactivity, because of a reduced blood flow reserve, renders the heart more vulnerable to ischaemia and yet requires it to work harder, even at rest, against an increased peripheral resistance. The resulting hypertrophy of the muscle of the left ventricle adds a further and important risk factor to those already existing as a consequence of inactivity. The high rate of mortality from coronary heart disease can be substantially reduced only if sudden unexpected coronary deaths, perhaps half of all coronary deaths, can be prevented (World Health Organization, 1985*b*). The data from Framingham show that a high resting pulse rate is strikingly associated with sudden cardiac death (Kannel & Thomas, 1982). In contrast to the rapid resting pulse rate associated with hypertension and left ventricular hypertrophy (Staessen *et al.* 1991), a slow resting pulse rate is a dependable indicator of physical fitness (Blomqvist & Saltin, 1983). This is consistent with physical fitness having a direct protective effect against sudden cardiac death.

Returning to consideration of diet as a preventive measure, it is difficult to see an effective preventive role for dietary modification. In wholly sedentary subjects, reduction in total dietary energy may help avoid obesity; maintaining energy balance inhibits the emergence of insulin resistance and hyperinsulinaemia as well as obesity. Reduction in dietary fats and carbohydrates may help to minimize the accumulation of excess lipid and glucose in the blood and tissues. On the other hand levels of anti-oxidant vitamins in the diet may delay the modification of lipids in lipoproteins, rendering them less liable to scavenger pathway uptake by macrophages. Salt restriction if sufficiently profound may assist some sedentary and salt sensitive subjects to reduce their blood pressure. However, there is no evidence that any dietary manipulation will affect total peripheral resistance, resting heart rate, sympathetic outflow or any other cardiodynamic function in a way that reduces the risk of sudden or premature cardiovascular death when the fundamental cause of the elevated risk is pathological inactivity.

Rose (1981), commenting on the population strategy and diet, showed that if 50 young men at average risk changed their diet to lower cholesterol by a mean 10% and continued that diet up to age 55, one heart attack would be prevented while 49 individuals would have eaten 'differently' every day for 40 years and got little or nothing from it. Rose described this as the prevention paradox: "a measure that brings large benefits to the community offers little to each participating individual". There has never been an adequate intervention study in normal individuals to determine whether or not there would be large benefits for the community but studies in high risk groups show that the dietary lowering of cholesterol is difficult to achieve and benefits are usually small and often absent (Ramsay *et al.* 1991). On the other hand Fentem (1992) and others (Eichner, 1983; Paffenbarger *et al.* 1993) have reviewed the evidence for exercise in prevention of disease and shown that exercise enhances not only cardiovascular reserves but also metabolic and psychological function. Thus unlike dietary restriction from which the healthy normal young individual is unlikely

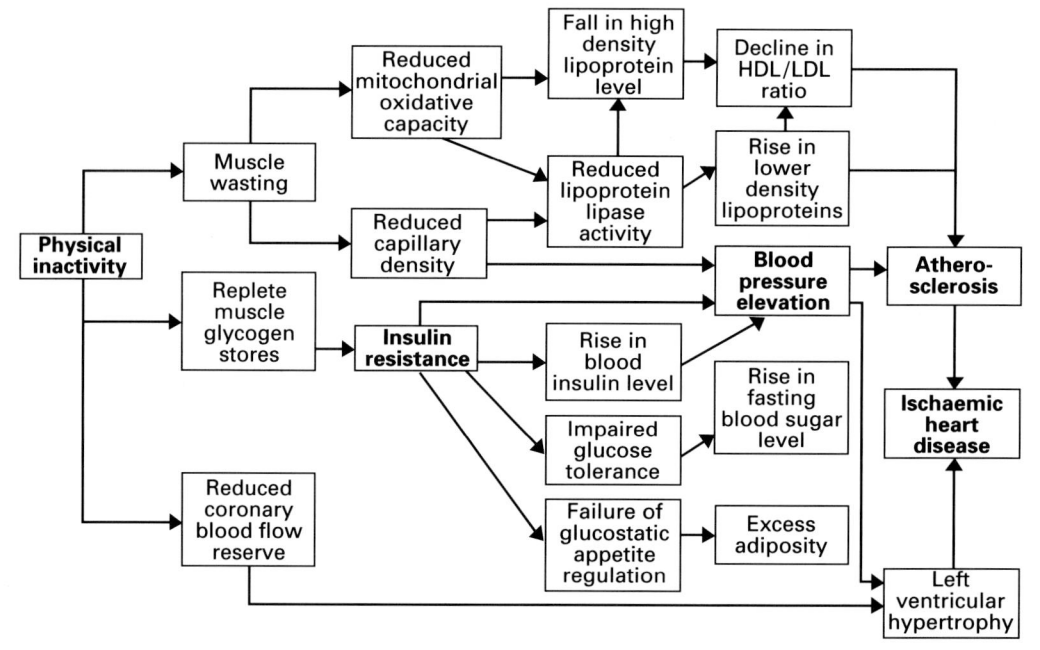

Fig. 1. Physical inactivity leads to underuse of both skeletal and ventricular muscle and failure to deplete glycogen stores in skeletal muscle. Both myocardium and skeletal muscle are left with reduced capacity to work while metabolic fuels tend to accumulate and skeletal myocytes become resistant to insulin-stimulated glucose uptake. The combination of reduced capillary density, insulin resistance and hyperinsulinaemia drive up resting blood pressure. Endothelial damage induced by pressure initiates atherosclerotic vascular disease which is exacerbated by an excess of blood fats, while the increased peripheral resistance to flow causes blood pressure elevation and hypertrophy of the cardiac muscle with increased vulnerability to ischaemic damage and electrical instability. Note that excess adiposity, fasting hyperglycaemia and dyslipidaemia are associated with the risk of ischaemic heart disease but not necessarily causally in this schema. The direct pathogenetic pathway leads from physical inactivity through insulin resistance to hypertension which directly damages the endothelium, initiating and sustaining the atherogenic process.

to see any potential benefit or to get much pleasure, regular vigorous physical activity maintained into middle age may contribute importantly to physical and even psychological wellbeing as well as to the longevity of each active individual. Even here, there is little case for a population approach since human genetic diversity and differing cultural pressures require the informed individual to identify his own determinants of energy balance and needs for physical training from which to devise an acceptable personal strategy for their long term management. However, since excess adiposity is the principal marker for energy balance failure and also, possibly, of failure to surpass the 'sedentary threshold', the monitor for achievement and maintenance of the required level of activity or training may be as simple as regular use of the bathroom scales.

REFERENCES

Ahrens, E. H., Blankenhorn, D. H. & Tsaltas, T. T. (1954). Effect on human serum lipids of substituting plant for animal fat in diet. *Proceedings of the Society for Experimental Biology and Medicine* **86**, 872–878.

Allen, E. M. (1925). *Treatment of Kidney Disease and Blood Pressure*; (Part I). Newark, NJ: Newark Printing Company Inc.

Ambard, L. & Beaujard, E. (1904). [Causes of arterial hypertension.] *Archives Générales de Médecine* **S.H.1**, 520–533.

Barker, D. J. P. (1992). *Fetal and Infant Origins of Adult Disease.* London: British Medical Journal.
Barker, D. J. P., Bull, A. R., Osmond, C. & Simmonds, S. J. (1990). Fetal and placental size and risk of hypertension in adult life. *British Medical Journal* **301**, 259–262.
Baron, A. D., Laakso, M., Brechtel, G. & Edelman, S. V. (1991). Mechanism of insulin resistance in insulin-dependent diabetes mellitus: a major role for reduced skeletal muscle blood flow. *Journal of Clinical Endocrinology and Metabolism* **73**, 637–643.
Baron, A. D., Laakso, M., Brechtel, G., Hoit, B., Watt, C. & Edelman, S. V. (1990). Reduced postprandial skeletal muscle blood flow contributes to glucose intolerance in human obesity. *Journal of Clinical Endocrinology and Metabolism* **70**, 1525–1533.
Beatty, O. L., Harper, R., Sheridan, B., Atkinson, A. B. & Bell, P. M. (1993). Insulin resistance in offspring of hypertensive parents. *British Medical Journal* **307**, 92–96.
Berenson, G. S., Blonde, C. V., Farris, R. P., Foster, T. A., Frank, G. C., Srinivasan, S. R., Voors, A. W. & Webber, L. S. (1979). Cardiovascular disease risk factor variables during the first year of life. *American Journal of Diseases of Childhood* **133**, 1049–1057.
Berenson, G. S., Srinivasan, S. R., Hunter, S. M., Nicklas, T. A., Freedman, D. S., Shear, C. L. & Webber, L. S. (1989). Risk factors in early life as predictors of adult heart disease: the Bogalusa Heart Study. *American Journal of the Medical Sciences* **298**, 141–151.
Björntorp, P. (1987). Adipose tissue distribution, plasma insulin, and cardiovascular disease. *Diabète et Métabolisme* **13**, 381–385.
Björntorp, P. (1990). 'Portal' adipose tissue as a generator of risk factors for cardiovascular disease and diabetes. *Arteriosclerosis* **10**, 493–496.
Björntorp, P. & Sjöström, L. (1978). Carbohydrate storage in man: speculations and some quantitative considerations. *Metabolism* **27**, Suppl. II, 1853–1865.
Blomqvist, C. G. & Saltin, B. (1983). Cardiovascular adaptations to physical training. *Annual Review of Physiology* **45**, 169–189.
Bouchard, C., Tremblay, A., Després, J.-P., Nadeau, A., Lupien, P. J., Thériault, G., Dussault, J., Moorjani, S., Pinault, S. & Fournier, G. (1990). The response to long-term overfeeding in identical twins. *New England Journal of Medicine* **322**, 1477–1482.
Brown, M. S. & Goldstein, J. L. (1984). How LDL receptors influence cholesterol and atherosclerosis. *Scientific American* **251** (5), 58–66.
Brown, M. S. & Goldstein, J. L. (1986). A receptor-mediated pathway for cholesterol homeostasis. *Science* **232**, 34–47.
Brown, M. S. & Goldstein, J. L. (1991). The hyperlipoproteinemias and other disorders of lipid metabolism. In *Harrison's Principles of Internal Medicine*, 12th edn, pp. 1814–1825 [J. D. Wilson, E. Braunwald, K. J. Isselbacher, R. G. Petersdorf, J. B. Martin, A. S. Fauci & R. K. Root, editors]. New York: McGraw-Hill.
Dahl, L. K. (1972). Salt and hypertension. *American Journal of Clinical Nutrition* **25**, 232–244.
Dahl, L. K. & Love, R. A. (1954). Evidence for a relationship between sodium chloride intake and human essential hypertension. *Archives of Internal Medicine* **94**, 525–531.
Dahl, L. K., Knudsen, K. D., Heine, M. A. & Leitl, G. J. (1968). Effects of chronic excess salt ingestion: modification of experimental hypertension in the rat by variations in the diet. *Circulation Research* **22**, 11–18.
Dahl, L. K., Silver, L. & Christie, R. (1958). Role of salt in the fall of blood pressure accompanying reduction of obesity. *New England Journal of Medicine* **258**, 1186–1192.
Davies, M. (1992). Atherosclerosis – can regression be achieved? *Cardiology in Practice* **10**, 4–5.
DeFronzo, R. A. (1988). The triumvirate: β-cell, muscle, liver: a collusion responsible for NIDDM. *Diabetes* **37**, 667–687.
DeFronzo, R. A., Cooke, C. R., Andres, R., Faloona, G. R. & Davis, P. J. (1975). The effect of insulin on renal handling of sodium, potassium, calcium, and phosphate in man. *Journal of Clinical Investigation* **55**, 845–855.
DeFronzo, R. A., Tobin, J. D. & Andres, R. (1979). Glucose clamp technique: a method for quantifying insulin secretion and resistance. *American Journal of Physiology* **237**, E214–E223.
De Wardener, H. E. (1990). The primary role of the kidney and salt intake in the aetiology of essential hypertension. *Clinical Science* **79**, 193–200, 289–297.
Dietschy, J. M., Turley, S. D. & Spady, D. K. (1993). Role of liver in the maintenance of cholesterol and low density lipoprotein homeostasis in different animal species, including humans—review. *Journal of Lipid Research* **34**, 1637–1659.
Dustan, H. P. (1985). Obesity and hypertension. *Annals of Internal Medicine* **103**, 1047–1049.
Eichner, E. R. (1983). Exercise and heart disease: epidemiology of the 'exercise hypothesis'. *American Journal of Medicine* **75**, 1008–1023.
Ekblom, B. (1971). Physical training in normal boys in adolescence. *Acta Paediatrica Scandinavica* Suppl. 217, 60–62.
Eriksson, J., Franssila-Kallunki, A., Ekstrand, A., Saloranta, C., Widén, E., Schalin, C. & Groop, L. (1989). Early metabolic defects in persons at increased risk for non-insulin dependent diabetes mellitus. *New England Journal of Medicine* **321**, 337–343.
Faggiotto, A. & Ross, R. (1984). Studies of hypercholesterolemia in the non-human primate. II. Fatty streak conversion to fibrous plaque. *Arteriosclerosis* **4**, 341–356.

Faggiotto, A., Ross, R. & Harker, L. (1984). Studies of hypercholesterolemia in the non-human primate. I. Changes that lead to fatty streak formation. *Arteriosclerosis* **4**, 323–340.
Fentem, P. H. (1992). Exercise in prevention of disease. *British Medical Bulletin* **48**, 630–650.
Flatt, J. P. (1987). Dietary fat, carbohydrate balance, and weight maintenance: effects of exercise. *American Journal of Clinical Nutrition* **45**, 296–306.
Fomon, S. J., Haschke, F., Ziegler, E. E. & Nelson, S. E. (1982). Body composition of reference children from birth to age 10 years. *American Journal of Clinical Nutrition* **35**, 1169–1175.
Frank, G. C., Berenson, G. S. & Webber, L. S. (1978). Dietary studies and the relationship of diet to cardiovascular disease risk factor variables in 10-year-old children: the Bogalusa Heart Study. *American Journal of Clinical Nutrition* **31**, 328–340.
Fraser, G. E., Phillips, R. L. & Harris, R. (1983). Physical fitness and blood pressure in school children. *Circulation* **67**, 405–411.
Freedman, D. S., Srinivasan, S. R., Burke, G. L., Shear, C. L., Smoak, C. G., Harsha, D. W., Webber, L. S. & Berenson, G. S. (1987). Relation of body fat distribution to hyperinsulinemia in children and adolescents: the Bogalusa Heart Study. *American Journal of Clinical Nutrition* **46**, 403–410.
Frohlich, E. D., Messerli, F. H., Reisin, E. & Dunn, F. G. (1983). The problems of obesity and hypertension. *Hypertension* **5** Suppl., III71–III78.
Garvey, W. T., Olefsky, J. M. & Marshall, S. (1986). Insulin induces progressive insulin resistance in cultured rat adipocytes: sequential effects at receptor and multiple post receptor sites. *Diabetes* **35**, 258–267.
Geleijnse, J. M., Grobbee, D. E. & Hofman, A. (1990). Sodium and potassium intake and blood pressure change in childhood. *British Medical Journal* **300**, 899–902.
Goldstein, J. L., Ho, Y. K., Basu, S. K. & Brown, M. S. (1979). Binding site on macrophages that mediates uptake and degradation of acetylated low density lipoprotein, producing massive cholesterol deposition. *Proceedings of the National Academy of Sciences, USA* **76**, 333–337.
Gordon, T. (1970). The Framingham Diet Study: diet and the regulation of serum cholesterol. In *The Framingham Study: an Epidemiological Investigation of Cardiovascular Disease,* Section 24 [W. B. Kannel and T. Gordon, editors]. Washington, DC: Government Printing Office.
Grobbee, D. E., van Hooft, I. M. S. & de Man, S. A. (1990). Determinants of blood pressure in the first decades of life. *Journal of Cardiovascular Pharmacology* **16**, Suppl. 7, S71–S74.
Guyton, A. C. (1989). Dominant role of the kidneys and accessory role of whole-body autoregulation in the pathogenesis of hypertension. *American Journal of Hypertension* **2**, 575–585.
Hagberg, J. M., Ehsani, A. S., Goldring, D., Hernandez, A., Sinacore, D. R. & Holloszy, J. O. (1984). Effect of weight training on blood pressure and hemodynamics in hypertensive adolescents. *Journal of Pediatrics* **104**, 147–151.
Havlik, R. J., Hubert, H. B., Fabsitz, R. R. & Feinleib, M. (1983). Weight and hypertension. *Annals of Internal Medicine* **98**, 855–859.
Hegsted, D. M., McGandy, R. B., Myers, M. L. & Stare, F. J. (1965). Quantitative effects of dietary fat on serum cholesterol in man. *American Journal of Clinical Nutrition* **17**, 281–295.
Henriksen, T., Mahoney, E. M. & Steinberg, D. (1981). Enhanced macrophage degradation of low density lipoprotein previously incubated with cultured endothelial cells: recognition by receptors for acetylated low density lipoproteins. *Proceedings of the National Academy of Sciences, USA* **78**, 6499–6503.
Hessler, J. R., Robertson, A. L. & Chisholm, G. M. (1979). LDL induced cytotoxicity and its inhibition by HDL in human vascular smooth muscle and endothelial cells in culture. *Atherosclerosis* **32**, 213–229.
Hofman, A., Walter, H. J., Connelly, P. A. & Vaughan, R. D. (1987). Blood pressure and physical fitness in children. *Hypertension* **9**, 188–191.
Hubert, H. B. (1984). The nature of the relationship between obesity and cardiovascular disease. *International Journal of Cardiology* **6**, 268–274.
Inadera, H., Ito, S., Ishikana, Y., Shinomiya, M., Shirai, K., Saito, Y. & Yoshida, S. (1993). Visceral fat deposition is seen in patients with insulinoma. *Diabetologia* **36**, 91–92.
Ingjer, F. & Brodal, P. (1978). Capillary supply of skeletal muscle fibers in untrained and endurance-trained women. *European Journal of Applied Physiology* **38**, 291–299.
Ingle, D. J. (1949). A simple means of producing obesity in the rat. *Proceedings of the Society for Experimental Biology and Medicine* **72**, 604–605.
Jarrett, R. J. (1988). Is insulin atherogenic? *Diabetologia* **31**, 71–75.
Jarrett, R. J. (1992). In defence of insulin: a critique of syndrome X. *Lancet* **340**, 469–471.
Jennings, G., Nelson, L., Nestel, P., Esler, M., Korner, P., Burton, D. & Bazelmans, J. (1986). The effects of changes in physical activity on major cardiovascular risk factors, hemodynamics, sympathetic function, and glucose utilization in man: a controlled study of four levels of activity. *Circulation* **73**, 30–40.
Julius, S., Gudbrandsson, T., Jamerson, K., Shahab, S. T. & Andersson, O. (1991). The hemodynamic link between insulin resistance and hypertension. *Journal of Hypertension* **9**, 983–986.
Kahn, H. A., Medalie, J. H., Neufeld, H. N., Riss, E., Balogh, M. & Groen, J. J. (1969). Serum cholesterol: its distribution and association with dietary and other variables in a survey of 10,000 men. *Israel Journal of Medical Sciences* **5**, 1117–1127.
Kannel, W. B. & Thomas, H. E. (1982). Sudden coronary death: The Framingham Study. *Annals of the New York Academy of Sciences* **382**, 3–20.

Kaplan, N. M. (1989). The deadly quartet: upper-body obesity, glucose intolerance, hypertriglyceridemia, and hypertension. *Archives of Internal Medicine* **149**, 1514–1520.

Kempner, W. (1948). Treatment of hypertensive vascular disease with rice diet. *American Journal of Medicine* **4**, 545–577.

Keys, A., Anderson, J. T. & Grande, F. (1957). Prediction of serum-cholesterol responses of man to changes in fats in the diet. *Lancet* **ii**, 959–966.

Kiens, B. & Lithell, H. (1989). Lipoprotein metabolism influenced by training-induced changes in human skeletal muscle. *Journal of Clinical Investigation* **83**, 558–564.

Kinsell, L. W., Partridge, J., Boling, L., Margen, S. & Michaels, G. (1952). Dietary modification of serum cholesterol and phospholipide levels. *Journal of Clinical Endocrinology and Metabolism* **12**, 909–913.

Kohner, E. M. (1993). Diabetic retinopathy. *British Medical Journal* **307**, 1195–1199.

Korner, P. I. (1994). Some thoughts on pathogenesis, therapy and prevention of hypertension. *Blood Pressure* **3**, 7–17.

Kramsch, D. M., Aspen, A. J., Abramowitz, B. M., Kreimendahl, T. & Hood, W. B. (1981). Reduction of coronary atherosclerosis by moderate conditioning exercise in monkeys on an atherogenic diet. *New England Journal of Medicine* **305**, 1483–1489.

Krotkiewski, M., Mandroukas, K., Sjöström, L., Sullivan, L., Wetterqvist, H. & Björntorp, P. (1979). Effects of long-term physical training on body fat, metabolism, and blood pressure in obesity. *Metabolism* **28**, 650–658.

Lapidus, L., Bengtsson, C., Larsson, B., Pennert, K., Rybo, E. & Sjöström, L. (1984). Distribution of adipose tissue and risk of cardiovascular disease and death: a 12 year follow up of participants in the population study of women in Gothenburg, Sweden. *British Medical Journal* **289**, 1257–1261.

Law, M. R., Frost, C. D. & Wald, N. J. (1991). By how much does dietary salt reduction lower blood pressure? I. Analysis of observational data among populations. *British Medical Journal* **302**, 811–815.

Lorenzi, M. (1992). Glucose toxicity in the vascular complications of diabetes: the cellular perspective. *Diabetes & Metabolism Review* **8**, 85–103.

McNamara, D. J., Kolb, R., Parker, T. S., Batwin, H., Samuel, P., Brown, C. D. & Ahrens, E. H. (1987). Heterogeneity of cholesterol homeostasis in man: responses to changes in dietary fat quality and cholesterol quantity. *Journal of Clinical Investigation* **79**, 1729–1739.

Martin, M. J., Hulley, S. B., Browner, W. S., Kuller, L. H. & Wentworth, D. (1986). Serum cholesterol, blood pressure, and mortality: implications from a cohort of 361 662 men. *Lancet* **ii**, 933–936.

Mayer, J. (1953). Glucostatic mechanism of regulation of food intake. *New England Journal of Medicine* **249**, 13–16.

Mayer, J., Marshall, N. B., Vitale, J. J., Christensen, J. H., Mashayekhi, M. B. & Stare, F. J. (1954). Exercise, food intake and body weight in normal rats and genetically obese adult mice. *American Journal of Physiology* **177**, 544–548.

Mayer, J., Roy, P. & Mitra, K. P. (1956). Relation between calorie intake, body weight, and physical work: studies in an industrial male population in West Bengal. *American Journal of Clinical Nutrition* **4**, 169–175.

Messerli, F. H. (1982). Cardiovascular effects of obesity and hypertension. *Lancet* **i**, 1165–1168.

Miller, J. H. & Bogdonoff, M. D. (1954). Antidiuresis associated with administration of insulin. *Journal of Applied Physiology* **6**, 509–512.

Miller, J. Z., Weinberger, M. H., Daugherty, S. A., Fineberg, N. S., Christian, J. C. & Grim, C. E. (1987). Heterogeneity of blood pressure response to dietary sodium restriction in normotensive adults. *Journal of Chronic Diseases* **40**, 245–250.

Modan, M., Halkin, H., Fuchs, Z., Lusky, A., Chetrit, A., Segal, P., Eshkol, A., Almog, S. & Shefi, M. (1987). Hyperinsulinemia: a link between glucose intolerance, obesity, hypertension, dyslipoproteinemia, elevated serum uric acid and internal cation imbalance. *Diabète et Métabolisme* **13**, 375–380.

Moller, D. E. & Flier, J. S. (1991). Insulin resistance—mechanisms, syndromes, and implications. *New England Journal of Medicine* **325**, 938–948.

Morris, J. N., Marr, J. W., Heady, J. A., Mills, G. L. & Pilkington, T. R. E. (1963). Diet and plasma cholesterol in 99 bank men. *British Medical Journal* **i**, 571–576.

Newman, W. P., Freedman, D. S., Voors, A. W., Gard, P. D., Srinivasan, S. R., Cresanta, J. L., Williamson, G. D., Webber, L. S. & Berenson, G. S. (1986). Relation of serum lipoprotein levels and systolic blood pressure to early atherosclerosis: the Bogalusa Heart Study. *New England Journal of Medicine* **314**, 138–144.

Nichols, A. B., Ravenscroft, C., Lamphiear, D. E. & Ostrander, L. D. (1976). Independence of serum lipid levels and dietary habits: the Tecumseh Study. *Journal of the American Medical Association* **236**, 1948–1953.

Ornish, D., Brown, S. E., Scherwitz, L. W., Billings, J. H., Armstrong, W. T., Ports, T. A., McLanahan, S. M., Kirkeeide, R. L., Brand, R. J. & Gould, K. L. (1990). Can lifestyle changes reverse coronary heart disease? *Lancet* **336**, 129–133.

Paffenbarger, R. S., Hyde, R. T., Wing, A. L., Lee, I.-M., Jung, D. L. & Kampert, J. B. (1993). The association of changes in physical-activity level and other lifestyle characteristics with mortality among men. *New England Journal of Medicine* **328**, 538–545.

Pilbeam, D. (1984). The descent of hominoids and hominids. *Scientific American* **250** (3), 60–69.

Quinn, D., Shiraj, K. & Jackson, R. L. (1982). Lipoprotein lipase: mechanism of action and role in lipoprotein metabolism. *Progress in Lipid Research* **22**, 35–78.

Ramsay, L. E., Yeo, W. W. & Jackson, P. R. (1991). Dietary reduction of serum cholesterol concentration: time to think again. *British Medical Journal* **303**, 953–957.
Randle, P. J., Garland, P. B., Hales, C. N. & Newsholme, E. A. (1963). The glucose fatty-acid cycle: its role in insulin sensitivity and the metabolic disturbances of diabetes mellitus. *Lancet* **i**, 785–789.
Rauramaa, R., Salonen, J. T., Kukkonen-Harjula, K., Seppänen, K., Seppälä, E., Vapaatalo, H. & Huttunen, J. K. (1984). Effects of mild physical exercise on serum lipoproteins and metabolites of arachidonic acid: a controlled randomised trial in middle aged men. *British Medical Journal* **288**, 603–606.
Ravussin, E., Lillioja, S., Knowler, W. C., Christin, L., Freymond, D., Abbott, W. G. H., Boyce, V., Howard, B. V. & Bogardus, C. (1988). Reduced rate of energy expenditure as a risk factor for body-weight gain. *New England Journal of Medicine* **318**, 467–472.
Reaven, G. M. (1988). Role of insulin resistance in human disease. *Diabetes* **37**, 1595–1607.
Reichl, D., Myant, N. B., Brown, M. S. & Goldstein, J. L. (1978). Biologically active low density lipoprotein in human peripheral lymph. *Journal of Clinical Investigation* **61**, 64–71.
Richter, E. A., Hansen, S. A. & Hansen, B. F. (1988a). Mechanisms limiting glycogen storage in muscle during prolonged insulin stimulation. *American Journal of Physiology* **255**, E621–E628.
Richter, E. A., Hansen, B. F. & Hansen, S. A. (1988b). Glucose-induced insulin resistance of skeletal muscle glucose transport and uptake. *Biochemical Journal* **252**, 733–737.
Roberts, S. B., Savage, J., Coward, W. A., Chew, B. & Lucas, A. (1988). Energy expenditure and intake in infants born to lean and overweight mothers. *New England Journal of Medicine* **318**, 461–466.
Romijn, J. A., Klein, S., Coyle, E. F., Sidossis, L. S. & Wolfe, R. R. (1993). Strenuous endurance training increases lipolysis and triglyceride-fatty acid cycling at rest. *Journal of Applied Physiology* **75**, 108–113.
Rose, G. (1981). Strategy of prevention: lessons from cardiovascular disease. *British Medical Journal* **282**, 1847–1851.
Rosenthal, M., Haskell, W. L., Solomon, R., Widstrom, A. & Reaven, G. M. (1983). Demonstration of a relationship between level of physical training and insulin stimulated glucose utilization in normal humans. *Diabetes* **32**, 408–411.
Ross, R. (1986). The pathogenesis of atherosclerosis – an update. *New England Journal of Medicine* **314**, 488–500.
Ross, R. & Glomset, J. A. (1976). The pathogenesis of atherosclerosis. *New England Journal of Medicine* **295**, 369–377, 420–425.
Schwartz, S. M. & Benditt, E. P. (1977). Aortic endothelial cell replication. I. Effects of age and hypertension in the rat. *Circulation Research* **41**, 248–255.
Shulman, G. I., Rothman, D. L., Jue, T., Stein, P., DeFronzo, R. A. & Shulman, R. G. (1990). Quantitation of muscle glycogen synthesis in normal subjects and subjects with non-insulin-dependent diabetes by ^{13}C nuclear magnetic resonance spectroscopy. *New England Journal of Medicine* **322**, 223–228.
Slater, E. (1991). Insulin resistance and hypertension. *Hypertension* **18** (3), Suppl., 108–114.
Soman, V. R., Koivisto, V. A., Deibert, D., Felig, P. & DeFronzo, R. A. (1979). Increased insulin sensitivity and insulin binding to monocytes after physical training. *New England Journal of Medicine* **301**, 1200–1204.
Somers, V. K., Conway, J., Johnston, J. & Sleight, P. (1991). Effects of endurance training on baroreflex sensitivity and blood pressure in borderline hypertension. *Lancet* **337**, 1363–1368.
Sørensen, T. I. A., Price, R. A., Stunkard, A. J. & Schulsinger, F. (1989). Genetics of obesity in adult adoptees and their biological siblings. *British Medical Journal* **298**, 87–90.
Sporik, R., Johnstone, J. H. & Cogswell, J. J. (1991). Longitudinal study of cholesterol values in 68 children from birth to 11 years of age. *Archives of Disease in Childhood* **66**, 134–137.
Staessen, J., Bulpitt, C. J., Thijs, L., Fagard, R., Joossens, J. V., Van Hoof, R. & Amery, A. (1991). Pulse rate and sodium intake interact to determine blood pressure. *American Journal of Hypertension* **4**, 107–112.
Stamler, R., Stamler, J., Riedlinger, W. F., Algera, G. & Roberts, R. H. (1978). Weight and blood pressure: findings in hypertension screening of 1 million Americans. *Journal of the American Medical Association* **240**, 1607–1610.
Steinberg, D. (1983). Lipoproteins and atherosclerosis: a look back and a look ahead. *Arteriosclerosis* **3**, 283–301.
Stout, R. W. (1968). Insulin-stimulated lipogenesis in arterial tissue in relation to diabetes and atheroma. *Lancet* **ii**, 702–703.
Stout, R. W. (1991). Insulin as a mitogenic factor: role in the pathogenesis of cardiovascular disease. *American Journal of Medicine* **90**, Suppl. 2A, 62S–65S.
Stout, R. W. & Vallance-Owen, J. (1969). Insulin and atheroma. *Lancet* **i**, 1078–1080.
Strong, J. P. & McGill, H. C. (1969). The pediatric aspects of atherosclerosis. *Journal of Atherosclerosis Research* **9**, 251–265.
Stunkard, A. J., Foch, T. T. & Hrubec, Z. (1986). A twin study of human obesity. *Journal of the American Medical Association* **256**, 51–54.
Sundram, K., Hayes, K. C. & Siru, O. H. (1994). Dietary palmitic acid results in lower serum cholesterol than does a lauric-myristic acid combination in normolipemic humans. *American Journal of Clinical Nutrition* **59**, 841–846.
Task Force on Blood Pressure Control in Children. (1987). Report of the second task force on blood pressure control in children – 1987. *Pediatrics* **79**, 1–25.
Texon, M. (1980). *Hemodynamic Basis of Atherosclerosis*. New York, Hemisphere.

Texon, M. (1986). The hemodynamic basis of atherosclerosis. Further observations: the linear lesion. *Bulletin of the New York Academy of Medicine* **62**, 875–880.

Tran, Z. V. & Weltman, A. (1985). Differential effects of exercise on serum lipids and lipoprotein levels seen with changes in body weight – a meta analysis. *Journal of the American Medical Association* **254**, 919–924.

Van Hooft, I. M. S., Hofman, A., Grobbee, D. E. & Valkenburg, H. A. (1988). Change in blood pressure in offspring of parents with high or low blood pressure: the Dutch hypertension and offspring study. *Journal of Hypertension* **6**, Suppl. 4, S594–S596.

Virchow, R. (1858). *Die Cellularpathologie in ihrer Begründung auf physiologische und pathologische Gewebelehre.* Berlin: Hirschwald.

Voors, A. W., Webber, L. S. & Berenson, G. S. (1979). Time course studies of blood pressure in children – the Bogalusa heart study. *American Journal of Epidemiology* **109**, 320–334.

Webber, L. S., Voors, A. W., Srinivasan, S. R., Frerichs, R. R. & Berenson, G. S. (1979). Occurrence in children of multiple risk factors for coronary artery disease: the Bogalusa Heart Study. *Preventive Medicine* **8**, 407–418.

Weintraub, M. S., Rosen, Y., Otto, R., Eisenberg, S. & Breslow, J. L. (1989). Physical exercise conditioning in the absence of weight loss reduces fasting and postprandial triglyceride-rich lipoprotein levels. *Circulation* **79**, 1007–1014.

Wong, N. D., Hei, T. K., Qaqundah, P. Y., Davidson, D. M., Bassin, S. L. & Gold, K. V. (1992). Television viewing and pediatric hypercholesterolemia. *Pediatrics* **90**, 75–79.

Wood, P. D., Haskell, W. L., Blair, S. N., Williams, P. T., Krauss, R. M., Lindgren, F. T., Albers, J. J., Ho, P. H. & Farquhar, J. W. (1983). Increased exercise level and plasma lipoprotein concentrations: a one-year, randomized, controlled study in sedentary, middle-aged men. *Metabolism* **32**, 31–39.

Wood, P. D., Stefanick, M. L., Dreon, D. M., Frey-Hewitt, B., Garay, S. C., Williams, P. T., Superko, H. R., Fortmann, S. P., Albers, J. J., Vranizan, K. M., Ellsworth, N. M., Terry, R. B. & Haskell, W. L. (1988). Changes in plasma lipids and lipoproteins in overweight men during weight loss through dieting as compared with exercise. *New England Journal of Medicine* **319**, 1173–1179.

World Health Organization (1985a). *Blood Pressure Studies in Children.* Technical Report Series No. 715. Geneva: WHO.

World Health Organization (1985b). *Sudden Cardiac Death.* Technical Report Series No. 726. Geneva: WHO.

Young, A. A., Bogardus, C., Wolfe-Lopez, D. & Mott, D. M. (1988). Muscle glycogen synthesis and disposition of infused glucose in humans with reduced rates of insulin-mediated carbohydrate storage. *Diabetes* **37**, 303–308.

NUTRITION, PHYSICAL ACTIVITY AND BONE HEALTH IN WOMEN

J. H. WILSON*

British Olympic Medical Centre, Northwick Park Hospital, Harrow, Middlesex HA1 3UJ

CONTENTS

INTRODUCTION	68
NUTRITIONAL NEEDS OF THE ATHLETE	70
ENERGY EXPENDITURE	70
THE PROBLEMS OF NUTRITIONAL ASSESSMENT	70
ENERGY INTAKE IN FEMALE ATHLETES	70
Sports in which low weight is of no direct benefit	72
Sports in which low weight may improve performance	72
Sports which are judged on aesthetic appeal	72
Sports in which the athlete competes in a weight category	73
THE RELATIONSHIP BETWEEN AMENORRHOEA AND LOW ENERGY INTAKE	73
ENERGY INTAKE OF AMENORRHOEIC ATHLETES COMPARED TO EUMENORRHOEIC SUBJECTS	73
CONSTITUENTS OF THE DIET	75
VEGETARIANISM	75
VITAMINS AND TRACE MINERALS	76
EATING DISORDERS	76
EATING DISORDERS IN THE GENERAL POPULATION	77
EATING DISORDERS IN ATHLETES	77
PREVALENCE OF EATING DISORDERS IN DIFFERENT ATHLETIC GROUPS	78
Sports in which low weight is not an advantage	78
Sports in which low weight may improve performance	78
Sports which are judged on aesthetic appeal	79
Sports in which the athlete competes in a weight category	79
IS THERE A RELATIONSHIP BETWEEN DISORDERED EATING IN ATHLETES AND MENSTRUAL DYSFUNCTION?	79
BONE METABOLISM IN ATHLETES	80
THE EFFECT OF EXERCISE ON BONE	80
THE EFFECT OF MENSTRUAL IRREGULARITY ON BONE	81
MECHANISM OF REDUCED BONE DENSITY IN AMENORRHOEA	81
CALCIUM INTAKE AND BONE DENSITY IN ATHLETES	82

* Present address: Clinical Lecturer in Rheumatology, Royal Oldham Hospital, Rochdale Road, Oldham, OL1 2JH.

CONSEQUENCES OF LOW BONE DENSITY IN ATHLETES	82
STRESS FRACTURES	82
MUSCULOSKELETAL INJURIES	83
LONG TERM CONSEQUENCES	83
PREVENTION, ASSESSMENT AND MANAGEMENT OF ATHLETIC AMENORRHOEA	84
EDUCATION	84
IDENTIFICATION OF 'AT RISK' ATHLETES	84
OPTIONS FOR TREATMENT	84
CONCLUSION	86
REFERENCES	86

INTRODUCTION

During the last two decades there has been a marked increase in the number of women undertaking physical activity at both competitive and recreational levels. In particular, the number of women involved in endurance aerobic exercise, such as running, has increased rapidly. Regular physical activity is recognized to have many beneficial health related effects, not least of which is a reduction in the rate of hip fracture due to osteoporosis in later life (Cooper *et al.* 1988). For this reason, exercise is now promoted as having a role in maintaining bone mineral density (BMD) and preventing the development of osteoporosis. However, there is evidence that intensive exercise at a young age may paradoxically reduce BMD.

Such changes in BMD have been found to be linked to profound alterations in the hypothalamic–pituitary–ovarian pathway in athletes that train intensively. Oligomenorrhoea, amenorrhoea, short luteal phases and late menarche have all been demonstrated with increased frequency in such athletes. Whereas secondary amenorrhoea occurs in only 2–5% of the general population it is common in sports in which thinness may be an added advantage to performance, such as running, gymnastics, ice skating and dancing. In the Great Britain National squads of these sports in 1988, 50–100% of women had some form of menstrual dysfunction (Wolman & Harries, 1989). Although oligomenorrhoea and amenorrhoea due to exercise were initially thought to be relatively benign, it is now known that the hypo-oestrogenic state is associated with reduced bone density. Some regions of the skeleton, for instance the vertebral bodies, may be as much as 25% lower in BMD in amenorrhoeic women compared to their eumenorrhoeic peers. In the short term, musculoskeletal injuries are more common, and in the long term concern has arisen that there may be a risk of early osteoporosis and fracture.

Menstrual dysfunction in athletes is probably multifactorial (Myburgh *et al.* 1992; Fig. 1); severity of training (Feicht *et al.* 1978; Sanborn *et al.* 1982; Drinkwater *et al.* 1984, 1986), low body fat (Frisch & McArthur, 1974; Sanborn *et al.* 1982), low weight for height (Shangold & Levine, 1982; Marcus *et al.* 1985; Drinkwater *et al.* 1986; Glass *et al.* 1987; Wolman & Harries, 1989; Harber *et al.* 1991; Myerson *et al.* 1992) and chronological age (Drinkwater *et al.* 1986; Glass *et al.* 1987; Snead *et al.* 1992) have all been suggested as playing a part. Amenorrhoeic athletes are also more likely to have had a later menarche (Frisch *et al.* 1981; Drinkwater *et al.* 1986; Casey *et al.* 1991), to have started training either before or soon after menarche (Frisch *et al.* 1981; Drinkwater *et al.* 1986; Snead *et al.* 1992) and to have had a previous episode of irregular periods (Shangold & Levine, 1982; Lloyd *et al.* 1987). But recent evidence suggests that inadequate nutritional intake and abnormal dietary habits may play a pivotal role in the development of menstrual disorders. Athletes

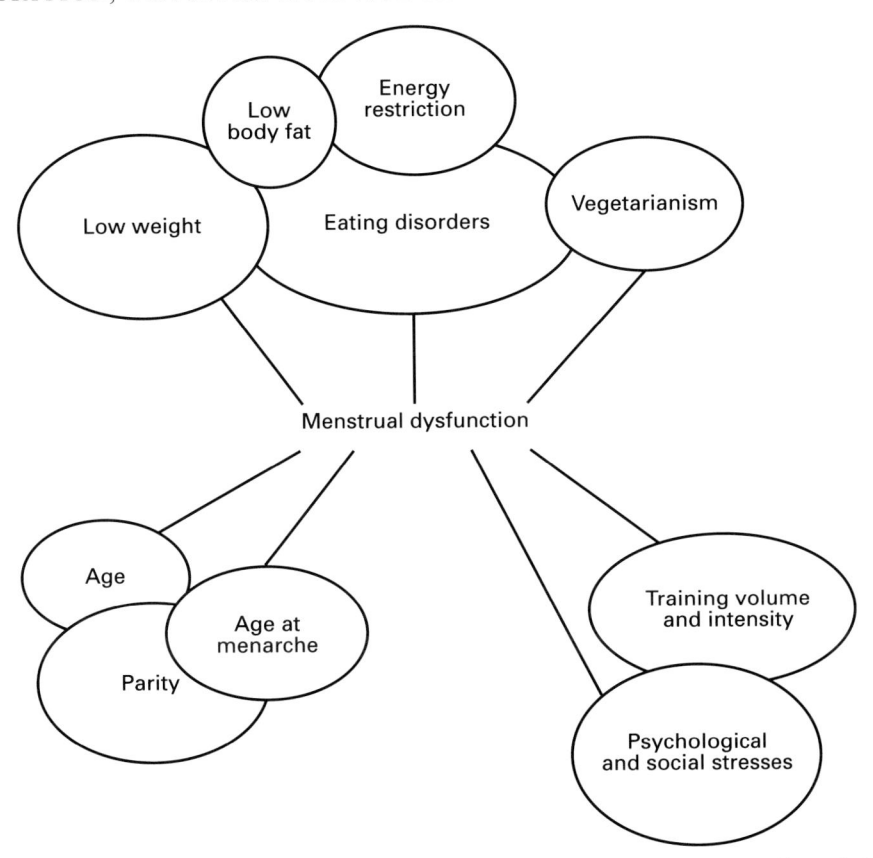

Fig. 1. Non-pathological factors likely to contribute to the development of menstrual irregularity in athletes. Those women with the greatest number of risk factors are most likely to develop oligomenorrhoea.

strive hard to attain the perfect shape and this combines with social and cultural pressures to reward a thin, sylph-like figure. In their efforts to achieve the ideal they may restrict food intake or use abnormal methods for weight reduction despite their high energy requirements. In some, frank eating disorders develop. The consequence of this can be a suppression of normal menstrual function. Attention has therefore focused on the proposed relationship between disordered eating habits, calorie restriction, menstrual irregularity and reduced bone density in female athletes. Early identification of those at risk may reduce the incidence of injury and low bone density.

The mechanisms by which any of the above risk factors induce hypothalamic suppression and hence menstrual dysfunction are unclear. One possibility is that activation of the adrenal axis during exercise inhibits the pulsatile release of gonadotrophin releasing hormone. Consistent with this hypothesis is the finding that cortisol levels are mildly elevated in amenorrhoeic women (Ding et al. 1988). Other putative mediators are thyroxine, endorphins and catecholoestrogens, all of which show changes with exercise, but no evidence is currently available to support a direct link. Of course, in any case of primary or secondary amenorrhoea pathological causes must be excluded and the diagnosis of exercise induced amenorrhoea depends on this.

This review assesses the evidence for nutritional disorders in female athletes and the relationship between these and menstrual abnormalities. The impact of menstrual dysfunction on bone metabolism is reviewed and the potential for intervention is discussed. Readers are referred to reviews by Loucks (1990), De Souza & Metzger (1991) and Loucks *et al.* (1992) for more detailed data on the hormonal aspects of athletic amenorrhoea.

NUTRITIONAL NEEDS OF THE ATHLETE
ENERGY EXPENDITURE

Energy expenditure can be extremely high during competition and training (McArdle *et al.* 1986) and in some sports, such as cycle racing and triathlon, athletes may approach the limit for working capacity on a sustained basis. Energy expenditure of cyclists during 10 days of competitive cycling (total race distance 1600 km) has been estimated to be as high as 4110 (SD 504) kcal d^{-1} (Snyder *et al.* 1991) and during the Tour de France cyclists may maintain a daily expenditure of 3·5–5·5 times the basal metabolic rate. Although races such as the Tour de France exemplify the enormous energy requirements of some athletes, more moderate levels of competition and training also make high demands. From McArdle *et al.* (1986) it can be estimated that a 50 kg female long distance runner completing a marathon in 2 h 37 min (6 min mile^{-1} pace) has an energy expenditure of 1978 kcal for that race alone.

Nutrition of the correct type and quantity is therefore of great importance to the athlete. In order to maintain energy balance it would be expected that energy intakes would be much higher in athletes than in sedentary peers. Yet recent literature suggests that many athetes, particularly females, do not adequately balance their energy demands with nutritional intake.

THE PROBLEMS OF NUTRITIONAL ASSESSMENT

During the last 10–15 years a number of studies have estimated the calorie intake of female athletes for a variety of sports (Table 1). These surveys have generally used three or seven day dietary records, in some cases backed up by 24 h recall. There are doubts of the accuracy of such records, particularly because of the tendency to underestimate intake (Gersovitz *et al.* 1978). Although some have demonstrated good correlation between measured food intake and dietary recall (Stunkard & Waxman, 1981; Karvetti & Knutts, 1992), Schoeller *et al.* (1990) reported large differences between 24-h recall methods and the double labelled water method in elite athletes. Similar disparities have also been shown in non-athletic subjects (Bandini *et al.* 1990).

There are further difficulties in interpreting nutritional surveys in athletes. Much of the data is presented as total calorie intake without adjustment for weight, which makes comparison across sports difficult. Calorie intake is more meaningful when expressed as a function of the biologically active tissue, which is lean body mass (Webb, 1981). Calorie intake may also vary widely according to time of competitive season. Short & Short (1983) found that, in female University team swimmers, intake was 77% higher during preseason training than at the end of season. Training intensity and competitive level may vary between studies and even between groups used within each study. A single three-day dietary record may not therefore be representative of long term energy intake in athletes.

ENERGY INTAKE IN FEMALE ATHLETES

Despite these drawbacks, it appears that some female athletes have lower energy intakes than might be expected from their level of activity. Highly active women have greater energy requirements than sedentary women, yet many studies report very similar energy

Table 1. *Estimated energy intakes of female athletes from various sports assessed from 24 hour, 2 day, 3 day and 7 day records/recall. Mean ± SD or (range)*

Activity	n	Calorie intake (kcal day^{-1})	Reference
Aerobic sports			
Running	19	1973 ± 145	Snead *et al.* 1992
Running	5	1690 ± 272	Wilmore *et al.* 1992
Running	9	1611 ± 76	Wilmore *et al.* 1990
Running	9	1988 ± 439	Mulligan & Butterfield, 1990
Running	7	1973 ± 350	Mulligan & Butterfield 1990
Running	9	1933 ± 382	Myerson *et al.* 1991
Running	103	1603 ± 488	Pate *et al.* 1990
Running	9	1611 ± 76	Schulz *et al.* 1992
Running	20	2488 ± 302	Bergen-Cico & Short, 1992
Swimming	10	1894 ± 634	Risser *et al.* 1990
Swimming	9	3988 (5874 – 2267)	Short & Short, 1983
Swimming	7	2248 (3532 – 1516)	Short & Short, 1983
Swimming	18	1892 ± 446	Benson *et al.* 1990
Rowing	24	2340 (3580 – 1260)	Short & Short, 1983
Anaerobic/Aerobic sports			
Basketball	9	3240 (3879 – 1872)	Short & Short, 1983
Basketball	9	1797 ± 870	Risser *et al.* 1990
Volleyball	11	2446 (3199 – 1144)	Short & Short, 1983
Volleyball	12	1610 ± 574	Risser *et al.* 1990
Lacrosse	7	2219 (3059 – 1438)	Short & Short, 1983
'Appearance' sports			
Gymnastics	12	1544 ± 398	Benson *et al.* 1990
Ice skating	23	1174 ± 454	Rucinski, 1989
Dancers	10	1431 ± 500·2	Frusztajer *et al.* 1990
Dancers	9	1909 (2909 – 898)	Short & Short, 1983
Dancers	16	1701 ± 448	Williford *et al.* 1989
Dancers	19	1405 ± 379	Hergenroeder *et al.* 1991

Table 2. *Energy intake of regularly menstruating non-athletes estimated from 1-day to 7-day dietary recall/records. Mean ± SD*

(The data presented in this Table are from subjects used as controls in the studies presented in Table 1)

Energy intake of young female non-athletes

Age	n	Calorie intake (kcal day^{-1})	Reference
20·5 ± 4·3	10	1672 ± 641	Frusztajer *et al.* 1990
30·6 ± 5·6	5	1744 ± 367	Mulligan & Butterfield 1990
19·8 ± 1·4	13	1750 ± 368	Risser *et al.* 1990
29·6 ± 2·2	8–9	1852 ± 161	Snead *et al.* 1992
25·0 ± 4·6	5	1763 ± 420	Wilmore *et al.* 1992
26·3 ± 1·5	6	1779 ± 365	Myerson *et al.* 1991
13·5 ± 1·2	34	1849 ± 391	Benson *et al.* 1990

intakes in the two groups. The mean calorie intake of all the runners in the papers presented in Table 1 is 1874 kcal d^{-1}. Non-athletic women, of a similar age group, presented in Table 2 have an estimated mean intake of 1772 kcal d^{-1}; only 102 kcal d^{-1} less. To evaluate this further, energy balance studies have compared total energy expenditure and estimated

Table 3. *Categorization of sports according to whether low body mass is likely to improve performance or increase marks obtained during performance*

Low weight not of specific benefit to performance	Low weight likely to improve performance	Performance judged on aesthetic appeal	Competing in weight categories
Heavyweight rowing	Running	Gymnastics	Judo
Ball games e.g.:	Middle	Rhythmic gymnastics	Martial arts
Lacrosse	Long	Figure skating	Rowing, lightweight
Hockey	Ultra	Ice dance	Wrestling
Basketball	Crosscountry	Dancing	Weight lifting
Netball	Orienteering	Competitive	
Tennis	Race walking	Ballet	
Golf	Jumping	Bodybuilding	
Sprinting	Pole vault	Synchro swimming	
Field events (throwing)	Rowing, cox	Diving	
Contact sports	Jockeys, flat-racing		
Swimming	Cycling		
Water polo	Triathlon		
Skiing	Climbing, competitive		
Speed skating	Windsurfing, Olympic		
Luge, bobsleigh	Sailing, some classes		

energy intake in female runners. Energy deficits reported in such literature vary from -221 to -788 kcal d^{-1} (Myerson *et al.* 1991; Bergen-Cico & Short, 1992; Schulz *et al.* 1992; Wilmore *et al.* 1992).

This suggests that female athletes may be restricting nutritional intake, most noticeably in endurance and aesthetic sports. However, the desired body weight patterns vary by sport (see Table 3), and it is possible that the demands of individual sports influence the energy intake and nutritional habits of the athlete.

Sports in which low weight is of no direct benefit

In sports such as hockey, volleyball, lacrosse, basketball and heavyweight rowing, athletes may lose fat to improve fitness but performance is unlikely to be improved by low weight. Overall energy intake appears to be higher in these activities and there are few reports of restricted eating or abnormal eating habits.

Young swimmers may be an exception to this. A study of 487 female and 468 male swimmers (age 9–18) found that girls were particularly likely to misperceive themselves as overweight, and although only 16·5% of swimmers were actually classified as overweight, 41·7% had attempted to lose weight. Of the girls who were actively dieting, 17·9% were underweight and 60·5% were average weight (Dummer *et al.* 1987). The decision to lose weight was based on misperceptions of body image induced by peer group comments rather than on parental or coach advice.

Sports in which low weight may improve performance

Sports which involve movement against gravity favour those with a low body mass. For example, running performance is improved by weight loss (Wilmore & Costill, 1987; Brandon & Boileau, 1992) and vertical jump power is negatively correlated with body weight in skaters (Delistraty *et al.* 1992). Both athletes and coaches may aim for a weight that is inappropriate for an individual athlete. Top distance runners are usually very slim and this combines with cultural influences to increase the desire for leanness. Anecdotal reports abound of the adolescent athlete who finds that a small amount of weight loss results in improved performance, reinforcing the desire to lose yet more weight and precipitating the onset of severe food restriction or abnormal weight loss habits.

Middle and long distance runners have high energy expenditure yet estimated nutritional intakes have been very low (Schulz et al. 1992; Wilmore et al. 1992). Pate et al. (1990) found that energy intakes in 103 female runners (age range 25–34 years) and 74 sedentary controls were remarkably similar even when adjusted for body size (27·5 (SD 0·9) v. 26·0 (SD 1·3) kcal kg^{-1} daily). Baer (1993) also reported almost identical intakes in eumenorrhoeic runners and sedentary controls (1944 (SD 75) v. 1950 (SD 56) kcal d^{-1}).

Not all authors agree that there is nutritional restriction in runners. Mixed results were obtained by Mulligan & Butterfield (1990) who found that intake was highest in a group of moderately or very active runners compared to non-runners (37·5, 37·0 and 32·2 kcal kg^{-1} daily respectively) although when body weight was accounted for it was evident that the very active runners may have been relatively energy restricted (49·0, 43·5 and 43·7 kcal (kg fat free mass)$^{-1}$ daily in moderately active, very active and sedentary respectively). Relatively high energy intakes (2397 (SD 104) kcal d^{-1}) have been recorded in Olympic marathon hopefuls (Deuster et al. 1986) and Blair et al. (1981) estimated a calorie intake of 2386 kcal d^{-1} in runners and 1871 kcal d^{-1} in controls.

Sports which are judged on aesthetic appeal

In this group, which includes gymnastics, dancing and ice skating, weight is lost in a bid for higher marks. Research supports this concept; women placed in a National gymnast team had markedly lower body fat levels than non-placers (Falls & Humphrey, 1978) and elite gymnasts have much lower calorie intakes per unit body weight than subelite gymnasts (van Erp-Baart et al. 1989). Dancers also have a low intake; mean calorie intake was similar in 16 female dancers and 11 controls (1702 (SD 448) v. 1603 (SD 405) kcal d^{-1} respectively; Williford et al. 1989). And in a study of adolescents (mean age 12·9 years), Benson et al. (1990) found that intakes were similar in swimmers, gymnasts and non-athletic controls (39·2 (SD 13·4), 39·5 (SD 10·6), 39·4 (SD 13·3) kcal kg^{-1} daily respectively).

Female bodybuilders also fall into this category because they are judged on muscle definition which is improved when subcutaneous fat is low. However, they also compete in weight categories and in terms of eating habits they are better considered with the next group.

Sports in which the athlete competes in a weight category

In for instance judo, karate, tae kwon do and weight lifting, athletes are classed according to their body weight. Rowing also comprises a light weight category in which women must be 59 kg or under. Some athletes have a natural weight above that of the weight category in which they wish to compete. They therefore need to reduce weight before competitions (Short & Short, 1983; McCargar et al. 1993; Walberg-Rankin et al. 1993). Although many are able to do this over several months, others use starvation diet methods or resort to illicit practices such as purgation or the use of diuretics (Walberg & Johnston, 1991).

Table 4. *Energy intake of female athletes estimated from 1-day to 7-day dietary recall and records*

(Comparison of energy intake between eumenorrhoeic (EU) and amenorrhoeic (AM) athletes)

EU (kcal day^{-1})	n	AM (kcal day^{-1})	n	Reference
Runners				
2490	9	1582*	8	Kaiserauer et al. 1989
2250 ± 141	17	1730 ± 152*	11	Nelson et al. 1986
2489 ± 132	33	2151 ± 236	12	Deuster et al. 1986
1937 ± 383	9	1732 ± 236	6	Myerson et al. 1991
1965 ± 98	14	1623 ± 145	14	Drinkwater et al. 1984
1715 ± 281	6	1272 ± 136	11	Marcus et al. 1985
1971 ± 145	19	2046 ± 115	12/13	Snead et al. 1992
1690 ± 227	5	1781 ± 283	8	Wilmore et al. 1992
1817 ± 523	38	1832 ± 463	38	Watkin et al. 1991
1944 ± 45	10	1627 ± 75*	10	Baer, 1993
Dancers				
1405 ± 379	19	1116 ± 365*	19	Hergenroeder et al. 1991

Mean ± SD. Comparison between women with amenorrhoea and regular menstrual cycles: * $P < 0.05$.

THE RELATIONSHIP BETWEEN AMENORRHOEA AND LOW ENERGY INTAKE

ENERGY INTAKE OF AMENORRHOEIC ATHLETES COMPARED TO EUMENORRHOEIC SUBJECTS

Several recent reports have suggested that the calorie intake of amenorrhoeic athletes may be even lower than that of eumenorrhoeic subjects and that restriction of energy intake may be causally related to menstrual dysfunction. Some workers have also shown that amenorrhoeic athletes weigh less than eumenorrhoeic athletes (Galle et al. 1983; Marcus et al. 1985; Drinkwater et al. 1986; Harber et al. 1991) and have a lower percentage body fat (Sanborn et al. 1982; Glass et al. 1987) although there is much overlap between the two groups.

Nelson et al. (1986) used a three-day dietary record to compare the intakes of 11 amenorrhoeic and 17 eumenorrhoeic highly trained runners. They reported the mean energy intake of amenorrhoeic runners to be 520 kcal d^{-1} less than eumenorrhoeics (see Table 4). The difference in calorie intake was highly significant ($P < 0.005$) when expressed as a function of lean body mass. Similar findings were reported by Kaiserauer et al. (1989) although the intake of eumenorrhoeic sedentary controls was very similar to that of the amenorrhoeic women. Baer (1993) also found lower intakes ($P < 0.05$) in amenorrhoeic runners compared to both eumenorrhoeic runners and controls (1627 (SD 75), 1944 (SD 45), 1950 (SD 56) kcal d^{-1} respectively). And in adolescent dancers, Hergenroeder et al. (1991) estimated intakes of 1405 and 1116 kcal d^{-1} ($P = 0.03$) in eumenorrhoeic and amenorrhoeic subjects respectively using a 24-h recall and food frequency form.

Four out of a further seven studies on runners have also shown lower total calorie intakes in the amenorrhoeic athletes but these differences have not been statistically significant, largely because of the wide variance of the mean (see Table 4). In one of these studies (Drinkwater et al. 1984) energy expenditure was estimated to be higher in the amenorrhoeic subjects compared to the eumenorrhoeics and the two groups were therefore not directly comparable. In only two studies has calorie intake been shown to be higher in amenorrhoeic athletes compared to eumenorrhoeic athletes (Snead et al. 1992; Wilmore et al. 1992). Most of these reports are based on small sample numbers but Watkin et al.

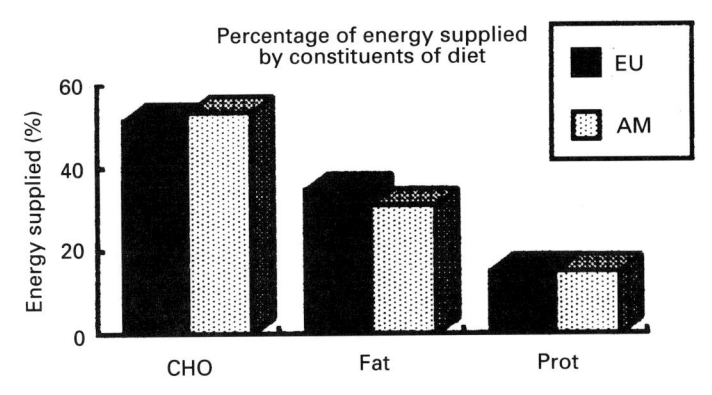

Fig. 2. Percentage of total energy supplied by main constituents of diet in amenorrhoeic and eumenorrhoeic female runners. Data accumulated from Nelson et al. (1986), Myerson et al. (1991), Snead et al. (1992), Kaiserauer et al. (1989), Drinkwater et al. (1984), Deuster et al. (1986), Watkin et al. (1991) and Baer (1993). AM = amenorrhoeic, EU = eumenorrhoeic.

(1991) surveyed 38 matched pairs of eumenorrhoeic and oligomenorrhoeic ultra-marathon runners. Similar energy intakes were found in both groups; however, of the 38 oligomenorrhoeic athletes only 53% were amenorrhoeic, the remainder having a cycle length of 35–90 d. These oligomenorrhoeic athletes may represent a subset with differing eating habits which may have skewed the results.

It is unclear whether short term energy restriction has such profound effects on menstrual function as long periods of reduced intake. Walberg-Rankin et al. (1993) surveyed six bodybuilders for two months which included a competition. Weight loss of 0·7–10·8% during the month prior to competition was achieved in most by gradual energy restriction. Mean energy intake in the month before the competition varied between 1536 (SD 28) and 1839 (SD 33·8) kcal d^{-1}. Immediately after the competition, intake rose to 3237 (SD 59) kcal d^{-1}. Of the three women not on the oral contraceptive, two reported missing one menstrual cycle during the past year although none had prolonged menstrual irregularity. Other workers have shown that it is possible to reduce weight slowly, by alternating constituents of the diet rather than energy restriction, without any physiological disturbance (McCargar et al. 1993).

Chronic nutritional restriction therefore appears to be a causative factor in the development of amenorrhoea, in addition to other physiological, psychological or training stresses, but data are scant for sports other than running, gymnastics and dancing. Even within this subgroup of athletes, opinion is divided. Some authors have suggested that amenorrhoeic runners are metabolically more efficient and do not require a high energy intake (Mulligan & Butterfield, 1990). Others have suggested that in runners the calorie deficit is overestimated and is more likely to be due to either consistent under-reporting of food intake or to restricted eating on the days of the dietary record (Wilmore et al. 1992). Under-reporting does occur in some subsets of women more than others: Bandini et al. (1990) demonstrated energy deficits of 562 and 1455 kcal d^{-1} in non-obese and obese subjects respectively and Prentice et al. (1986) found a difference of 837 kcal d^{-1} in obese subjects but no deficit in lean subjects. The suggestion in both of these studies is that obese women under-report food intake. It is possible that, with disturbances in body image, amenorrhoeic women behave in a similar way. In addition, self-reporting of physical activity is only moderately accurate (Klesges et al. 1990) with up to 300% overestimate of aerobic activity in some subjects. Studies relying on self-reporting of activity may therefore exaggerate any energy deficit that is present.

CONSTITUENTS OF THE DIET

There may be other nutritional factors contributing to the development of menstrual disturbance. Deuster *et al.* (1986) suggested that the total calorie intake may not be as important as the balance between the constituents of the diet and in particular that a low intake of fat may play some role in this relationship. In eight papers in which the constituents of the diet are shown, fat contributed 27·1–35·7% of the calories in amenorrhoeic athletes and 31·3–37·9% in eumenorrhoeic athletes (Fig. 2). However, the difference in fat intake was significant in only two papers and although the overall trend is for fat to provide a lower percentage of calories in amenorrhoeic athletes these differences are remarkably small. It seems unlikely that such small differences in fat intake could have profound effects on the pituitary–ovarian axis but this is another area which requires further work.

VEGETARIANISM

Diet may influence sex hormone production in other ways. In small studies in non-athletes, a vegetarian weight reduction diet has been associated with low oestrogen and progesterone levels, short luteal phases (Schweiger *et al.* 1987) and anovulation (Pirke *et al.* 1986). These changes were related to severity of weight loss and were not seen in non-vegetarian diets. Other authors have also noted an increased frequency of menstrual irregularity in vegetarian compared to non-vegetarian women (Lloyd *et al.* 1987; Pedersen *et al.* 1991) and some have shown a negative correlation between dietary fibre and sex hormone levels (Barbosa *et al.* 1990).

In athletes, Lloyd *et al.* (1991) found that oligomenorrhoeic collegiate women consumed greater amounts of dietary fibre than their eumenorrhoeic counterparts and suggested dietary habits as a cause of their menstrual irregularity. Several studies of small numbers of amenorrhoeic athletes have suggested that vegetarianism is more common than in eumenorrhoeic athletes (Brooks *et al.* 1984; Slavin *et al.* 1984; Kaiserauer *et al.* 1989) and this may be a fruitful area for further longitudinal research.

VITAMINS AND TRACE MINERALS

Low intake of a number of nutrients is common in athletes, particularly those with amenorrhoea. Low protein intake (Kaiserauer *et al.* 1989) and low intakes of some vitamins and trace minerals (Deuster *et al.* 1986; Kaiserauer *et al.* 1989; Rucinski, 1989; Benson *et al.* 1990; Myerson *et al.* 1991; Bergen-Cico & Short, 1992) have all been demonstrated. In most, the low intakes are due to an overall reduced food intake rather than a specific deficiency of one nutrient. Whether these deficiencies are linked to menstrual disturbance is not yet known.

EATING DISORDERS

Analysis of calorie intake alone may present only half the picture and for this reason it may have been difficult to link directly energy intake with amenorrhoea. Current literature suggests a high incidence of anorexic or bulimic types of behaviour in some athletes, particularly runners and dancers. Preoccupation with weight and food, a desire to maintain a low body weight or body fat and intensive exercise are all hallmarks of anorectic behaviour. Both anorexia and bulimia are associated with menstrual disturbance and it may be that disordered eating rather than energy intake *per se* is the main contributor to the high incidence of amenorrhoea. Alternatively these eating habits may be normal

Table 5. *Adapted from the Diagnostic and Statistical Manual of Mental Disorders (American Psychiatric Association, 1987)*

DSM-III-R* diagnostic criteria for anorexia nervosa	DSM-III-R* diagnostic criteria for bulimia nervosa
1. Refusal to maintain weight over a minimal normal weight for age and height, i.e. weight loss leading to maintenance of body weight 15% below that expected or failure to make expected weight gain during a period of growth leading to body weight 15% below that expected	1. Recurrent episodes of binge eating (rapid consumption of a large amount of food in a discrete period of time, usually < 2 h)
2. Intense fear of gaining weight or becoming fat, even though underweight	2. A feeling of lack of control over eating behaviour during the eating binges
3. Disturbance in the way in which one's body weight, size, or shape is experienced, e.g. the person claims to 'feel fat' even when emaciated or believes that one area of the body is 'too fat', even when obviously underweight	3. Regularly engaging in self-induced vomiting, the use of laxatives or diuretics, strict dieting or fasting, or vigorous exercise in order to prevent weight gain
4. In females, absence of at least three consecutive menstrual cycles when otherwise expected to occur (primary or secondary amenorrhoea)	4. A minimum of two binge eating episodes a week for at least three months

amongst elite athletes in whom the correct timing and constituents of the diet may be very different from the rest of the population. It is likely too that within any group of athletes there are some who use exercise to become thin.

EATING DISORDERS IN THE GENERAL POPULATION

Estimates of eating disorders, using strict DSM III criteria of the American Psychiatric Association (1987; see Table 5), vary from 1·0 to 4·8% of female student populations (Schotte & Stunkard, 1987; Drewnowski *et al.* 1988; Kurtzmann *et al.* 1989). Using eating habits questionnaires in 2544 high school girls, Johnson & Whitaker (1992) found that 10·2% frequently had episodes of binge eating, 4·0% frequently self-induced vomiting and 1·6 and 5·1% regularly used laxatives or diet pills respectively. When questioned on dieting habits 43·5% had dieted for less than 28 days and 18·5% had dieted for longer than this. An alarming 7·4% had fasted for longer than three days. Similarly high values for the incidence of dieting (61% black girls, 77% white girls), fasting (33% of dieters) and purging (up to 18%) in high school girls were found by Emmons (1992) in a study of 1269 male and female high school students. Thus although the estimates for strictly defined eating disorders in the general population may be relatively low, abnormal eating patterns and weight control practices are common in this age group of girls. Indeed, in a large study of 603 elite female athletes and 522 controls, Sundgot-Borgen & Larsen (1993) found similar numbers with eating disorders (12 and 11% respectively) or who were at risk of developing an eating disorder (22 and 26%). It is against this background that the eating habits and prevalence of these disorders in athletes must be considered.

EATING DISORDERS IN ATHLETES

Many athletes have strange but not pathological eating habits and this makes it difficult to estimate the prevalence of true eating disorders in the athletic population. In addition to this, eating disorders are, by their nature, secretive activities and many will not admit to such a problem. Roommates and relatives may be entirely unaware that an eating disorder exists. Studies that rely on questionnaires may therefore underestimate this incidence.

Most studies on the athletic population do not use strict DSM criteria and comparison with the normal population is therefore difficult. Research on athletes has usually entailed the Eating Disorders Inventory, the Bulimia Test – Revised or the Eating Attitudes Test (EAT) in which the athlete responds to statements such as "I vomit after I have eaten", "I am terrified about being overweight", "I think about burning up calories when I exercise". A progressive rating of replies from 'always' to 'never' enables the degree of disordered eating to be assessed. Athletes may under-report symptoms and behaviours associated with eating disorders and it has been suggested that both clinical interviews and questionnaires may provide a better estimate of incidence (Sundgot-Borgen, 1993).

It has been suggested that intensive exercise itself may precipitate eating disorders. Gleaves *et al.* (1992) tested the hypothesis that weight loss due to running leads to body image disturbance and places runners at increased risk of bulimic disorders. They found no difference in measures of depression, bulimia or body image disturbance between the running group and normal controls whereas bulimia nervosa patients differed from both groups in most measures. However, Davis *et al.* (1990) identified two groups of women within a cohort of 112 regularly exercising females: 'dieters' (57%) and 'non-dieters' (43%). The dieters expressed weight and diet concerns equivalent to eating-disordered patients and much more frequently than the non-dieters. They also engaged in more frequent and more intense exercise than the non-dieters. Davis *et al.* concluded that "regular participation in a fitness programme may be causally related to excessive concern with weight and dieting". It is possible that, within any group of athletes, those with a tendency towards body image preoccupation are precipitated into a full eating disorder by the combined sociocultural influences and the demands of sport.

PREVALENCE OF EATING DISORDERS IN DIFFERENT ATHLETIC GROUPS

Sports in which low weight is not an advantage

In a study of elite female Norwegian athletes (Sundgot-Borgen & Larsen, 1993), women competing in ball games and power sports (sprinting, discus, shot put etc.) had a very similar prevalence of abnormal weight control habits in comparison with non-athletic controls (8, 6, and 7% respectively). Downhill skiers, golfers and others involved in technical sports also had a lower prevalence of such habits than those in 'appearance' and weight dependent sports (10, 16 and 17% respectively). Estimates of eating disorders in young swimmers are high but are also no different from those of the non-athletic population of the same age (Dummer *et al.* 1987; Benson *et al.* 1990; Rosenvinge & Vig, 1993).

Sports in which low weight may improve performance

A large survey was undertaken by *Runner's World* magazine in the USA; of the 1908 female runners that responded, 38% reported binge eating at least once a month and in 6% this occurred at least three times a week. Purging was common: 26% had purged at least once and 4% purged at least three times a week (Brownell *et al.* 1988). These results are similar in many ways to those of high school girls in the USA but it is possible that athletes with significant eating problems failed to return the questionnaire. Prussin & Harvey (1991) support a high incidence of bulimic behaviour in female runners: 19% of their cohort ($n = 174$) fulfilled DSM-IIIR criteria for bulimia.

Some of the variation in results may be due to the wide range of abilities in large questionnaire based studies. Weight control habits in highly competitive athletes may be more common than in lower achievers (Mulligan & Butterfield, 1990). In one study, 14% of 125 female marathon runners were symptomatic for anorexia and elite runners were more likely to have high EAT scores ($P < 0.05$). Eighty per cent of the runners with both the physical and psychological features of anorexia nervosa were reported to be highly successful competitors (Weight & Noakes, 1987).

Sports which are judged on aesthetic appeal

Estimates of eating disorders are high in a number of studies in this group of athletes. In 42 female college gymnasts assessed using the Michigan State University Weight Control Survey, all were actively dieting and 62% were using at least one form of pathological weight control (Rosen & Hough, 1988). Two-thirds of the sample had been told by their coaches that they were too heavy, following which 75% resorted to weight control methods of which self-induced vomiting, diet pills and fasting were most frequently used. In a larger study of 218 female gymnasts (mean age 19·4 years), 48·6% used exercise specifically to burn calories, 49·1% had fasted or used a strict diet in the last year and 58·7% ate uncontrollably at least two or three times a month.

A lower incidence of disordered eating has been found in prepubertal gymnasts (Benson *et al.* 1990). Body fat stores are lower before puberty and the typical 'ideal' gymnast figure more closely resembles that of the prepubertal girl than a fully mature figure. It becomes harder to achieve this as the athlete ages, which may account for the increased frequency of disordered eating in the older age group.

Rucinski (1989) found 48% of female ice skaters to have EAT scores in the range of clinical anorexia nervosa and also found that energy intake was negatively correlated with EAT scores. In dancers, self-reported anorexia or bulimia ranges from 11% in highly select dancers to 33% in regional and national companies and University dancers (Brooks-Gunn *et al.* 1987; Evers, 1987; Hamilton *et al.* 1988).

Sports in which the athlete competes in a weight category

Female weight lifters (including a subset who competed in body building) have beeen found to score highly on the Drive for Thinness subscale of the Eating Disorders Inventory, and of those that competed 67% were terrified of becoming fat, 50% experienced uncontrollable urges to eat and 42% had a history of anorexia (Walberg & Johnston, 1991). Data from male athletes in these sports suggest that they might be at high risk of rapid weight reduction measures in order to 'make weight' (Short & Short, 1983). Methods employed by male wrestlers to induce weight loss include laxative abuse, self-induced vomiting and rapid dehydration methods (saunas, sweat suits, diuretics). Similar problems have been found in American majorettes (Humphries & Gruber, 1986).

IS THERE A RELATIONSHIP BETWEEN DISORDERED EATING IN ATHLETES AND MENSTRUAL DYSFUNCTION?

In the study of eating habits in high school girls by Johnson & Whitaker (1992), the relative risk for secondary amenorrhoea was 4·17 for frequent binge purging and 2·59 for weight fluctuation due to weight control measures. Amenorrhoea is sometimes the first sign in anorexia nervosa and the DSM III (American Psychiatric Association, 1987) criteria for this disorder include amenorrhoea (Table 5). In bulimia nervosa endocrine disturbance often results in menstrual irregularity and sometimes amenorrhoea, although it is not a criterion for diagnosis (Table 5). Eating disorders are therefore likely candidates for a key role in the development of amenorrhoea in athletes.

During a study on diet in long distance runners, Gadpaille *et al.* (1987) noted a greater preoccupation with diet in amenorrhoeic subjects compared to cyclic women. Analysis of their group of 13 amenorrhoeic and 19 regularly menstruating runners showed that while nearly two thirds of amenorrhoeic runners had evidence of an eating disorder as defined by DSM III criteria, no eumenorrhoeic athlete was affected. A quarter of the amenorrhoeic runners had evidence of a bipolar disorder or major depression and there was a high incidence of first or second degree relatives with a major affective disorder. None of the cyclic women had a depressive disorder and very few had relatives affected. They suggest that athletic amenorrhoea may be either a variant of anorexia nervosa or may be a defence against one of the other disorders, particularly depression (running is known to ameliorate symptoms of depression; Harris, 1986). Higher EAT scores and subscales of the score have also been found in amenorrhoeic runners compared to eumenorrhoeic runners (Myerson *et al.* 1991).

Prolonged amenorrhoea has been strongly associated with dieting in dancers. Brooks-Gunn *et al.* (1987) found that 50% of amenorrhoeic dancers reported anorexia nervosa as opposed to only 13% of dancers with normal periods. The amenorrhoeic dancers weighed less and were leaner ($P < 0.05$) than their eumenorrhoeic peers and had higher scores on the EAT26 scale ($P < 0.01$). Although some workers have found no differences between the incidence of either overt anorexia and bulimia or altered eating behaviour or attitudes in amenorrhoeic athletes (Myburgh *et al.* 1992; Snead *et al.* 1992), the majority of workers have shown a trend towards some form of dysfunctional eating or weight control in those with menstrual abnormalities (Walberg & Johnston, 1991; Wilmore *et al.* 1992).

Although there have been few large scale studies on amenorrhoeic athletes, disordered eating is strongly implicated in the development of menstrual irregularity and this relationship is starting to be taken seriously. In 1990, the National Collegiate Athletic Association of the United States released a videotape series with written information on anorexia nervosa and bulimia. This was aimed at all levels of sport from administrators to athletes and included lists of specific warning signs for both conditions (Wilmore, 1991). The Sports Council of Great Britain and most British bodies governing sports do not yet produce information that is readily available to all levels of sport with the exception of the British Amateur Athletic Federation. A booklet, produced by the International Amateur Athletic Federation and aimed at coaches, is available through their bookshop.

BONE METABOLISM IN ATHLETES

THE EFFECT OF EXERCISE ON BONE

Julius Wolff (1892) first suggested that bone responds to mechanical stress to increase strength at areas of high strain. Animal models support the concept that dynamic loading of bone results in increased bone strength at predicted sites (Lanyon, 1992). In athletes, BMD of the femur, pelvis, tibia and os calcis have all been shown to be high in athletes undertaking weight-bearing sport (Heinrich *et al.* 1990; Risser *et al.* 1990; Wolman *et al.* 1991; Slemenda & Johnston, 1993; Wilson *et al.* 1994*b*), and in tennis players BMD is higher in the dominant arm than on the non-dominant side (Dalén *et al.* 1985; Pirnay *et al.* 1987). In rowers, BMD is high in the lumbar spine which correlates well with their greater back strength (Wolman *et al.* 1990) and bodybuilders have higher lumbar BMD than runners (Heinrich *et al.* 1990). Indeed, BMD of spine and femur has been shown to be highly correlated with muscle strength (Pocock *et al.* 1989; Conroy *et al.* 1993; Eickoff *et al.* 1993). Bone density has also been shown to increase during a training programme (Margulies *et al.* 1986) and in small cross-sectional studies higher bone density has been demonstrated in those who weight train compared to those who are aerobically trained (Davee *et al.* 1990; Heinrich *et al.* 1990).

The beneficial effect of regular exercise on BMD occurs not only at a young age, but also appears to continue at least until the menopause (Jacobson *et al.* 1984; Wilson *et al.* 1994*b*) and possibly beyond (Talmage *et al.* 1986). In addition, physical activity may be a powerful tool in reducing hip fractures associated with older age (Cooper *et al.* 1988) although some of this reduction may be due to improvements in muscle strength, coordination and balance rather than greater BMD. To date most prospective studies using exercise programmes have measured changes in BMD of spine or radius rather than hip. Therefore there is no evidence yet which demonstrates improvements in hip BMD with exercise. These studies are awaited but it is hoped that exercise prescription may be useful in reducing the risk of osteroporosis later in life.

THE EFFECT OF MENSTRUAL IRREGULARITY ON BONE

Early studies on BMD in athletes with amenorrhoea showed reduced BMD compared to sedentary controls or eumenorrhoeic athletes (Cann *et al.* 1984; Drinkwater *et al.* 1984; Marcus *et al.* 1985). Many others have now confirmed these findings especially in the lumbar spine (Nelson *et al.* 1986; Lindberg *et al.* 1987; Wolman *et al.* 1990; Wilson *et al.* 1994*b*). In the largest study of its kind, Drinkwater *et al.* (1990) demonstrated a linear relationship between vertebral BMD and menstrual history in 97 active women. Those who had a long history of oligomenorrhoea had a mean BMD which was 17% lower than those who had always had regular periods. This linear relationship was also demonstrated in our laboratory. In a study of 50 runners we found that mean lumbar BMD in amenorrhoeic runners was 16% below that of eumenorrhoeic subjects and 6% lower than an oligomenorrhoeic group. In a few of the amenorrhoeic group, BMD was as much as three standard deviations below the mean of age matched European data (Wilson *et al.* 1994*b*).

There is less agreement about the effect of amenorrhoea on weight bearing bones in which the mechanical stresses of exercise may offset bone loss. Reduction in femoral shaft BMD in amenorrhoeic active women has been noted by one author (Drinkwater *et al.* 1990) but Wolman *et al.* (1991) found no difference in femoral shaft BMD between amenorrhoeic and eumenorrhoeic runners, rowers and dancers. Femoral neck and total leg BMD have also been shown to be well maintained despite amenorrhoea (Drinkwater *et al.* 1990; Myerson *et al.* 1992; Snead *et al.* 1992). However, in our recent study of 24 amenorrhoeic national and international standard runners BMD was low in the proximal femur despite the positive osteogenic effects of running (Wilson *et al.* 1994*b*). In the neck of the femur BMD was 0·56 SD below the age matched European mean and 16% lower than in eumenorrhoeic runners. If the differential in BMD between these athletes and the normal population persists into later life their risk of hip fracture may be increased as much as five times (Cummings *et al.* 1993).

MECHANISM OF REDUCED BONE DENSITY IN AMENORRHOEA

Sex hormones are important in bone turnover; bone resorption is reduced in the presence of oestrogen and progesterone is a trophic hormone for bone (Prior, 1990). In athletic amenorrhoea levels of follicle stimulating hormone and luteinizing hormone are low due to suppression of hypothalamic gonadotrophin releasing hormone. In turn oestrogen production in the ovary is reduced and prolonged low levels may result in increased bone resorption with loss of bone minerals. Bone biopsy may reveal the resorption surface to be greatly increased while the bone formation surface remains normal (Warren et al. 1990).

CALCIUM INTAKE AND BONE DENSITY IN ATHLETES

Bone mass is higher in children and adolescents with a high calcium intake (Chan, 1991; Sentipal et al. 1991) and this may result in increased bone mass in later life (Matković et al. 1979). Work on healthy premenopausal women supports a role for dietary calcium in the development of bone, particularly when associated with exercise (Kanders et al. 1988; Halioua & Anderson, 1989). The synergic effect of calcium and exercise on bone has also been demonstrated in animals (Lanyon et al. 1986).

A positive linear relationship between trabecular BMD in the lumbar spine and calcium intake in athletes has been demonstrated by Wolman et al. (1992), a finding that was independent of menstrual status and which has not been shown in other studies (Nelson et al. 1986; Grimston et al. 1990; Heinrich et al. 1990). These differences may be due to the methods used, particularly in assessing calcium intake. Alternatively the relationship between calcium intake and bone mineral content may not be linear. Kanders et al. (1988) showed a positive relationship between calcium intake and vertebral BMD in normal healthy eumenorrhoeic women but not above a daily intake of 800–1000 mg.

Low calcium intakes have been reported in many athletes (Rucinski, 1989; Benson et al. 1990; Pate et al. 1990; Bergen-Cico & Short, 1992; Delistraty et al. 1992; Frederick & Hawkins, 1992; Stensland & Sobal, 1992) particularly in amenorrhoeic women. Low oestrogen levels are associated with decreased intestinal absorption of calcium and increased urinary loss (Nordin & Heaney, 1990), so dietary calcium requirements may be even higher in amenorrhoeic athletes. Both Marcus et al. (1985) and Nelson et al. (1986) found that 55% of amenorrhoeic athletes failed to meet the recommended daily allowance for calcium compared to 35–40% of cyclic women. Kaiserauer et al. (1989) also noted a lower intake in amenorrhoeic compared to eumenorrhoeic runners (600 mg $v.$ 1200 mg d^{-1}, $P < 0.05$). Yet it remains unclear how much of the variance in BMD in amenorrhoeic athletes is due to low calcium intake.

CONSEQUENCES OF LOW BONE DENSITY IN ATHLETES

STRESS FRACTURES

Lloyd et al. (1986) reviewed the medical records of 207 collegiate athletes and found X-ray confirmed fractures (type of fracture not defined) in 9% of regularly menstruating women and 24% of women with irregular or absent menses. For dancers, two papers report a relationship between bone injuries or stress fractures and amenorrhoea (Warren et al. 1986; Benson et al. 1989) and a survey of 240 female athletes showed a higher incidence of stress fractures in those with fewer than five menses per year (49%) compared to those with 10 or more menses per year (29%) (Barrow & Saha, 1988). However, two reports have refuted a relationship between menstrual history and stress fractures although both contained small numbers of subjects (Frusztajer et al. 1990; Grimston et al. 1990).

It remains unclear whether the increased rate of stress fractures in amenorrhoeic athletes is related to low BMD. Femoral bone density has been shown to be low in young male military recruits with femoral stress fractures (Pouilles et al. 1989) but Carbon et al. (1990) assessed elite female runners with and without stress fractures and found no difference in the femoral BMD between the two groups. Others have also described a lack of association between tibial BMD and stress fractures in military recruits (Milgrom et al. 1989).

MUSCULOSKELETAL INJURIES

Stress fractures may not be the only injury more prevalent in amenorrhoeic athletes. Participants in a 10 km race who responded to a questionnaire were more likely to have taken time off training owing to any form of musculoskeletal injury if they had irregular menses (Lloyd et al. 1986). In Benson's study (Benson et al. 1989) of 49 female dancers, those with abnormal menses had more 'bone injuries' (mean = 15·0) than normally menstruating dancers (mean = 5·0; $P < 0.05$). Additionally, dancers with a low body mass index (< 19.0 kg m^{-2}) had a greater duration of low grade musculoskeletal injury (mean = 24·1 d) than those with a higher body mass index (mean = 11·6 d; $P < 0.05$).

More severe bone injury also occurs in amenorrhoeic athletes. In dancers, scoliosis was found to be more common in those with delayed menarche and in whom anorectic behaviour was more prevalent (Warren et al. 1986). Warren et al. (1990) also described a 20 year old ballet dancer with long standing anorexia nervosa, primary amenorrhoea and low BMD, who suffered femoral head collapse. Recently we have reported an osteoporotic fracture in the neck of the humerus of a 30 year old marathon runner with a history of anorexia and low bone density (Wilson & Wolman, 1994).

LONG TERM CONSEQUENCES

Amenorrhoeic athletes may be at risk of premature osteoporosis and fractures but as yet there is little long term information on the natural history of bone metabolism in this condition. Bone mass peaks at approximately 30–35 years of age and from the fourth decade onwards there is a progressive decline in bone density of 0·5–1·5% per year (Riggs & Melton, 1986; Riggs et al. 1986; Slemenda et al. 1987) which increases (3–8% per year) in the immediate postmenopausal years (Riggs & Melton, 1986). The lifetime risk of hip fracture has been related to bone mass; the higher the bone mass at any age, the lower the lifetime risk of fracture (Melton et al. 1988). Eumenorrhoeic athletes, who have a high peak bone mass, are likely to have reduced risk of osteoporotic fracture whereas amenorrhoeic athletes may never achieve their maximum potential bone mass.

BMD may increase when menstruation returns in these athletes. Drinkwater et al. (1986) followed nine athletes over a 15·5 month period. Seven of the women had regained menses and two had remained amenorrhoeic. Lumbar BMD increased 6·3% in the former amenorrhoeic women whilst decreasing a further 0·3% in those who had remained amenorrhoeic. Small increases in BMD were also seen in the radius. These results are very similar to those of Lindberg et al. (1987) who retested seven amenorrhoeic runners at 15 months. Four had recovered menses and showed an improvement of 6·5% in lumbar BMD. The other three remained amenorrhoeic and showed no improvement in BMD. In our laboratory we have followed 26 oligomenorrhoeic runners over an 18 month period. At one year those who had either regained menstruation or who were on hormone replacement therapy ($n = 14$) had increased BMD by 4·1% in the lumbar spine while in those who remained amenorrhoeic ($n = 12$), BMD further declined by 1·7%. Similar trends were seen in the proximal femur (J. H. Wilson, unpublished data).

These short term studies all suggest that small improvements in BMD may occur when menstruation returns. It is unknown whether such increases are maintained over several years. Recently we have compared 13 perimenopausal athletic women (mean age 42·9 years) with a history of menstrual irregularity with 37 similar athletic women (mean age 45·8 years) who had continual eumenorrhoea (Wilson et al. 1994a). Bone density was lower in the first group in both the proximal femur and the lumbar spine but this only reached significance in the spine ($P < 0.05$). More importantly, BMD in the proximal femur and lumbar spine of the previously amenorrhoeic women was not significantly different from the mean for European age matched women and in the neck of femur it was higher ($P < 0.02$). This suggests that prolonged menstrual irregularity does have long term effects on BMD but these may be offset by continued exercise.

PREVENTION, ASSESSMENT AND MANAGEMENT OF ATHLETIC AMENORRHOEA

The available evidence suggests that the onset of menstrual irregularity is associated with one or more of the factors in Fig. 1. Because they are so interlinked it has proved impossible to determine which are of greatest importance. However, alterations in eating habits such as food restriction and purging would seem to lie at the centre of the problem for many and better assessment and management of these behaviours may reduce the incidence. Prevention of eating disorders requires a number of approaches.

EDUCATION

Staff at all levels need to have greater awareness of the potential for eating disorders within their own sport. Coaches need advice on correct assessment of body composition and the optimum for performance for each athlete. Individualization of programmes for weight maintenance (and reduction if necessary) will reduce the likelihood that ill-advised comments will precipitate abnormal eating habits. The emphasis should move away from the very slender look in aesthetic sports although this also requires a change in general cultural attitudes. Finally, the athletes themselves must be educated. Unfortunately many athletes will continue to see role models, who are very thin, undertaking high levels of training and competition. This discourages them from weight gain or changes in training. In these cases, reduction of other risk factors and alterations in diet constituents may be sufficient for menstrual function to return (Myburgh *et al.* 1992).

IDENTIFICATION OF 'AT RISK' ATHLETES

A high index of suspicion may aid the detection of these problems. The preparticipation examination or routine medical for women and girls is an opportunity to screen for menstrual and nutritional disorders. Eating habit questionnaires specifically designed for athletes will aid identification of disturbed body image and eating attitudes. All sports specialists and primary care physicians should seek such information when female athletes present with problems such as stress fractures, recurrent injuries, weight loss or fatigue. Suggestions for assessment by primary care physicians, sports physicians or others involved in the care of athletes is given in Fig. 3.

OPTIONS FOR TREATMENT

Drinkwater *et al.* (1986) and Lindberg *et al.* (1987) demonstrated an increase in BMD associated with resumption of menses but in all cases this occurred owing to a reduction in training volume and intensity with a concomitant increase in weight. Not all athletes are willing to alter training habits in order to resume menstruation. In many, menstruation would be a nuisance and in some it might interfere with performance. In such athletes treatment to prevent further bone mineral loss or to improve low bone density may be an option.

NUTRITION, PHYSICAL ACTIVITY AND BONE HEALTH IN WOMEN

Athlete with menstrual dysfunction
│
Pathological causes?
 ├── Yes ── Appropriate referral
 └── No ── Assess risk factors
   ├── Nutrition (*sports nutritionist, parents*)
   │   Dietary analysis
   │   Inadequate intake
   │   Disordered eating
   │   Vegetarian
   │   │
   │   Clinical symptoms?
   │   ├── Yes ── Referral
   │   │   *Sports Psychologist*
   │   │   *Psychiatrist*
   │   └── No
   ├── Body composition (*exercise physiologist, coach*)
   │   Optimal weight and body fat (%)
   │   Power:weight
   │   Training volume/intensity
   │   │
   │   Weight too low?
   │   ├── Yes ── Monitor:
   │   │   food intake
   │   │   weight
   │   │   performance
   │   │   Reassign targets
   │   └── No
   └── Menstrual history (*GP*)
       Menarche
       Parity
       Previous dysfunction
       │
       Reduce other risk factors:
       Competition and training stresses
       Social stresses

Continuing menstrual dysfunction?
 ├── Yes ── Specialist referral:
 │   *endocrinologist*
 │   *gynaecologist*
 │   Bone density assessment
 │   Treatment if necessary
 └── No ── Continue to monitor

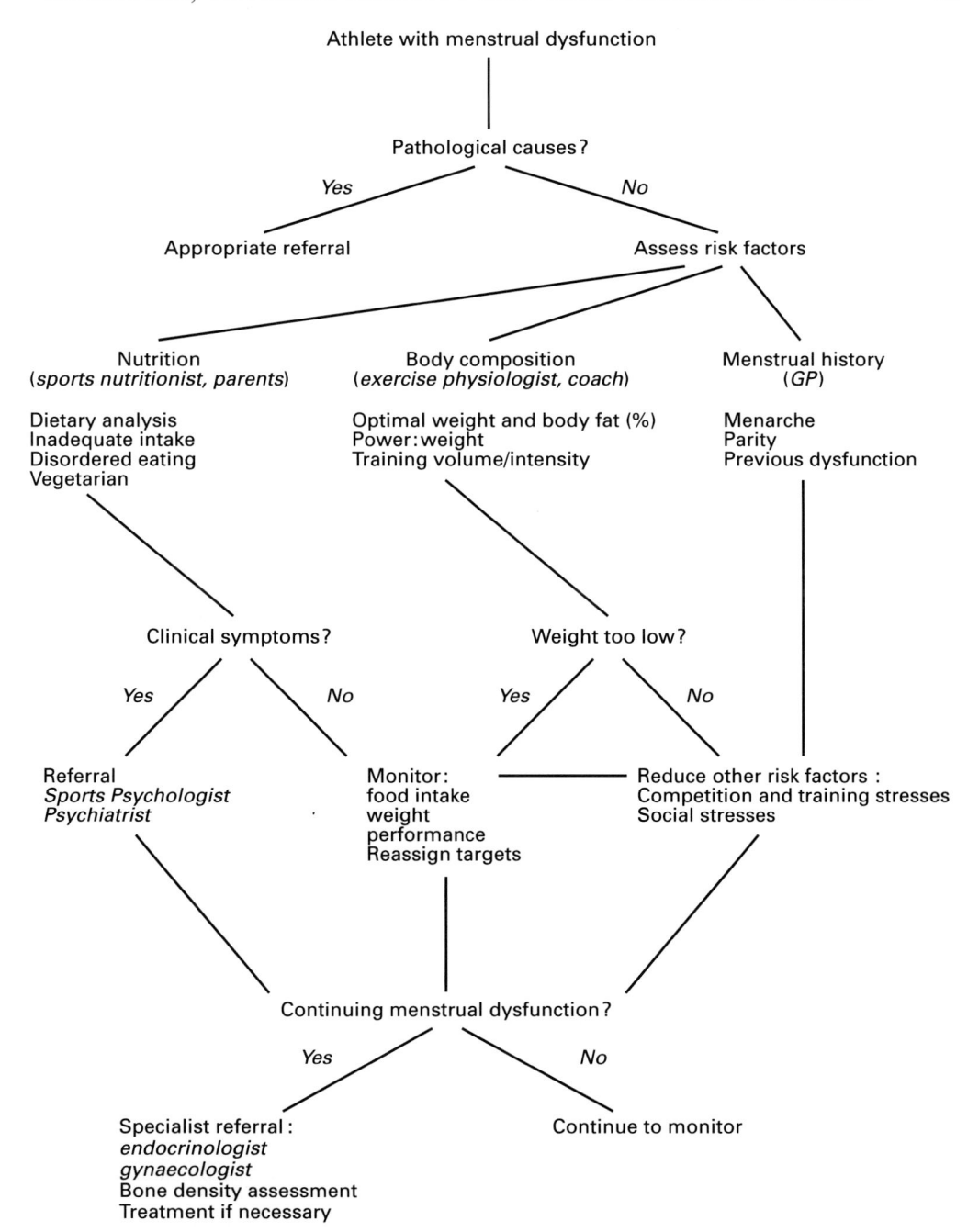

Fig. 3. Flow diagram for assessment of athletes with menstrual dysfunction by clinicians. Management of athletes requires team approach and suggestions for personnel are given at relevant stages, if not already part of decision making. No stages are mutually exclusive.

Treatment regimes used in women with secondary amenorrhoea include calcium supplementation, hormone replacement therapy, and intranasal calcitonin. Readers are referred to Prior *et al.* (1992) for further discussion of the treatment of this condition which is beyond the scope of this paper. Management of the athlete with eating disorders requires a team approach with appropriate specialist referral.

CONCLUSION

Amenorrhoea in athletes is a common occurrence and is associated with low levels of sex hormones. The major consequence of this disorder is a reduction in bone density compared to eumenorrhoeic peers. In some, BMD may be so low that fractures occur and there may be a risk of early osteoporosis. However, long term studies on large numbers of amenorrhoeic athletes are required to determine the natural history of BMD in such women and the response to treatment. Multicentre studies and trials may be able to provide these answers.

Prevention of menstrual disturbance and careful management should it occur will help to prevent such dramatic consequences but, unfortunately, the aetiology of athletic amenorrhoea is complex. Abnormalities in eating are implicated but further large scale studies are required to ascertain the true incidence of these disorders and their relationship to menstrual dysfunction in athletes. All those involved in the care and support of young athletes should be aware of the prevalence of eating disorders and should place nutrition high on the agenda for optimum performance.

I thank Dr Jonathon Reeve, Consultant Physician at Addenbrooke's Hospital, Cambridge and Dr Mark Harries, Consultant Physician and Honorary Medical Director of the British Olympic Medical Centre, for their advice and editorial comments.

REFERENCES

Baer, J. T. (1993). Endocrine parameters in amenorrheic and eumenorrheic adolescent female runners. *International Journal of Sports Medicine* **14**, 191–195.

Bandini, L. G., Schoeller, D. A., Cyr, H. N. & Dietz, W. H. (1990). Validity of reported energy intake in obese and nonobese adolescents. *American Journal of Clinical Nutrition* **52**, 421–425.

Barbosa, J. C., Shultz, T. D., Filley, S. J. & Nieman, D. C. (1990). The relationship among adiposity, diet, and hormone concentrations in vegetarian and nonvegetarian postmenopausal women. *American Journal of Clinical Nutrition* **51**, 798–803.

Barrow, G. W. & Saha, S. (1988). Menstrual irregularity and stress fractures in collegiate female distance runners. *American Journal of Sports Medicine* **16**, 209–214.

Benson, J. E., Allemann, Y., Theintz, G. E. & Howald, H. (1990). Eating problems and calorie intake levels in Swiss adolescent athletes. *International Journal of Sports Medicine* **11**, 249–252.

Benson, J. E., Geiger, C. J., Eisermann, P. A. & Wardlaw, G. M. (1989). Relationship between nutrient intake, body mass index, menstrual function, and ballet injury. *Journal of the American Dietetic Association* **89**, 58–63.

Bergen-Cico, D. K. & Short, S. H. (1992). Dietary intakes, energy expenditures, and anthropometric characteristics of adolescent female cross-country runners. *Journal of the American Dietetic Association* **92**, 611–612.

Blair, S. N., Ellsworth, N. M., Haskell, W. L., Stern, M. P., Farquhar, J. W. & Wood, P. D. (1981). Comparison of nutrient intake in middle-aged men and women runners and controls. *Medicine and Science in Sports and Exercise* **13**, 310–315.

Brandon, L. J. & Boileau, R. A. (1992). Influence of metabolic, mechanical and physique variables on middle distance running. *Journal of Sports Medicine and Physical Fitness* **32**, 1–9.

Brooks, S. M., Sanborn, C. F., Albrecht, B. H. & Wagner, W. W. (1984). Diet in athletic amenorrhoea. *Lancet* **i**, 559–560.

Brooks-Gunn, J., Warren, M. P. & Hamilton, L. H. (1987). The relation of eating problems and amenorrhoea in ballet dancers. *Medicine and Science in Sports and Exercise* **19**, 41–44.

Brownell, K. D., Rodin, J. & Wilmore, J. H. (1988). Eat, drink, and be worried? *Runner's World* Aug. 28, 28–34.

Cann, C. E., Martin, M. C., Genant, H. K. & Jaffe, R. B. (1984). Decreased spinal mineral content in amenorrheic women. *Journal of the American Medical Association* **251**, 626–629.

Carbon, R., Sambrook, P. N., Deakin, V., Fricker, P., Eisman, J. A., Kelly, P., Maguire, K. & Yeates, M. G. (1990). Bone density of elite female athletes with stress fractures. *Medical Journal of Australia* **153**, 373–376.

Casey, M. J., Foster, C., Thompson, N. N., Jones, E. C. & Snyder, A. C. (1991). Menstrual function in elite speed skaters. *Sports Training, Medicine and Rehabilitation* **2**, 69–76.

Chan, G. M. (1991). Dietary calcium and bone mineral status of children and adolescents. *American Journal of Diseases of Children* **145**, 631–634.

Conroy, B. P., Kraemer, W. J., Maresh, C. M., Fleck, S. J., Stone, M. H., Fry, A. C., Miller, P. D. & Dalsky, G. P. (1993). Bone mineral density in elite junior Olympic weightlifters. *Medicine and Science in Sports and Exercise* **25**, 1103–1109.

Cooper, C., Barker, D. J. P. & Wickham, C. (1988). Physical activity, muscle strength and calcium intake in fracture of the proximal femur in Britain. *British Medical Journal* **297**, 1443–1446.

Cummings, S. R., Black, D. M., Nevitt, M. C., Browner, W., Cauley, J., Ensrud, K., Genant, H. K., Palermo, L., Scott, J. & Vogt, T. M. (1993). Bone density at various sites for prediction of hip fractures. *Lancet* **341**, 72–75.

Dalén, N., Låftman, P., Ohlsén, H. & Strömberg, L. (1985). The effect of athletic activity on bone mass in human diaphyseal bone. *Orthopedics* **8**, 1139–1141.

Davee, A. M., Rosen, C. J. & Adler, R. A. (1990). Exercise patterns and trabecular bone density in college women. *Journal of Bone and Mineral Research* **5**, 245–250.

Davis, C., Fox, J., Cowles, M., Hastings, P. & Schwass, K. (1990). The functional role of exercise in the development of weight and diet concerns in women. *Journal of Psychosomatic Research* **34**, 563–574.

Delistraty, D. A., Reisman, E. J. & Snipes, M. (1992). A physiological and nutritional profile of young female figure skaters. *Journal of Sports Medicine and Physical Fitness* **32**, 149–155.

De Souza, M. J. & Metzger, D. A. (1991). Reproductive dysfunction in amenorrheic athletes and anorexic patients: a review. *Medicine and Science in Sports and Exercise* **23**, 995–1007.

Deuster, P. A., Kyle, S. B., Moser, P. B., Vigersky, R. A., Singh, A. & Schoomaker, E. B. (1986). Nutritional intakes and status of highly trained amenorrheic and eumenorrheic women runners. *Fertility and Sterility* **46**, 636–643.

Ding, J.-H., Sheckter, C. B., Drinkwater, B. L., Soules, M. R. & Bremner, W. J. (1988). High serum cortisol levels in exercise-associated amenorrhea. *Annals of Internal Medicine* **108**, 530–534.

Drewnowski, A., Hopkins, S. A. & Kessler, R. C. (1988). The prevalence of bulimia nervosa in the US college student population. *American Journal of Public Health* **78**, 1322–1325.

Drinkwater, B. L., Bruemner, B. & Chesnut, C. H. (1990). Menstrual history as a determinant of current bone density in young athletes. *Journal of the American Medical Association* **263**, 545–548.

Drinkwater, B. L., Nilson, K., Chesnut, C. H., Bremner, W. J., Shainholtz, S. & Southworth, M. B. (1984). Bone mineral content of amenorrheic and eumenorrheic athletes. *New England Journal of Medicine* **311**, 277–281.

Drinkwater, B. L., Nilson, K., Ott, S. & Chesnut, C. H. (1986). Bone mineral density after resumption of menses in amenorrheic athletes. *Journal of the American Medical Association* **256**, 380–382.

Dummer, G. M., Rosen, L. W., Heusner, W. W., Roberts, P. J. & Counsilman, J. E. (1987). Pathogenic weight-control behaviors of young competitive swimmers. *Physician and Sportsmedicine* **15**, 75–84.

Eickoff, J., Molczyk, L., Galagher, J. C. & De Jong, S. (1993). Influence of isotonic, isometric and isokinetic muscle strength on bone density of the spine and femur in young women. *Bone and Mineral* **20**, 201–209.

Emmons, L. (1992). Dieting and purging behavior in black and white high school students. *Journal of the American Dietetic Association* **92**, 306–312.

Evers, C. L. (1987). Dietary intake and symptoms of anorexia nervosa in female university dancers. *Journal of the American Dietetic Association* **87**, 66–68.

Falls, H. B. & Humphrey, L. D. (1978). Body type and composition differences between placers and non-placers in an AIAW gymnastics meet. *Research Quarterly for Exercise and Sport* **49**, 38–43.

Feicht, C. B., Johnson, T. S., Martin, B. J., Sparkes, K. E. & Wagner, W. W. (1978). Secondary amenorrhoea in athletes. *Lancet* **ii**, 1145–1146.

Frederick, L. & Hawkins, S. T. (1992). A comparison of nutrition knowledge and attitudes, dietary practices, and bone densities of postmenopausal women, female college athletes, and non-athletic college women. *Journal of the American Dietetic Association* **92**, 299–305.

Frisch, R. E., Gotz-Welbergen, A. V., McArthur, J. W., Albright, T., Witschi, J., Bullen, B., Birnholz, J., Reed, R. B. & Hermann, H. (1981). Delayed menarche and amenorrhea of college athletes in relation to age of onset of training. *Journal of the American Medical Association* **246**, 1559–1563.

Frisch, R. E. & McArthur, J. W. (1974). Menstrual cycles: fatness as a determinant of minimum weight for height necessary for their maintenance or onset. *Science* **185**, 949–951.

Frusztajer, N. T., Dhuper, S., Warren, M. P., Brooks-Gunn, J. & Fox, R. P. (1990). Nutrition and the incidence of stress fractures in ballet dancers. *American Journal of Clinical Nutrition* **51**, 779–783.

Gadpaille, W. J., Sanborn, C. F. & Wagner, W. W. (1987). Athletic amenorrhea, major affective disorders, and eating disorders. *American Journal of Psychiatry* **144**, 939–942.

Galle, P. C., Freeman, E. W., Galle, M. G., Huggins, G. R. & Sondheimer, S. J. (1983). Physiologic and psychologic profiles in a survey of women runners. *Fertility and Sterility* **39**, 633–639.

Gersovitz, M., Madden, J. P. & Smiciklas-Wright, H. (1978). Validity of the 24-hr dietary recall and seven-day record for group comparisons. *Journal of the American Dietetic Association* **73**, 48–55.

Glass, A. R., Deuster, P. A., Kyle, S. B., Yahiro, J. A., Vigersky, R. A. & Schoomaker, E. B. (1987). Amenorrhea in Olympic marathon runners. *Fertility and Sterility* **48**, 740–745.

Gleaves, D. H., Williamson, D. A. & Fuller, R. D. (1992). Bulimia nervosa symptomatology and body image disturbance associated with distance running and weight loss. *British Journal of Sports Medicine* **26**, 157–160.

Grimston, S. K., Engsberg, J. R., Kloiber, R. M. & Hanley, D. A. (1990). Menstrual, calcium, and training history: relationship to bone health in female runners. *Clinics in Sports Medicine* **2**, 119–128.

Halioua, L. & Anderson, J. J. B. (1989). Lifetime calcium intake and physical activity habits: independent and combined effects on the radial bone of healthy premenopausal Caucasian women. *American Journal of Clinical Nutrition* **49**, 534–541.

Hamilton, L. H., Brooks-Gunn, J., Warren, M. P. & Hamilton, W. G. (1988). The role of selectivity in the pathogenesis of eating problems in ballet dancers. *Medicine and Science in Sports and Exercise* **20**, 560–565.

Harber, V. J., Webber, C. E., Sutton, J. R. & MacDougall, J. D. (1991). The effect of amenorrhea on calcaneal bone density and total bone turnover in runners. *International Journal of Sports Medicine* **12**, 505–508.

Harris, D. V. (1986). The psychology of the female runner. In *Female Endurance Athletes* (conference, 1984), pp. 59–74 (B. L. Drinkwater, editor). Champaign, IL: Human Kinetics.

Heinrich, C. H., Going, S. B., Pamenter, R. W., Perry, C. D., Boyden, T. W. & Lohman, T. G. (1990). Bone mineral content of cyclically menstruating female resistance and endurance trained athletes. *Medicine and Science in Sports and Exercise* **22**, 558–563.

Hergenroeder, A. C., Fiorotto, M. L. & Klish, W. J. (1991). Body composition in ballet dancers measured by total body electrical conductivity. *Medicine and Science in Sports and Exercise* **23**, 528–533.

Humphries, L. L. & Gruber, J. J. (1986). Nutrition behaviors of university majorettes. *Physician and Sportsmedicine* **14**, 91–98.

Jacobson, P. C., Beaver, W., Grubb, S. A., Taft, T. N. & Talmage, R. V. (1984). Bone density in women: college athletes and older athletic women. *Journal of Orthopedic Research* **2**, 328–332.

Johnson, J. & Whitaker, A. H. (1992). Adolescent smoking, weight changes, and binge–purge behavior: associations with secondary amenorrhea. *Journal of the American Dietetic Association* **82**, 47–54.

Kaiserauer, S., Snyder, A. C., Sleeper, M. & Zierath, J. (1989). Nutritional, physiological, and menstrual status of distance runners. *Medicine and Science in Sports and Exercise* **21**, 120–125.

Kanders, B., Dempster, D. W. & Lindsay, R. (1988). Interaction of calcium nutrition and physical activity on bone mass in young women. *Journal of Bone and Mineral Research* **3**, 145–149.

Karvetti, R.-L. & Knuts, L.-R. (1992). Validity of the estimated food diary: comparison of 2-day recorded and observed food and nutrient intakes. *Journal of the American Dietetic Association* **92**, 580–584.

Klesges, R. C., Eck, L. H., Mellon, M. W., Fulliton, W., Somes, G. W. & Hanson, C. L. (1990). The accuracy of self-reports of physical activity. *Medicine and Science in Sports and Exercise* **22**, 690–697.

Kurtzmann, F. D., Yager, J., Landswerk, E., Wiesmeier, E. & Bodurka, D. C. (1989). Eating disorders among selected female student populations at UCLA. *Journal of the American Dietetic Association* **89**, 45–53.

Lanyon, L. E. (1992). The success and failure of the adaptive response to functional load-bearing in averting bone fracture. *Bone* **13**, Suppl. 1, S17–S21.

Lanyon, L. E., Rubin, C. T. & Baust, G. (1986). Modulation of bone loss during calcium insufficiency by controlled dynamic loading. *Calcified Tissue International* **38**, 209–216.

Lindberg, J. S., Powell, M. R., Hunt, M. M., Ducey, D. E. & Wade, C. E. (1987). Increased vertebral bone mineral in response to reduced exercise in amenorrheic runners. *Western Journal of Medicine* **146**, 39–42.

Lloyd, T., Buchanan, J. R., Bitzer, S., Waldman, C. J., Myers, C. & Ford, B. G. (1987). Interrelationships of diet, athletic activity, menstrual status, and bone density in collegiate women. *American Journal of Clinical Nutrition* **46**, 681–684.

Lloyd, T., Schaeffer, J. M., Walker, M. A. & Demers, L. M. (1991). Urinary hormonal concentrations and spinal bone densities of premenopausal vegetarian and nonvegetarian women. *American Journal of Clinical Nutrition* **54**, 1005–1010.

Lloyd, T., Triantafyllou, S. J., Baker, E. R., Houts, P. S., Whiteside, J. A., Kalenak, A. & Stumpf, P. G. (1986). Women athletes with menstrual irregularity have increased musculoskeletal injuries. *Medicine and Science in Sports and Exercise* **18**, 374–379.

Loucks, A. B. (1990). Effects of exercise training on the menstrual cycle: existence and mechanisms. *Medicine and Science in Sports and Exercise* **22**, 275–280.

Loucks, A. B., Vaitukaitis, J., Cameron, J. L., Rogol, A. D., Skrinar, G., Warren, M. P., Kendrick, J. & Limacher, M. C. (1992). The reproductive system and exercise in women. *Medicine and Science in Sports and Exercise* **24**, Suppl., S288–S293.

McArdle, W. D., Katch, F. I. & Katch, V. L. (1986). *Exercise Physiology: Energy, Nutrition and Human Performance*, 2nd edn, pp. 642–649. Philadelphia, PA: Lea & Febiger.

McCargar, L. J., Simmons, D., Craton, N., Taunton, J. E. & Birmingham, C. L. (1993). Physiological effects of weight cycling in female lightweight rowers. *Canadian Journal of Applied Physiology* **18**, 291–303.

Marcus, R., Cann, C., Madvig, P., Minkoff, J., Goddard, M., Bayer, M., Martin, M., Gaudiani, L., Haskell, W. & Genant, H. (1985). Menstrual function and bone mass in elite women distance runners: endocrine and metabolic features. *Annals of Internal Medicine* **102**, 158–163.

Margulies, J. Y., Simkin, A., Leichter, I., Bivas, A., Steinberg, R., Giladi, M., Stein, M., Kashtan, H. & Milgrom, C. (1986). Effect of intense physical activity on the bone mineral content in the lower limbs of young adults. *Journal of Bone and Joint Surgery* **68A**, 1090–1093.

Matković, V., Kostial, K., Šimonović, I., Buzina, R., Brodarec, A. & Nordin, B. E. C. (1979). Bone status and fracture rates in two regions of Yugoslavia. *American Journal of Clinical Nutrition* **32**, 540–549.

Melton, L. J., Kan, S. H., Wahner, H. W. & Riggs, B. L. (1988). Lifetime fracture risk: an approach to hip fracture risk assessment based on bone mineral density and age. *Clinical Epidemiology* **41**, 985–994.

Milgrom, C., Giladi, M., Simkin, A., Rand, N., Kedem, R., Kashtan, H., Stein, M. & Gomori, M. (1989). The area moment inertia of the tibia: a risk factor for stress fractures. *Journal of Biomechanics* **22**, 1243–1248.

Mulligan, K. & Butterfield, G. E. (1990). Discrepancies between energy intake and expenditure in physically active women. *British Journal of Nutrition* **64**, 23–36.

Myburg, K. H., Watkin, V. A. & Noakes, T. D. (1992). Are risk factors for menstrual dysfunction cumulative? *Physician and Sportsmedicine* **20**, 114–125.

Myerson, M., Gutin, B., Warren, M. P., May, M. T., Contento, I., Lee, M., Pi-Sunyer, F. X., Pierson, R. N. & Brooks-Gunn, J. (1991). Resting metabolic rate and energy balance in amenorrheic and eumenorrheic runners. *Medicine and Science in Sports and Exercise* **23**, 15–22.

Myerson, M., Gutin, B., Warren, M. P., Wang, J., Lichtman, S. & Pierson, R. N. (1992). Total body bone density in amenorrheic runners. *Obstetrics and Gynecology* **79**, 973–978.

Nelson, M. E., Fisher, E. C., Catsos, P. D., Meredith, C. N., Turksoy, R. N. & Evans, W. J. (1986). Diet and bone status in amenorrheic runners. *American Journal of Clinical Nutrition* **43**, 910–916.

Nordin, B. E. C. & Heaney, R. P. (1990). Calcium supplementation of the diet: justified by present evidence. *British Medical Journal* **300**, 1056–1060.

Pate, R. R., Sargent, R. G., Baldwin, C. & Burgess, M. L. (1990). Dietary intake of women runners. *International Journal of Sports Medicine* **11**, 461–466.

Pedersen, A. B., Bartholomew, M. J., Dolence, L. A., Aljadir, L. P., Netteburg, K. L. & Lloyd, T. (1991). Menstrual differences due to vegetarian and nonvegetarian diets. *American Journal of Clinical Nutrition* **53**, 879–885.

Pirke, K. M., Schweiger, U., Laessle, R., Dickhaut, B., Schweiger, M. & Waechtler, M. (1986). Dieting influences the menstrual cycle: vegetarian versus nonvegetarian diet. *Fertility and Sterility* **46**, 1083–1088.

Pirnay, F., Bodeux, M., Crielaard, J. M. & Franchimont, P. (1987). Bone mineral content and physical activity. *International Journal of Sports Medicine* **8**, 331–335.

Pocock, N., Eisman, J., Gwinn, T., Sambrook, P., Kelly, P., Freund, J. & Yeates, M. (1989). Muscle strength, physical fitness, and weight but not age predict femoral neck bone mass. *Journal of Bone and Mineral Research* **4**, 441–448.

Pouilles, J. M., Bernard, J., Tremollières, F., Louvet, J. P. & Ribot, C. (1989). Femoral bone density in young male adults with stress fractures. *Bone* **10**, 105–108.

Prentice, A. M., Black, A. E., Coward, W. A., Davies, H. L., Goldberg, G. R., Murgatroyd, P. R., Ashford, J., Sawyer, M. & Whitehead, R. G. (1986). High levels of energy expenditure in obese women. *British Medical Journal* **292**, 983–987.

Prior, J. C. (1990). Progesterone as a bone-trophic hormone. *Endocrine Reviews* **11**, 386–398.

Prior, J. C., Vigna, Y. M. & McKay, D. W. (1992). Reproduction for the athletic woman. New understandings of physiology and management. *Sports Medicine* **14**, 190–199.

Prussin, R. A. & Harvey, P. D. (1991). Depression, dietary restraint and binge-eating in female runners. *Addictive Behaviour* **16**, 295–301.

Riggs, B. L. & Melton, L. J. (1986). Involutional osteoporosis. *New England Journal of Medicine* **314**, 1676–1686.

Riggs, B. L., Wahner, H. W., Melton, L. J., Richelson, L. S., Judd, H. L. & Offord, K. P. (1986). Rates of bone loss in the appendicular and axial skeletons of women. Evidence of substantial vertebral bone loss before menopause. *Journal of Clinical Investigation* **77**, 1487–1491.

Risser, W. L., Lee, E. J., LeBlanc, A., Poindexter, H. B. W., Risser, J. M. H. & Schneider, V. (1990). Bone density in eumenorrheic female college athletes. *Medicine and Science in Sports and Exercise* **22**, 570–574.

Rosen, L. W. & Hough, D. O. (1988). Pathogenic weight-control behaviors of female college gymnasts. *Physician and Sportsmedicine* **16**, 141–144.

Rosenvinge, J. H. & Vig, C. (1993). Eating disorders and associated symptoms among adolescent swimmers. *Scandinavian Journal of Medicine and Science in Sports* **3**, 164–169.

Rucinski, A. (1989). Relationship of body image and dietary intake of competitive ice-skaters. *Journal of the American Dietetic Association* **89**, 98–100.

Sanborn, C. F., Martin, B. J. & Wagner, W. W. (1982). Is athletic amenorrhea specific to runners? *American Journal of Obstetrics and Gynecology* **143**, 859–861.

Schoeller, D. A., Bandini, L. G. & Dietz, W. H. (1990). Inaccuracies in self-reported intake identified by comparison with the doubly labelled water method. *Canadian Journal of Physiology and Pharmacology* **68**, 941–949.

Schotte, D. E. & Stunkard, A. J. (1987). Bulimia versus bulimic behaviors on a college campus. *Journal of the American Medical Association* **258**, 1213–1215.

Schulz, L. O., Alger, S., Harper, I., Wilmore, J. H. & Ravussin, E. (1992). Energy expenditure of elite female runners measured by respiratory chamber and doubly labeled water. *Journal of Applied Physiology* **72**, 23–28.

Schweiger, U., Laessle, R., Pfister, H., Hoehl, C., Schwingenschloegel, M., Schweiger, M. & Pirke, K.-M. (1987). Diet-induced menstrual irregularities: effects of age and weight loss. *Fertility and Sterility*. **48**, 746–751.

Sentipal, J. M., Wardlaw, G. M., Mahan, J. & Matkovic, V. (1991). Influence of calcium intake and growth indexes on vertebral bone mineral density in young females. *American Journal of Clinical Nutrition* **54**, 425–428.

Shangold, M. M. & Levine, H. S. (1982). The effect of marathon training upon menstrual function. *American Journal of Obstetrics and Gynecology* **143**, 862–869.

Short, S. H. & Short, W. R. (1983). Four-year study of university athletes' dietary intake. *Journal of the American Dietetic Association* **82**, 632–645.

Slavin, J., Lutter, J. & Cushman, S. (1984). Amenorrhoea in vegetarian athletes. *Lancet* **i**, 1474–1475.

Slemenda, C., Hui, S. L., Longcope, C. & Johnston, C. C. (1987). Sex steroids and bone mass. A study of changes about the time of menopause. *Journal of Clinical Investigation* **80**, 1261–1269.

Slemenda, C. W. & Johnston, C. C. (1993). High intensity activities in young women: site specific bone mass effects among female figure skaters. *Bone and Mineral* **20**, 125–132.

Snead, D. B., Weltman, A., Weltman, J. Y., Evans, W. S., Veldhuis, J. D., Varma, M. M., Teates, C. D., Dowling, E. A. & Rogol, A. D. (1992). Reproductive hormones and bone mineral density in women runners. *Journal of Applied Physiology* **72**, 2149–2156.

Snyder, A. C., Seifert, J. G. & Welsh, R. (1991). Energy intake and expenditure during a cycling series (abstract). *Medicine and Science in Sports and Exercise* **23**, Suppl. 1, S470.

Stensland, S. H. & Sobal, J. (1992). Dietary practices of ballet, jazz, and modern dancers. *Journal of the American Dietetic Association* **92**, 319–324.

Stunkard, A. J. & Waxman, M. (1981). Accuracy of self-reports of food intake. *Journal of the American Dietetic Association* **79**, 547–551.

Sundgot-Borgen, J. (1993). Prevalence of eating disorders in elite female athletes. *International Journal of Sport Nutrition* **3**, 29–40.

Sundgot-Borgen, J. & Larsen, S. (1993). Pathogenic weight control methods and self-reported eating disorders in female elite athletes and controls. *Scandinavian Journal of Medicine and Science in Sports* **3**, 150–155.

Talmage, R. V., Stinnett, S. S., Landwehr, J. T., Vincent, L. M. & McCartney, W. H. (1986). Age-related loss of bone mineral density in non-athletic and athletic women. *Bone and Mineral* **1**, 115–125.

van Erp-Baart, A. M. J., Saris, W. H. M., Binkhorst, R. A., Vos, J. A. & Elvers, J. W. H. (1989). Nationwide survey on nutritional habits in elite athletes. I. Energy, carbohydrate, protein, and fat intake. *International Journal of Sports Medicine* **10**, Suppl. 1, S3–S10.

Walberg, J. L. & Johnston, C. S. (1991). Menstrual function and eating behavior in female recreational weight lifters and competitive body builders. *Medicine and Science in Sports and Exercise* **23**, 30–36.

Walberg-Rankin, J., Edmonds, C. E. & Gwazdauskas, F. C. (1993). Diet and weight changes of female bodybuilders before and after competition. *International Journal of Sport Nutrition* **3**, 87–102.

Warren, M. P., Brooks-Gunn, J., Hamilton, L. H., Warren, L. F. & Hamilton, W. G. (1986). Scoliosis and fractures in young ballet dancers: relation to delayed menarche and secondary amenorrhea. *New England Journal of Medicine* **314**, 1348–1353.

Warren, M. P., Shane, E., Lee, M. J., Lindsay, R., Dempster, D. W., Warren, L. F. & Hamilton, W. G. (1990). Femoral head collapse associated with anorexia nervosa in a 20-year-old ballet dancer. *Clinical Orthopaedics and Related Research* no. 251, 171–176.

Watkin, V. A., Myburg, K. H. & Noakes, T. D. (1991). Low nutrient intake does not cause the menstrual cycle interval disturbance seen in some ultramarathon runners. *Clinical Journal of Sport Medicine* **1**, 154–161.

Webb, P. (1981). Energy expenditure and fat free mass in men and women. *American Journal of Clinical Nutrition* **34**, 1816–1826.

Weight, L. M. & Noakes, T. D. (1987). Is running an analog of anorexia? A survey of the incidence of eating disorders in female distance runners. *Medicine and Science in Sports and Exercise* **19**, 213–217.

Williford, H. N., Olson, M. S., Blessing, D. L., Barksdale, J. & Keith, R. E. (1989). Iron and dietary status of long-term female exercisers (abstract). *Medicine and Science in Sports and Exercise* **23**, Suppl. 4, S79.

Wilmore, J. H. (1991). Eating and weight disorders in the female athlete. *International Journal of Sport Nutrition* **1**, 104–117.

Wilmore, J. H. & Costill, D. L. (1987). *Training for Sport and Activity: the Physiological Basis of the Conditioning Process*, 3rd edn. Boston, MA: Allyn & Bacon.

Wilmore, J. H., Wambsgans, K. C., Brenner, M., Broeder, C. E., Paijmans, I., Volpe, J. A. & Wilmore, K. M. (1992). Is there energy conservation in amenorrheic compared with eumenorrheic distance runners? *Journal of Applied Physiology* **72**, 15–22.

Wilson, J. H., Harries, M. G. & Reeve, J. (1994*a*). Bone mineral density in premenopausal veteran female athletes and the influence of menstrual irregularity. *Clinical Science* **86**(2), 2p.

Wilson, J. H., Reeve, J. & Harries, M. G. (1994*b*). Determinants of bone mineral density in female athletes. *Bone* **15**, 000–000.

Wilson, J. H. & Wolman, R. L. (1994). Osteoporosis and fracture complications in an amenorrhoeic athlete. *British Journal of Rheumatology* **33**, 480–481.

Wolff, J. (1892). In *Das Gesetz der Transformation der Knocken*. Berlin: Hirschwald.

Wolman, R. L., Clark, P., McNally, E., Harries, M. & Reeve, J. (1990). Menstrual state and exercise as determinants of spinal trabecular bone density in female athletes. *British Medical Journal* **301**, 516–518.

Wolman, R. L., Clark, P., McNally, E., Harries, M. G. & Reeve, J. (1992). Dietary calcium as a statistical determinant of spinal trabecular bone density in amenorrhoeic and oestrogen-replete athletes. *Bone and Mineral* **17**, 415–423.

Wolman, R. L., Faulmann, L., Clark, P., Hesp, R. & Harries, M. G. (1991). Different training patterns and bone mineral density of the femoral shaft in elite, female athletes. *Annals of the Rheumatic Diseases* **50**, 487–489.

Wolman, R. L. & Harries, M. G. (1989). Menstrual abnormalities in elite athletes. *Clinical Sports Medicine* **1**, 95–100.

HUMAN BIOAVAILABILITY OF VITAMINS

Members of EC Flair Concerted Action No. 10: 'Measurement of micronutrient absorption and status'†

Compiled by: C. J. BATES[1] AND H. HESEKER

[1] MRC Dunn Nutrition Unit, Milton Road, Cambridge CB4 1XJ, UK

CONTENTS

INTRODUCTION	95
VITAMIN A AND CAROTENOIDS	96
FOOD SOURCES	96
ABSORPTION AND TRANSPORT	96
CONVERSION OF CAROTENOIDS TO RETINOL	97
CONCLUSIONS	98
VITAMIN D	98
FOOD SOURCES	99
ABSORPTION	99
PHYSIOLOGICAL CONDITION	100
CONCLUSIONS	100
VITAMIN E	100
FOOD SOURCES	100
ABSORPTION, TRANSPORT AND METABOLISM	101
FACTORS INFLUENCING BIOAVAILABILITY	101
CONCLUSION	102
VITAMIN K	102
FOOD SOURCES AND HUMAN REQUIREMENTS	103
ABSORPTION MECHANISMS AND BLOOD TRANSPORT	103
FUTURE RESEARCH	103
THIAMIN	104
FOOD SOURCES	104
ABSORPTION AND EXCRETION	104
FACTORS AFFECTING ABSORPTION AND UTILIZATION	105
Food processing	105
Antagonists	105
Alcohol, diseases and drugs	105

† Contributors: H. van den Berg, Zeist, Netherlands (introduction (van den Berg, 1993) and vitamin D); C. J. Bates, Cambridge, UK (vitamin K, riboflavin and vitamin C); R. Bitsch, Jena, Germany (vitamin B_6); P. Finglas, Norwich, UK (thiamin); H. Heseker and J. Zempleni, Giessen, Germany (niacin); M. Jagerstad and K. Wigerty, Lund, Sweden (folate); G. Maiani, M. Serafini and A. Ferro-Luzzi, Rome, Italy (biotin and pantothenate); K. Pietrzik and J. Dierkes, Bonn, Germany (vitamin B_{12}); A. Sheehy, Cork, Eire (vitamin E); C. E. West and T. van Vliet, Wageningen/Zeist, Netherlands (vitamin A and carotenoids).

　　　　Smoking and age 105
　　　　Other nutrient interactions 106
　　CONCLUSIONS. 106
RIBOFLAVIN 106
　　FOOD SOURCES 106
　　RELEASE AND ABSORPTION 106
　　EXCRETION 107
　　ANTAGONISTS AND HOMEOSTATIC INFLUENCES 107
　　CONCLUSION 107
VITAMIN B_6 107
　　FOOD PROCESSING; BOUND B_6 108
　　INTERACTIONS WITH NUTRIENTS AND DRUGS. 109
　　CONCLUSION 109
NIACIN 109
　　FOOD SOURCES 109
　　RELEASE AND ABSORPTION 110
　　EXCRETION 110
　　SYNTHESIS FROM TRYPTOPHAN 110
　　METABOLISM AND TRANSPORT BETWEEN TISSUES 110
　　CONCLUSIONS. 111
FOLATE 111
　　ABSORPTION OF SYNTHETIC MONO- AND POLY-GLUTAMYL FOLATES . . 111
　　AVAILABILITY OF FOOD FOLATES 112
　　MILK FOLATE BINDING PROTEIN. 112
　　DIETARY FIBRE AND MATRIX EFFECTS 112
　　EXCRETION 113
　　CONCLUSIONS AND FUTURE RESEARCH PRIORITIES 113
VITAMIN B_{12} 113
　　VITAMIN B_{12} IN FOOD 113
　　ABSORPTION 114
　　RETENTION 114
　　CONCLUSION 114
BIOTIN 114
　　ABSORPTION AND EXCRETION 115
　　DIETARY SOURCES 115
　　REQUIREMENTS AND NUTRITIONAL STATUS 115
　　CONCLUSION 116
PANTOTHENIC ACID 116
　　RELEASE AND ABSORPTION 116
　　EXCRETION 116
　　NUTRITIONAL STATUS AND INDUCED DEFICIENCY. 116

CONCLUSION 116
VITAMIN C 117
    FOOD SOURCES 117
    INTESTINAL ABSORPTION 117
    TURNOVER AND EXCRETION AND FACTORS WHICH AFFECT AVAILABILITY 118
    TRANSPORT AT OTHER SITES 118
    CONTRASTS BETWEEN GROUPS OF PEOPLE 118
    CONCLUSION 119
REFERENCES 119

INTRODUCTION

Although vitamins are usually considered to be well absorbed and readily available compared with mineral nutrients, there are situations where bioavailability is limited. Absorption processes are complex, and not completely understood, and interactions with food and physiological processes are critical.

The percentage absorption of individual vitamins from food in the human intestine varies between 20 and 98%. Active transport in the duodenum and ileum has been shown for retinol and the water-soluble vitamins except B_6 and pantothenate. This is characterized as saturable, sodium dependent, high affinity and low capacity, and is therefore most important at low intraluminal concentrations (Matthews, 1974; Rose, 1988; Bowman *et al.* 1989). It was missed in some of the earlier studies of transport mechanisms which had to rely on relatively insensitive detection methods. At high concentrations, passive diffusion generally predominates.

Water-soluble vitamins may occur in protein bound forms which require release by proteinase or phosphatase action before absorption. Stomach acidity may be critical, and pathology may influence absorption.

Food intake usually enhances vitamin absorption by stimulating enzyme and bile acid secretion, and by increasing transit time (Jusko & Levy, 1975). In contrast, some foods contain inhibitors of vitamin utilization, which reduce solubility or release. Proteinase inhibitors in some raw fruits and vegetables can affect vitamin absorption, as can substances which reduce bile acid reabsorption (Kern *et al.* 1978).

Food processing may have either beneficial or adverse effects. Beneficial effects include the removal of antivitamin factors, or interconversions which enhance availability. Changes in fat content, its saturation level, and the use of fat substitutes may have profound effects on fat-soluble vitamin availability.

Measurement of vitamin bioavailability is a complex problem. Simple pharmacokinetic parameters, such as area under or maximum height of the curve of plasma concentration following a test dose, may not be suitable, if pre-existing stores and postabsorption events interfere, as they often do. Animal based studies using growth rates or other dose–effect relationships, urinary excretion or balance studies (for fat-soluble vitamins) can yield very useful information, but they may not be ideal to investigate bioavailability in humans, because of interspecies differences. The use of stable isotope labelled vitamins in man seems promising, but is still at an early stage of development.

Vitamin absorption does not appear to be controlled by the body pool size, as is the case for some metal ions, and only vitamins A and B_{12} are stored (in the liver) in amounts greatly in excess of the immediate functional requirement. Excess amounts of water-soluble vitamins are generally removed by urinary excretion; those of some fat-soluble vitamins by biliary excretion. Turnover rates (metabolic degradation) largely determine minimum daily requirements; these have not yet been determined precisely for humans. Adaptation to varying intake levels is poorly understood and deserves further study. The effects of age and sex are poorly characterized: nutrient density requirements may be different for old people than for children or young adults; pregnancy, lactation and preterm birth clearly have major effects on requirements and, possibly, on availability. Metabolic diseases and drug use may likewise affect vitamin availability (van den Berg, 1991).

In conclusion, the bioavailability of vitamins owes its complexity to the fact that enhancers or inhibitors in foods, the different chemical form of vitamers, and the physiological state of the subject, all affect availability. Quantitative data, especially for vitamins in foods and diets, are scarce and, as a result, requirement estimates are imprecise. New standardized methods of estimation over physiologically relevant intake ranges are needed. Bioavailability needs to be included in the evaluation of commercial food products and in their design. In view of current health concerns that focus on the antioxidant vitamins for their relevance to degenerative diseases, and on the special micronutrient requirements of pregnant and lactating women, the elderly, adolescents, dieters, preterm infants, and other high risk groups, this challenge assumes considerable practical significance. Please refer to van den Berg (1993) for a more detailed discussion.

Each vitamin will now be considered in more detail.

VITAMIN A AND CAROTENOIDS

Vitamin A is obtained either preformed, or from provitamin A carotenoids. There are two forms: retinol (vitamin A_1), the most common, and 3-dehydroretinol (vitamin A_2) with about half the activity. Most dietary vitamin A occurs as esters, readily hydrolysed in the gut. Other active compounds are retinal (convertible to retinol), and retinoic acid and its β-glucuronides which replace retinol in functions such as growth and differentiation, but not vision. Retinyl esters are less polar and retinoic acid and glucuronides more polar than vitamin A. Hydroxycarotenoids, e.g. lutein, are more polar than carotenoids such as β-carotene (Craft & Soares, 1992). Carotenoids may protect against cancer and may modify the immune response, independently of retinol. Disease prevalence seems frequently related to carotenoid intake but not to serum retinol levels. Antioxidant and other functions of carotenoids have been considered (Burton, 1989). Retinol, retinal and retinoic acid are toxic in excess (Bendich & Langseth, 1989); carotenoids are generally regarded as non-toxic.

FOOD SOURCES

Rich sources of retinol are animal products such as milk, butter, cheese, egg yolk, liver and marine fishes; dehydroretinol is common in freshwater fishes. Traces of retinal, retinoic acid and the glycosides of retinol and retinoic acid also occur in foods. Carotenoids occur in dark green leafy vegetables and yellow and orange coloured fruits and vegetables, but less in animal products. In meat, fish and oils from animals and fish, 10% of the vitamin activity occurs as carotenoids, while in poultry, eggs and milk 30% occurs as carotenoids (Wu Leung et al. 1968).

ABSORPTION AND TRANSPORT

Retinyl esters are hydrolysed in the intestinal lumen by the same pancreatic enzyme which hydrolyses cholesteryl esters. Retinol is then absorbed into enterocytes by facilitated diffusion at physiological concentrations (< 150 nM) involving a cellular retinol binding protein (Said et al. 1989), or by passive diffusion at pharmacological levels. In enterocytes, retinol is esterified by microsomal enzymes, lecithin:retinol acyltransferase and acyl-CoA:retinol acyltransferase; the latter mainly at high retinol loads. Retinyl esters enter chylomicrons and are transported via the thoracic duct to the blood. Goodman et al. (1966) showed that one fifth of orally administered radioactive retinol was recovered in thoracic lymph, and Sauberlich et al. (1974) showed that half of an oral load is recovered in the liver.

Water-soluble retinoids such as retinoic acid and glycosides of retinol and retinoic acid travel from the intestine via the portal vein to the liver, where retinol and retinoic acid are glycosylated and then excreted into bile (Barua et al. 1989).

Carotenoids are absorbed from the small intestine into enterocytes by passive diffusion (Hollander & Ruble, 1978) and thence via chylomicrons to lymph. Less than one sixth of β-carotene is absorbed intact in humans (Goodman et al. 1966). Balance studies indicate that 50–75% of β-carotene may be utilized (Lala & Reddy, 1970) but breakdown may also occur. The initial increase in β-carotene level on feeding is in the chylomicrons and VLDL fraction (Johnson & Russell, 1992). The kinetics of the serum response to oral β-carotene is not simply dependent on dose (Henderson et al. 1989); indeed, when large doses are fed the serum level increases exponentially (Prince et al. 1991).

Study of the absorption of carotenoids other than β-carotene has recently been facilitated by high pressure liquid chromatography. Lutein and canthaxanthin are accumulated in the eye, while α- and β-carotene, lycopene, zeaxanthin and cryptoxanthin accumulate in various tissues (Kaplan et al. 1990; Stahl et al. 1992). Micozzi et al. (1992) found that feeding β-carotene reduced the concentration of lutein in serum while Henderson et al. (1989) found no effect of β-carotene on serum level of α-carotene, cryptoxanthin, lycopene and lutein.

Jensen et al. (1987) concluded that all-*trans* β-carotene is absorbed more readily in man that the 9-*cis* form because from mixtures the proportion of 9-*cis* in serum was less than one sixth of that in the mixture fed. However, Stahl et al. (1992) showed that the proportions of 9-, 13- and 15-*cis* isomers were much higher in liver and other tissues than in serum.

CONVERSION OF CAROTENOIDS TO RETINOL

Provitamin A carotenoids are converted to retinol primarily in the enterocytes, but also in tissues such as liver. The first FAO/WHO (1967) expert group on vitamin A requirements proposed that 6 μg β-carotene is equivalent to 1 μg retinol. The second FAO/WHO (1988) expert group proposed that the amount of β-carotene equivalent to 1 μg retinol is 4, 6 and 10 μg for intakes per meal of < 1000, 1000–4000 and > 4000 μg β-carotene respectively.

For vitamin A activity a carotenoid needs at least the unaltered β-ionone ring structure with an attached polyene side chain containing 11 carbons. Simpson & Chichester (1981) have suggested on the basis of structure alone that 50–60 carotenoids and apocarotenoids could have vitamin A activity.

The enzyme responsible for the formation of retinol from β-carotene in intestinal mucosa, carotene 15,15'-dioxygenase*, has been only partly purified (Goodman & Huang, 1965; Olson & Hayaishi, 1965). It is not known whether just one enzyme exists, or whether the cleavage of β-carotene is always central, yielding two molecules of retinal. In studies by

* (*EC* 1.13.11.21).

Goodman & Huang (1965) and Olson & Hayaishi (1965) in rats, central cleavage predominated. In *in vitro* studies, Gronowska-Senger & Wolf (1970) found retinal to be the sole product of cleavage, while Wang *et al.* (1991) identified β-apo-12′-carotenal (80%) and β-apo-10′-carotenal (11%) as the main products and retinal as only 5%. Few studies have been carried out *in vivo* in humans. Goodman *et al.* (1966) showed that of the radioactive β-carotene recovered in the lymph, most of that metabolized was retinyl esters, retinol and retinal. In a study using ^{13}C-labelled β-carotene in humans, Parker *et al.* (1992) recovered most of the label in retinyl esters. Such evidence suggests that β-carotene cleavage is central. Feeding β-carotene-rich foods increases serum retinol levels only when these are initially low (Lala & Reddy, 1970). No studies have been carried out in humans on the effect of vitamin A status on dioxygenase activity although studies in rats (Villard & Bates, 1986) as well as in hamsters (Van Vliet *et al.* 1992) have shown that diets low in vitamin A may increase dioxygenase activity. Low protein diets reduced dioxygenase activity in rats (Gronowska-Senger & Wolf, 1970). Failure to split β-carotene in man is rare but can lead to metabolic carotenaemia at low intakes of carotenoids (Monk, 1982) or to vitamin A deficiency if retinol intake is low (McLaren & Zekian, 1971).

Pure β-carotene is absorbed more readily than that in foods (Hume & Krebs, 1949). Thus the plasma response to pure β-carotene was about five times that of a similar amount in carrots (Brown *et al.* 1989; Micozzi *et al.* 1992). Absorption of carotenoids is markedly reduced when the intake of fat is low, and bile acids are necessary to prevent inhibition of the intraluminal phase of digestion ('maldigestion'). β-Carotene absorption in humans is reduced by pectin (Rock & Swendseid, 1992). Higher serum levels of β-carotene but not lycopene were found in women than in men (Kaplan *et al.* 1990). Carotene absorption is not affected by age (Maiani *et al.* 1989). Defects in absorption ('malabsorption') are seen in non-tropical sprue (Evans & Wollaeger, 1966) and cystic fibrosis (James *et al.* 1992). Intestinal parasites, however, do not interfere with carotene absorption. Absorption of carotenes in underweight children appears normal (Lala & Reddy, 1970). Acute respiratory infections, however, reduce absorption as measured by serum carotenoid levels (Heymann, 1936; Lala & Reddy, 1970).

CONCLUSIONS

The availability of preformed vitamin A is much greater than that of precursor carotenoids. There are important unsolved questions regarding the distribution of vitamin A, its delivery, turnover and toxicity at large intakes.

Further studies are also needed on the carotenoid content of foods, from different food matrices, with different levels of fat in the diet and in the presence of various parasites, on the extent of carotenoid absorption especially from different food matrices, on the extent of conversion of carotenoids to retinol and on the biological significance of the carotenoids.

VITAMIN D

Vitamin D is the generic name for a group of closely related secosteroids exhibiting qualitatively the biological activity of cholecalciferol (vitamin D_3). The international unit is equivalent to $1/40$ μg cholecalciferol. The fat-soluble vitamin D is only an essential nutrient under conditions of limited sun exposure of the skin, as the vitamin can be made endogenously in the skin epidermis from 7-dehydrocholesterol by action of ultraviolet light (DeLuca, 1988). 25-Hydroxyvitamin D (25-OHD), formed in the liver, and 1,25-dihydroxyvitamin D (1,25-DHD), produced by the kidneys, are the most important metabolites. Synthetic ergocalciferol (vitamin D_2) is assumed to have the same biological activity in man as the 'natural' cholecalciferol (vitamin D_3) from animal origin. Quantitative data on the human body pool are lacking. Adipose tissue and voluntary muscle are the principal storage sites of (non-hydroxylated) vitamin D in man (Mawer *et al.* 1972). The factors controlling the release of vitamin D from these stores are largely unknown but are most likely related to lipolysis and muscle protein breakdown, rather than vitamin D status *per se*. Liver and other tissues also contain some 25-OHD. In high dosages vitamin D can be toxic, especially in children (see Hathcock, 1985). Vitamin D toxicity seems attributable to high levels of 1,25-DHD (Vieth, 1990).

FOOD SOURCES

Only a few foods contain significant amounts of vitamin D, e.g. fatty fish and fish oils. Meat, especially liver, and dairy products contain up to about 100 international units/100 g. Human and bovine milk also contain 25-OHD (Hollis *et al.* 1981*a*), which has a higher biological activity than the non-hydroxylated vitamin D (Tanaka *et al.* 1973). The water-soluble vitamin D activity in milk is explained by the presence of protein bound 25-OHD (Hollis *et al.* 1981*b*). Meat products may also contain low amounts of 25-OHD which may contribute up to 50% of the vitamin D activity, and in many countries margarines, and sometimes milk products, are fortified with vitamin D (vitamin D_2 or D_3). Some plants have been reported to contain compounds with vitamin D activity (e.g. 1,25-DHD-glycoside in Solanaceae; Boland, 1986). Vitamin D is an acid- and heat-labile compound, but is generally stable during storage and processing in foods.

ABSORPTION

Vitamin D is, with the other fat-soluble vitamins, absorbed by emulsification into mixed micelles, uptake in the enterocyte, followed by incorporation in the chylomicrons and transport into the circulation *via* the lymphatic pathway (Hollander, 1981). Vitamin D esters, if present, are hydrolysed during solubilization in the mixed micelles. Vitamin D is efficiently absorbed in the proximal small intestine in the presence of fat (dependent on bile acid secretion). Long chain fatty acids facilitate the absorption of vitamin D (Holmberg *et al.* 1990).

The more polar hydroxylated metabolites (25-OHD) are at least partly absorbed *via* the portal venous system (Maislos *et al.* 1981). 25-OHD is more rapidly and efficiently absorbed than unhydroxylated vitamin D and is less dependent on bile salts (Sitrin & Bengoa, 1987). For both vitamin D and 25-OHD a clear dose–response relation between intake and serum 25-OHD content has been established (Whyte *et al.* 1979). A single large oral dose of vitamin D was found to give the same serum peak 25-OHD level as the same amount given in small doses over a longer period (Stamp, 1975).

Although bile is the main route for vitamin D excretion, the significance of a conservative enterohepatic circulation is still controversial and probably only functionally significant in conditions of a marginal vitamin D status (Gascon-Barré, 1986). A high fibre diet has been reported to lead to enhanced elimination of vitamin D (Batchelor & Compston, 1983). This may at least partly explain the low vitamin D status observed in vegetarians and macrobiotics (Dagnelie *et al.* 1990). Iron deficiency may result in impaired vitamin D absorption (Heldenberg *et al.* 1992).

Vitamin D bioavailability is mainly measured as the increase in 25-OHD after an oral load of vitamin D. Most of the available data on vitamin D availability relate to relative rather than absolute absorption. In animal experiments overall net absorption rates between 66 and 75% have been reported (Weber, 1981). For man, an average absorption of 50–80% from a mixed diet has been estimated (Lawson, 1980).

PHYSIOLOGICAL CONDITION

There is no evidence that vitamin D absorption, contrary to its postabsorptive utilization, is affected by vitamin D status (Stamp, 1975; Fraser, 1983). In conditions of primary or secondary hyperparathyroidism, increased metabolic inactivation of vitamin D has been reported (Clements *et al.* 1987). Although ageing may compromise vitamin D metabolism, especially at the level of 1,25-DHD production and receptor responsiveness, it does not seem to affect vitamin D absorption (Horst *et al.* 1990).

CONCLUSIONS

A large part of the vitamin D requirement is met by skin synthesis rather than by dietary ingestion. This is especially true under conditions with ample solar exposure. However, people living at latitudes above 50° need dietary sources, especially during the winter period, to maintain adequate vitamin D status (Webb *et al.* 1988). Also, groups with a limited capacity for endogenous vitamin D synthesis, such as people with heavily pigmented skins, or living at higher altitudes, or the elderly (with age related changes in skin thickness) are more dependent on dietary sources or supplements (Need *et al.* 1993). Only a limited number of foods contains vitamin D, sometimes present as 25-OHD. Vitamin D is normally efficiently absorbed in the proximal small intestine. For absorption, the presence of fat and normal gastrointestinal function (fat digestion) are prerequisites. Absorption of 25-OHD is more efficient than that of the non-hydroxylated form and less dependent on bile acid secretion, as the more polar 25-OHD is preferentially absorbed *via* the portal system instead of the lymph (in chylomicrons).

VITAMIN E

Vitamin E activity in foods is derived from two groups of fat-soluble compounds of plant origin called tocopherols and tocotrienols (Kasparek, 1980). α-Tocopherol is a derivative of 2-methyl-6-chromanol to which a saturated 16-carbon isoprenoid chain is attached at C(2), and which is methylated at C(5), C(7) and C(8). The other forms (β-, γ-, δ-) differ in the number and position of the methyl groups on the chromanol ring. Tocotrienols differ from tocopherols in that the side chain is unsaturated at C(3'), C(7') and C(11').

The biological activity of tocopherols is influenced by the configuration of both the chromanol ring and the side chain (Drevon, 1991). Because the most active form, RRR-α-tocopherol, is neither easily available nor stable, the International Unit is based on synthetic (all-rac) α-tocopheryl acetate, an equimolar mixture of all eight possible stereoisomers.

FOOD SOURCES

α-Tocopherol in the diet occurs mainly in the unesterified form. Its distribution in plants is influenced by species, variety, and stage of maturity as well as by harvesting, processing and storage procedures (Bauernfeind, 1980). The richest dietary sources of α-tocopherol are vegetable oils. Cereals are good sources, while supplies in fruits and vegetables are generally moderate to poor. Animal products are usually low in α-tocopherol, although the concentration can be increased by dietary supplementation. Some oils and cereals provide appreciable amounts of other tocopherols and tocotrienols.

ABSORPTION, TRANSPORT AND METABOLISM

α-Tocopherol is absorbed unchanged from the intestinal lumen, whereas tocopheryl esters are first hydrolysed by pancreatic esterase (Bjørneboe et al. 1990). Pancreatic juice and bile are essential for this process. The greatest capacity for absorption appears to be in the region between the upper and middle thirds of the small intestine (Gallo-Torres, 1980).

It is generally considered that the absorption of α-tocopherol and its esters is incomplete. However, precise absorption efficiencies are uncertain. Absorption of a single bolus of α-tocopherol by rats was calculated to be approximately 40% (Bjørneboe et al. 1986), while 65% absorption was observed when α-tocopheryl acetate was administered to rats as a slow, continuous infusion (Traber et al. 1986). In human studies, estimates of 24-h absorption efficiencies for α-tocopherol and α-tocopheryl acetate range from 21 to 86% (Gallo-Torres, 1980). Interpretation of these results is complicated by the limited sample numbers and the variety of experimental approaches used by different authors. Data are lacking on the efficiency of vitamin E absorption from foods.

There are no major differences in the absorption of α- and γ-tocopherol (Traber et al. 1986) but the latter is preferentially excreted in bile (Traber & Kayden, 1989), which accounts for its lower concentration in plasma despite its widespread distribution in the diet. The other tocopherols (β- and δ-) are poorly absorbed (Gallo-Torres, 1980).

α-Tocopherol enters the enterocyte by passive diffusion and is secreted into chylomicrons. Hydrolysis of chylomicrons in the circulation by lipoprotein lipase forms chylomicron remnants which are taken up by the liver, from which α-tocopherol is secreted in VLDL (Traber et al. 1988). Metabolism of VLDL results in the simultaneous delivery of α-tocopherol into LDL and HDL. The uptake of α-tocopherol by peripheral tissues may occur during the catabolism of chylomicrons and VLDL by lipoprotein lipase, via the LDL receptor or by uptake that is not dependent upon receptors (Drevon, 1991). In rats, uptake is most rapid in lung, liver, small intestine, plasma, kidney and red cells and slowest in brain, testes, adipose tissue and spinal cord (Ingold et al. 1987). The major stores of α-tocopherol in the body are adipose tissue, liver and muscle (Drevon, 1991). However, mobilization of adipose tissue α-tocopherol in response to dietary vitamin E deficiency is very slow (Bjørneboe et al. 1990).

Apart from oxidation and reduction, the metabolism of α-tocopherol is limited (Drevon, 1991). However, when large doses are ingested, a considerable amount of α-tocopherol is secreted in bile, which may account for the relative safety of vitamin E compared to vitamins A and D (Traber & Kayden, 1989). The principal route of elimination is via the faeces, arising from incomplete absorption, secretion from mucosal cells, desquamation and biliary excretion. The extent of faecal excretion may vary from 10 to 75% of the administered dose while urinary excretion represents about 1% of the dose (Gallo-Torres, 1980).

FACTORS INFLUENCING BIOAVAILABILITY

Vitamin E bioavailability is influenced by a variety of luminal and physiological factors. In animal bioassays, the choice of vehicle used to deliver the test compound is an important variable (Burton et al. 1988). Solubilization of the vitamin in medium chain compared to long chain tryglycerides enhances hydrolysis and absorption, possibly by influencing the formation of micelles (Gallo-Torres, 1980; Fukui et al. 1989). Administration in an oil solution, a mixed diet or in capsule form influences whether oxidation of the vitamin occurs in the intestinal chyme, while high intakes of polyunsaturated fatty acids may increase the oxidation of α-tocopherol *in vivo* (Bjørneboe *et al.* 1990).

Studies in rats have indicated that the bioavailability of vitamin E increases with age (Hollander & Dadufalza, 1989) but is reduced by high intakes of pectin (Schaus *et al.* 1985), wheat bran (Omaye & Chow, 1984; Kahlon *et al.* 1986) and alcohol (Bjørneboe *et al.* 1990). One group (Verdon & Blumberg, 1988) reported that the activity of hepatic α-tocopherol binding protein increases during vitamin E deficiency, suggesting that bioavailability is higher when vitamin E status is low. However, other workers (Burton *et al.* 1988) observed that the overall absorption and transport of RRR-α-tocopherol and RRR-α-tocopheryl acetate was similar in vitamin E replete and deficient rats.

The development of stable isotope methods represents an important advance in the study of vitamin E bioavailability. Studies using stable isotopes have shown that RRR- and all-rac α-tocopheryl acetate are absorbed equally well in both rats and humans under normal conditions (Burton *et al.* 1988). However, long term feeding of deuterium substituted RRR- and SRR-α-tocopheryl acetate to rats indicated that some tissues, especially brain, have a marked preference for RRR-α-tocopherol (Ingold *et al.* 1987). A similar, though less pronounced, preference is evident in human red blood cells and plasma (Traber *et al.* 1990). Discrimination may be due partly to the preferential secretion of RRR-α-tocopherol into VLDL. While these experiments confirm the greater bioavailability of the natural stereoisomer, they also suggest that bioassay duration and the tissues in which the symptoms of deficiency and cure are observed, can influence the biopotency estimate (Ingold *et al.* 1987, 1990). Consequently, the results of animal bioassays must be interpreted with caution. Extension of stable isotope studies in humans should provide more reliable data regarding the uptake and retention of compounds with vitamin E activity.

CONCLUSION

Vitamin E consists of eight naturally occurring fat-soluble compounds of which α-tocopherol has the highest biological activity. α-Tocopherol is absorbed unchanged from the intestine whereas tocopheryl esters are hydrolysed before absorption. The absorption process is incomplete but precise efficiencies remain to be established. α-Tocopherol is absorbed *via* the lymphatic pathway and transported in association with chylomicrons. At normal intakes, α-tocopherol is exported from the liver to the peripheral tissues in VLDL but pharmacological doses lead to some biliary losses. The major stores are adipose tissue, liver and muscle. Some tissues show a strong preference for natural (RRR-) rather than synthetic (all-rac) α-tocopherol. There is clearly a need for further research on vitamin E bioavailability, particularly from important food sources, because of its possible role in chronic disease prevention.

VITAMIN K

Vitamin K exists in two major forms: phylloquinone or vitamin K_1 from plants, and bacterial forms, the menaquinones which have variable side chain lengths, known collectively as vitamin K_2. A third water-soluble substance with vitamin K activity, which is purely synthetic, is menadione which lacks the hydrocarbon side chain. During the carboxylation* of glutamic acid residues in the proteins involved in blood clotting (Stenflo *et al.* 1974), in bone formation (Suttie, 1984), and other oxidation states, vitamin K epoxide and hydroquinone are formed transiently. More is known about the biological activity of

* involving vitamin K.

diet derived phylloquinone than of gut flora menaquinones. Phylloquinone is considered non-toxic in pharmacological amounts; menadione is potentially toxic and should be used with caution.

FOOD SOURCES AND HUMAN REQUIREMENTS

Rich sources of phylloquinone include leafy vegetables (e.g. broccoli, spinach, lettuce, Brussels sprouts) and liver (Suttie, 1984); typical concentrations range between ten and a few hundred micrograms per 100 g. Human requirements are 0·4–1 μg/kg body weight (Olson, 1987; Suttie et al. 1988). Absorption of menaquinones from the colon is limited, and their biological activity is uncertain (Shearer et al. 1974; Shearer, 1990).

Biochemical changes, including those of protein carboxylation, have been observed in adult human subjects receiving a diet low in vitamin K (Ferland et al. 1993). Nevertheless, a severe functional deficiency of vitamin K in adult humans is rarely achieved by diet alone, but requires antibiotics also, to interrupt the supply from the gut flora (Allison et al. 1987).

Young infants, especially if breast fed, are potentially at risk of developing functional vitamin K deficiency, particularly with respect to the blood clotting cascade (Shearer et al. 1982; Lane & Hathaway, 1985; Greer et al. 1988; Yang et al. 1989; Matsuda et al. 1991; Guillaumont et al. 1993; Schubiger et al. 1993; von Kries et al. 1993). There is evidence of considerable variation in the absorptive capacity of individual infants (Shinzawa et al. 1989).

ABSORPTION MECHANISMS AND BLOOD TRANSPORT

Phylloquinone is mainly absorbed in the proximal ileum, by an energy dependent pathway, into the lymphatic circulation (Blomstrand & Forsgren, 1968). Studies in rat gut segments indicated that phylloquinone was taken up most readily in the proximal ileum (Hollander, 1973) but that menaquinones of bacterial origin could be absorbed by the mammalian colon, at a rate sufficient to prevent a functional deficiency (Hollander et al. 1977). Since oestrogens can affect vitamin K absorption (Jolly et al. 1977), it is likely that the requirements differ somewhat between males and females.

Bjornsson et al. (1979, 1980) infused [^{3}H]phylloquinone into human subjects and made calculations of the half life and body pool size, the latter being remarkably small.* Shearer et al. (1974) obtained similar values: after an oral dose of 1 mg labelled phylloquinone a peak plasma level was observed after about 2 h, declining exponentially to baseline values after 48–72 h (Shearer et al. 1970). Vitamin K is probably transferred from the chylomicron remnants to the liver, is then incorporated into very low density lipoproteins and transported to other tissues *via* low density lipoproteins (Sadowski et al. 1988). No specific carrier protein for vitamin K has been identified, but apolipoprotein E seems especially important in plasma vitamin K transport, in relation to chylomicron remnant clearance (Saupe et al. 1994). Both phylloquinone and high molecular weight menaquinones from bacteria are found in the liver (Rietz et al. 1970).

FUTURE RESEARCH

Vitamin K research has, in the past, been hampered by the difficulty of measuring the very low concentrations that occur in blood and tissues. Developments in chromatographic and detection procedures are now beginning to overcome this problem, and the use of stable

* (50–100 μg).

isotopes should also facilitate future studies of bioavailability and disposition of vitamin K in humans. Clearly there is more to be learned about the availability of vitamin K from different sources, both exogenous (foods) and endogenous (bacterial). Studies of the changes in availability with age, and in different physiological and disease states, are also becoming feasible.

THIAMIN

Thiamin (vitamin B_1) is a water-soluble, B complex vitamin with an essential role in carbohydrate metabolism and neural function (Davis & Icke, 1983). Although thiamin can be synthesized by gut microflora, this is insufficient to contribute significantly to body requirements. Pure thiamin is well absorbed, but there is little information on the bioavailability of the vitamin from food sources, as coenzyme complexes.

FOOD SOURCES

Thiamin is widely distributed in most foods at relatively low concentrations. The richest sources are yeasts, cereals and meats, especially liver. In plant foods, thiamin occurs predominantly free. In animal foods, it occurs almost entirely (95–98%) in the phosphorylated forms (mono-, di- and triphosphates) with about 80% as the diphosphate (Gubler, 1991). In food or multivitamin supplements, thiamin hydrochloride or mononitrate is used.

ABSORPTION AND EXCRETION

Thiamin is readily released. Phosphorylated forms are cleaved in the intestine to the free vitamin. Studies using inverted jejunal sacs have shown that rat and human mucosae absorb thiamin at low concentrations ($< 2 \mu M$) by a saturable, active carrier mediated process with a K_m 0·16–0·38 μM, similar to the K_m of thiamin pyrophosphokinase. A close association between transport and phosphorylation has been suggested (Schaller & Höller, 1974). At higher thiamin concentrations (5–50 μM) absorption is *via* passive diffusion (Hoyumpa *et al.* 1975, 1982). Active absorption of thiamin across the mucosal cell, therefore, may not be associated with or dependent upon phosphorylation of thiamin. A specific carrier is implied, and supported by a thiamin binding protein in *Escherichia coli* and rat serum, associated with cellular thiamin transport (Matsuura *et al.* 1975). This binding protein may also be involved in the distribution of thiamin to critical tissues (Combs, 1992). Ethanol, taken orally or intravenously, inhibits intestinal uptake of thiamin.

Active absorption is greatest in the jejunum and ileum. The intestinal mucosa has a thiamin pyrophosphokinase activity with a K_m of the same order as that of carrier mediated absorption. However, the role of thiamin pyrophosphokinase is not fully understood. While most of the vitamin in the intestinal mucosa is phosphorylated, thiamin on the serosal side is mainly in the free form. Thus uptake is dependent either on thiamin phosphorylation/dephosphorylation, or on some other energy dependent mechanism, possibly activated by sodium (Basilico *et al.* 1979). Thiamin release on the serosal side is dependent on Na^+ and on the normal function of ATPase.

Thiamin preparations with greater lipid solubility (e.g. allithiamin derivatives such as thiamin propyldisulphide and 5-benzoylthiamin-O-monophosphate (benfotiamine)) are much more readily absorbed than thiamin hydrochloride, and result in higher blood and tissue levels (Baker & Frank, 1976). Allithiamin derivatives have been used to correct thiamin deficiencies, especially in alcoholics (Nose & Iwashima, 1965).

Thiamin is transported into red blood cells by diffusion (Hoyumpa, 1982), while in other tissues it proceeds by the active, carrier mediated process (Matsuura et al. 1975). Whole blood concentrations are 150–350 nmol/l, 90% of which is in the red cells and leukocytes. Thiamin has a high turnover rate and is not stored. Even a few days with inadequate intake can lead to signs of deficiency.

Excess thiamin is rapidly excreted in the urine, chiefly as the free vitamin and thiamin monophosphate. In addition, thiamine diphosphate, thiochrome, thiamin disulphide and other metabolites are excreted in small amounts. The relative proportion of metabolites to thiamin excreted increases with decreasing thiamin intake (Pearson, 1967).

FACTORS AFFECTING ABSORPTION AND UTILIZATION

Food processing

The vitamin is susceptible to destruction by several factors including neutral and alkaline pH, heat, oxidation and ionizing radiation. Thiamin is very stable at acidic pH but becomes unstable at pH > 7, especially when heated. The protein bound thiamin in animal tissue is more stable. In cereal gains, thiamin is unevenly distributed, being low in endosperm and high in germ. Milling and polishing of brown rice depletes thiamin. Most white rice and flour is re-enriched or fortified with the vitamin. Thiamin is destroyed by sulphites added to vegetables during blanching, by rupture of the methylene–thiazole bond (Gubler, 1991).

Antagonists

Several thiamin antagonists, which reduce absorption, are analogues that differ in either the thiazole (e.g. pyrithiamin), or pyrimidine (e.g. oxythiamin), rings. Although such compounds bind with thiamin dependent apoenzymes, they are inactive. Other antagonists include chick anticoccidial compounds (e.g. amprolium) which do not have a hydroxyether group and therefore cannot be phosphorylated. The sources and levels of them in foods, and their effect on absorption, are not known.

Thiamin destroying or inactivating enzymes (thiaminases) occur in fish, shellfish, ferns, tea, betel nuts and vegetables. Two types have been isolated, one from fresh fish, shellfish and ferns which catalyses a base exchange involving thiazole, the other a hydrolytic enzyme that hydrolyses the methylene–thiazole bond (cf. thiamine destruction by sulphites; Murata, 1965). Although thiaminases in raw fish are heat labile, this can still cause deficiency if consumption is high, as in Japan.

Heat stable antagonists are present in several plants (e.g. ferns, tea, nuts) and include hydroxypolyphenols (e.g. caffeic, chlorogenic and tannic acids) (Somogyi, 1971). They occur in fruits (blueberries, red currants), vegetables (Brussels sprouts, red cabbage), nuts, tea and coffee. Thiamin disulphide production indicates destruction of thiazole. Other thermostable inactivators include flavonoids (e.g. rutin, quercetin), found in brassicas, and haemins in animal products.

Alcohol, diseases and drugs

Excessive intake of alcohol inhibits intestinal ATPase involved in absorption of thiamin. Its utilization is impaired by liver dysfunction (hepatitis, cirrhosis). Drugs that cause nausea, induce diuresis, or increase intestinal motility decrease its availability.

Smoking and age

Older studies have shown that nicotine decreases tissue thiamin concentrations in the developing chick embryo (Kato, 1959) and excretion is markedly reduced in heavy smokers (Strauss & Scheer, 1939; Rafsky et al. 1947). Elderly persons may have increased requirements for B group vitamins, including thiamin, owing to a variety of factors including poor eating habits and malabsorption. Old rats require more thiamin/g food to maintain blood and tissue thiamin concentrations than younger ones (Mills et al. 1976). Thiamin transport across the intestine is lower in older animals (Nishino & Itokawa, 1977).

Other nutrient interactions

The amount of carbohydrate in the diet influences thiamin requirements. Other nutrients (e.g. Mg^{2+}, Ca^{2+}, B group vitamins) can affect thiamin metabolism and function in rats (Howard et al. 1974; Kimura & Itokawa, 1977; Nishino & Itokawa, 1977).

CONCLUSIONS

Both free and phosphorylated forms of thiamin are readily released from animal and plant foods. Phosphorylated forms are cleaved in the intestine to the free vitamin which is absorbed by an active, carrier mediated process at low concentrations, and by passive diffusion at higher concentrations. Thiaminases in foods can lead to loss of vitamin activity by destruction of the vitamin, or conversion to a form that is biologically unavailable. Metabolic demands for thiamin are increased by diets rich in carbohydrate. Alcohol inhibits absorption.

RIBOFLAVIN

Riboflavin or vitamin B_2, a water-soluble vitamin of the B group, is an essential dietary component for man, although some animal species can obtain sufficient from their intestinal flora *via* coprophagy (Prentice & Bates, 1980). Free riboflavin, and the free coenzymes riboflavin-5-phosphate (usually known as flavin mononucleotide), and the adenyl derivative of FMN (known as flavin adenine dinucleotide), are readily available for absorption and utilization by animals and man, whereas those flavin coenzymes which are linked to enzymes covalently, usually through the 8-methyl group, are not available.

FOOD SOURCES

Rich sources of riboflavin and its coenzymes include offal, yeast hydrolysate, milk, dairy products, and eggs. Because ungerminated grains, especially polished rice, are poor sources, those communities which rely heavily on simple rice dishes are at high risk of deficiency. Milk and eggs contain riboflavin bound to a carrier protein; hydrolysed products such as yeast hydrolysate and vitamin supplements contain free riboflavin. Other foods contain a major proportion of their riboflavin as coenzymes bound to enzymes, 60–90% occurring as FAD.

RELEASE AND ABSORPTION

Riboflavin phosphate and FAD are hydrolysed by various types of phosphatase in the lumen of the intestine (Akiyama et al. 1982) thus yielding free riboflavin, which is then absorbed in the upper ileum, mainly by a sodium dependent saturable transport mechanism (McCormick, 1972; Jusko & Levy, 1975; McCormick, 1989). After riboflavin phosphate is dephosphorylated to riboflavin in the intestinal lumen, free riboflavin is then rephosphorylated in the mucosal cells during transport, but is dephosphorylated again once it enters the bloodstream (Jusko & Levy, 1975). In humans, absorption is linear for single doses up to around 30 mg riboflavin, if given with a meal. Bile salts, and factors which increase the transit time, increase absorption (Jusko & Levy, 1975; Mayersohn *et al.* 1969). In riboflavin saturated subjects, 15–60% of an oral dose between 5 and 30 mg is rapidly excreted in the urine. Some synthetic fatty acid esters of riboflavin are more readily absorbed than riboflavin itself (Jusko & Levy, 1975). In everted jejunal segments of rat intestine, physiological amounts of riboflavin are transported by a carrier mediated saturable mechanism. At higher concentrations, diffusion predominates (Said *et al.* 1985; Daniel *et al.* 1983). The large bowel may absorb small amounts of riboflavin (Jusko & Levy, 1975), but it is much less efficient than the ileum in this respect.

An active transport system for efflux of riboflavin is present in the choroid plexus (Spector & Boose, 1979), and there are also specific transport systems in liver (Aw *et al.* 1983), kidney (Spector & Boose, 1982) and placenta (Dancis *et al.* 1985). A specific riboflavin binding protein which appears to play an important role in riboflavin transport from maternal to fetal circulation has been identified in pregnant rats and in cows, and preliminary indications suggest that it may also occur in humans (Murty & Adiga, 1982; White & Merrill, 1988).

EXCRETION

Amounts of riboflavin that are absorbed in excess of immediate needs, and in excess of the renal tubular reabsorption threshold, are rapidly excreted in the urine. Various hormonal, chemical and disease factors affect urinary excretion (Jusko & Levy, 1975). Loss through biliary excretion is a minor pathway in humans (Jusko & Levy, 1975); secretion into breast milk is significant, but is smaller in humans than the cow and rat.

ANTAGONISTS AND HOMEOSTATIC INFLUENCES

In animals, certain structural analogues of riboflavin, notably galactoflavin (7,8-dimethyl-10-(d-1′-dulcityl)-isoalloxazine), compete and give rise to functional deficiency (Warkany, 1975). Chlorpromazine, tricyclic antidepressants (Pinto *et al.* 1981), and certain antimalarial drugs (Dutta *et al.* 1985) have structural analogies with riboflavin, and may also compete either for intestinal absorption or for flavokinase. Boric acid in large amounts can cause deficiency, and other substances can alter availability in various ways (Jusko & Levy, 1975). Experimental uraemia can interfere with riboflavin absorption (Vaziri *et al.* 1985). In one human study, increasing efficiency of absorption occurred with age, in parallel with decreasing gut transit rate (Jusko & Levy, 1975), but no age changes were detected in an animal model (Said & Hollander, 1985), and this requires further study.

CONCLUSION

Non-covalently bound riboflavin and its coenzymes in food are rapidly and efficiently absorbed by a saturable mechanism, especially if given with a meal, and the excess over immediate needs is rapidly excreted in the urine. At low intakes, the body efficiently conserves tissue flavins, but certain xenobiotics and hormonal effects can influence riboflavin status, usually by altering flavokinase activity.

VITAMIN B_6

Vitamin B_6 is the generic name of three vitamers: pyridoxol, pyridoxal and pyridoxamine, which can be interconverted and occur either free or as phosphorylated forms in various plant and animal food products. Good sources of pyridoxine are vegetables, cereals, seeds (nuts) and meat products. Problems of vitamin B_6 bioavailability can be subdivided into external and internal factors, affecting vitamin B_6 or its metabolites and hence utilization.

External factors are food components and their environment that may impair the release of the vitamin, thus affecting digestibility and gastrointestinal absorption of B_6. Some of these food borne determinants are inherent; others can arise from food processing. B_6 bioavailability may also be affected by internal factors, including malabsorptive states, the retention capacity of organs and tissues, interactions with nutrients and non-nutrients, or inborn errors of B_6 metabolism. The absorption of free B_6 from the intestine is, in contrast to other B vitamins, unlimited and not subjected to saturation kinetics. High oral B_6 doses of more than 1 g/day may cause neuropathies.

FOOD PROCESSING; BOUND B_6

Heating of foods can cause B_6 decomposition or a loss of B_6 availability through formation of reaction products. The more reactive vitamers pyridoxal and pyridoxamine and their phosphates can bind to amino (NH_2-) or sulphydryl (SH-) groups of amino acid residues from proteins, forming aldimines or Schiff bases, which may be stabilized by chelating in presence of metal ions. Under reducing conditions a stable aldimine or ketimine is formed, e.g. pyridoxyl-ϵ-aminolysine.

The protein bound forms have a low availability in man. In rat bioassay studies pyridoxyl-ϵ-aminolysine was found to possess approximately 50% activity of the free form and for man may even have some antivitamin B_6 activity (Gregory & Kirk, 1977; Gregory, 1980). Cysteine bound pyridoxal was identified in heat sterilized milk. Partial destruction of B_6 during sterilization and the low availability of the remaining bound vitamin resulted in clinical deficiency symptoms in infants fed a commercially sterilized milk formula (Coursin, 1954; Davies et al. 1959). Similarly, pyridoxamine can react with carbonyl compounds of reducing sugars and in the presence of ascorbic acid the inactive 6-hydroxypyridoxine can be formed (Tadera et al. 1986).

A few years ago it was demonstrated by several authors that vitamin B_6 can react with equimolar amounts of glucose to form 5-O-β-glucopyranosylpyridoxine (PNG) first identified in rice bran (Gregory & Ink, 1987; Reynolds, 1988). Studies with rats using intrinsic and extrinsic labelling indicated that purified PNG was relatively well absorbed, but about 80% of the glucoside was rapidly excreted in urine. Only 20–30% bioavailability of the glucoside relative to pyridoxol was found (Gregory et al. 1991b; Gilbert & Gregory, 1992). In man, however, PNG was 58% utilized as determined with isotopes. When intravenously administered, the glucoside was only 30% available, suggesting a role of β-glucosidases of the intestinal mucosa, microflora or both. The glucoside also seems to alter metabolism and in vivo retention of free pyridoxol, and it may retard the utilization of other non-glycosylated B_6 vitamers (Gregory et al. 1991b; Gilbert & Gregory, 1992).

Glucosidic B_6 occurs only in plant material, and ranges from 8% to 50% of total B_6. Plant components with storage functions, such as root vegetables, exhibit the highest glucoside contents (Reynolds, 1988; Bitsch & Schramm, 1992). In some seeds, such as wheat, sunflower, lucerne and mung bean, two glucosidic forms are found. Higher conjugated glucosides were found in rice bran (Tadera et al. 1988). Glucosidic pyridoxol seems relatively stable, and 2 h autoclaving at 121 °C in acid was required for appreciable hydrolysis (Bitsch & Schramm, 1992).

Several authors estimated 70–80% bioavailability of B_6 in the average American diet, based on total (including glycosylated) B_6 content (Kabir et al. 1983; Reynolds, 1988). Although glucosidic B_6 was not detected in animals, a study on lactating women from a vegetarian population revealed small amounts in breast milk. The glucoside content was correlated between food and breast milk samples (Reynolds, 1988).

INTERACTIONS WITH NUTRIENTS AND DRUGS

In addition to naturally bound forms and those produced by food processing, there are also B_6 antagonists. Some can compete for absorption; others inactivate B_6 vitamers by complexing pyridoxal and pyridoxal phosphate, or by acting as structural analogues, e.g. isoniazid, penicillamine, cycloserine and carbonyl drugs such as benzerazide and carbidopa. Coadministration of pyridoxamine counteracts these effects. Theophylline used in asthma therapy is also thought to act as a pyridoxal phosphate antagonist (Merrill & Henderson, 1987). It is not clear in every case what is the principal site of interaction *in vivo*.

A close connection exists between vitamin B_6 and protein turnover. In animal experiments, the level and quality of dietary protein affected the B_6 retention capacity of liver and muscle (Sampson & Chung, 1991; Trumbo, 1991). The ratio, 0·015–0·02 mg vitamin B_6/g dietary protein, has been suggested as the basis of human vitamin B_6 recommended intake (Hansen *et al.* 1991).

Riboflavin dependent enzymes are involved in the conversion of pyridoxine to pyridoxal phosphate and in the oxidation of pyridoxal to 4-pyridoxic acid, so riboflavin status may affect the metabolic availability of B_6 vitamers, as was suggested in experiments with rats (Lakshmi & Bamji, 1979).

CONCLUSION

While B_6 vitamers in animal foods are well absorbed and bioavailable, a major part of B_6 in plant foods is glucosidically bound with a low bioavailability for man. More research is needed in order to clarify the contribution of plant food to human B_6 supplies.

NIACIN

The term niacin includes nicotinic acid, nicotinamide, $NAD(H)^+$, and $NADP(H)^+$. Humans are able to synthesize niacin from tryptophan (Henderson, 1983; Bender & Bender, 1986). Therefore the niacin content of foods is usually expressed as niacin equivalents (niacin + (tryptophan/60)).

FOOD SOURCES

Niacin is widely distributed in foods of plant and animal origin. In typical western diets the most important sources of preformed niacin are meat and meat products, cereals, dairy products, beverages and eggs (Deutsche Gesellschaft für Ernährung, 1988). Niacin is resistant to heat, light, storage, and oxygen. In foods, niacin occurs mainly in its coenzyme forms, e.g. $NAD(H)^+$. To some extent $NAD(H^+)$ is hydrolysed to nicotinamide during food processing. In cereals the main part of niacin is bound as niacytin. Niacytin is not a well defined compound, but is a mixture of different peptides, proteins and carbohydrates (Mason *et al.* 1973). In some cereals niacin is only partly available. In maize, at least 70% of the niacin is unavailable, unless it is treated with alkali (Carter & Carpenter, 1982; National Research Council, 1989). In coffee beans there are high concentrations of trigonellin (1-methyl nicotinic acid) from which nicotinic acid can be liberated by roasting

(Baessler et al. 1992). Therefore coffee can contribute significantly to niacin intake. Naturally occurring antagonists of niacin are rare, but some synthetic antagonists exist (Henderson, 1983). Biosynthesis of niacin by gut flora may occur, but subsequent absorption is uncertain. A lactating woman typically secretes 1·0–1·3 mg preformed niacin daily in 750 ml milk (National Research Council, 1989).

RELEASE AND ABSORPTION

NAD(H)$^+$ is hydrolysed by an intestinal pyrophosphatase to nicotinamide ribonucleotide and riboside, which accumulate in the lumen. Conversion to nicotinamide is the rate limiting step. There is no deamination to nicotinic acid (Henderson, 1983). Nicotinamide is rapidly absorbed in the stomach and small intestine (Bechgaard & Jespersen, 1977). At low concentrations niacin is absorbed by a sodium dependent facilitated diffusion, whereas higher concentrations are mainly absorbed by passive diffusion (Sadoogh-Abasian & Evered, 1980). Physiological doses of niacin are nearly completely absorbed. If high doses of niacin (e.g. 3 g) are ingested, 85% is found in urine (Friedrich, 1987). To achieve a slow elimination of niacin from blood plasma, nicotinic acid esters can be used. When 700 mg are given, 75% of the nicotinic acid is absorbed from such esters (Friedrich, 1987). Slow release forms of niacin have been tested for pharmacological benefits.

EXCRETION

Urinary metabolites in humans are very similar after oral loads of nicotinamide or nicotinic acid respectively (Mrochek et al. 1976). After doses in the 100–3000 mg range the main metabolites were N^1-methyl-2-pyridone-5-carboxamide, N^1-methylnicotinamide, and nicotinuric acid. The latter two increase with higher doses. Of a physiological dose of niacin, 40% is eliminated via urine within 24 h (Baessler et al. 1992). An enterohepatic cycle exists. Nicotinamide enters the gastrointestinal tract via bile; it is partly deaminated microbiologically to nicotinic acid. Nicotinamide and nicotinic acid are then re-absorbed and converted to NAD(H)$^+$.

SYNTHESIS FROM TRYPTOPHAN

Part of the tryptophan in the diet can be converted to niacin (Bender & Bender, 1986). Indeed, for an adult in nitrogen balance a typical protein intake can provide more than enough tryptophan to meet niacin requirements by endogenous synthesis of NAD, with no need for any preformed dietary niacin. Analysis of urinary metabolites revealed a mean yield of 1 mg niacin from c. 60 mg (range 39–86 mg) tryptophan (Horwitt et al. 1981; National Research Council, 1989). The rate of conversion is controlled by hormones (Wolf, 1971), and is elevated in pregnancy and by oral contraceptives (Horwitt et al. 1981). It is regulated by nicotinate-nucleotide pyrophosphorylase (carboxylating) (EC 2.4.2.19), but synthesis may be reduced by leucine at the kynureninase (EC 3.7.1.3) step. It depends on a sufficient supply of riboflavin and pyridoxine (Satyanarayana & Narasinga Rao, 1980). A typical tryptophan content of proteins is c. 1% (National Research Council, 1989).

METABOLISM AND TRANSPORT BETWEEN TISSUES

Absorbed nicotinamide is first accumulated by liver and erythrocytes, and is converted to NAD(H)$^+$.

CONCLUSIONS

Niacin is a rapidly and nearly completely absorbed vitamin and is easily transformed to its coenzyme forms NAD(H)$^+$ and NADP(H)$^+$. It is very stable under nearly all conditions and is ubiquitously distributed in foods from both animal and plant origin. Biosynthesis from dietary tryptophan also contributes to the supply.

FOLATE

Folate is the generic term for a B vitamin which is essential for one-carbon transfer reactions in many metabolic pathways, including purine and pyrimidine metabolism and amino acid interconversions. It exists primarily as reduced one-carbon substituted forms of pteroylpolyglutamates. Naturally occurring reduced folates are light and heat sensitive and destroyed by oxidation. Some is lost during cooking. Antioxidants, e.g. ascorbic acid, can prevent oxidation.

The most important food sources are liver, egg and green vegetables such as cabbage, Brussels sprouts, broccoli, lettuce and spinach. Yeast, bran, yogurt and orange juice also contain high levels. The polyglutamyl form of folate is *c.* 80 % of dietary folate and must be cleaved to monoglutamate before absorption. In milk, folate is bound to a specific folate binding protein.

ABSORPTION OF SYNTHETIC MONO- AND POLY-GLUTAMYL FOLATES

Monoglutamate folates are absorbed by an active energy dependent carrier mediated process at physiological concentrations, and by passive diffusion at higher concentrations (Selhub *et al.* 1984). Absorption takes place mainly in the jejunum and is markedly influenced by pH, with a maximum at pH 6·3 and sharp decline between 6·3 and 7·6 (Russell *et al.* 1979). The pKa and pH affect the proportions of ionic and non-ionic forms; non-ionic forms facilitate diffusion across the cell membrane.

The polyglutamate folates must be cleaved to monoglutamates before uptake in the intestinal epithelial cells (primarily in the jejunum) by a pteroylpolyglutamate hydrolase, or folate conjugase. Reisenauer *et al.* (1977) reported two separate folate conjugase activities in human jejunal mucosa, one soluble and intracellular and the other membrane bound and concentrated in the brush border. Human brush border conjugase is a zinc dependent exopeptidase with optimum activity at pH 6·5 catalysing stepwise hydrolysis of polyglutamyl folates. It plays a major role in the digestion of dietary folate, and zinc deficiency may affect folate bioavailability by impairing folate conjugase activity (see Canton & Cremin, 1990). Human intracellular folate conjugase is most active at pH 4·5, and its function in absorption is unclear. In contrast to man, the pancreas in chickens and rats provides a rich source of conjugase (Halstead, 1990), suggesting the lumen as the main site of hydrolysis in these species. Some studies indicated a lower (70–80 %) availability of polyglutamate compared with monoglutamate folate (Halstead, 1990). These human studies used a jejunal perfusion technique and labelled synthetic forms of folic acid. Using plasma response, Pietrzik (1993) showed that pteroylheptaglutamic acid was absorbed to 70 % of pteroylmonoglutamic acid following a single oral dose equivalent to 1000 μg pteroylmonoglutamic acid. By using stable isotopes, ^{2}H-monoglutamyl folate and ^{2}H-hexaglutamyl folates, Gregory *et al.* (1991*a*) showed the latter to be approximately half as available as the monoglutamate. Oxidized folates were better absorbed than reduced forms.

Oxidized folates in physiological amounts become reduced in passing through the intestinal mucosa. At higher concentrations, reduction occurs in the liver, which is the major tissue where folates are stored as polyglutamates. All intracellular folate occurs in polyglutamyl forms, whereas folates are transported in plasma in the monoglutamyl form.

AVAILABILITY OF FOOD FOLATES

Data on the absorption of dietary folate in man are based mainly on an assessment of bioavailability by urinary excretion after known intakes of synthetic folic acid and presaturation of tissues with repeated folate doses. After ingestion of banana, lima beans, liver and egg, uptake of folate exceeded 70% compared with synthetic folic acid. For the majority of food items, uptakes of 40–70% of that for pteroylmonoglutamic acid were observed. The lowest figures were seen for wheat germ, yeast, tomatoes and orange juice. Tamura & Stokstad (1973) concluded that bioavailability of folates from orange juice was low due to inhibition of intestinal conjugase by the low pH (3·2) of orange juice. However, they used unphysiologically high amounts of orange juice in their meals (2400 ml). When physiological quantities of orange juice were given to human subjects, the bioavailability of folate in orange juice was equal to that of pteroylmonoglutamate (Rhode et al. 1983). Yeast is rich in polyglutamates and several studies indicate poor absorption (Tamura et al. 1973; Babu & Srikantia, 1976).

Bioavailability data have also been obtained by a rat bioassay. This requires depletion of tissue folates by a folate deficient diet, followed by a test diet containing food folate for 1–2 weeks. The animals are then killed and folates measured in liver, kidney, plasma, erythrocytes or intestinal mucosa. With this model, availability of dietary folates generally exceeds 80%, with no values below 58%. The lowest figures were reported for wheat germ, cabbage and broccoli (Hoppner & Lampi, 1986; Abad & Gregory, 1987; Clifford et al. 1990). Foods rich in polyglutamates showed the lowest bioavailability. In contrast, yeast folate with a high proportion of polyglutamates was well utilized in this model (Hoppner & Lampi, 1986).

MILK FOLATE BINDING PROTEIN

In milk, the naturally occurring folate is protein bound (Ghitis, 1967; Wagner, 1985). The physiological role of these folate binding proteins is unclear. Ford et al. (1969) suggested that in milk these proteins may act in the mammary gland as a trapping agent for folate. After ingestion by the infant, they could prevent uptake by intestinal bacteria, and might promote the transport of folate across the mucosa. This latter proposal was supported by the work of Salter & Blakeborough (1988).

Bound folate may be absorbed in a manner different from free folate. While free monoglutamate is absorbed in the jejunum, protein bound folate is mainly absorbed in the ileum and at a much slower rate (Said et al. 1986) which may improve bioavailability.

DIETARY FIBRE AND MATRIX EFFECTS

A negative effect of wheat bran on availability of polyglutamyl folate has been reported (Keagy et al. 1988) both in rats and humans. Ion exchange may occur between the folates and wheat bran which reaches the colon faster than other types of fibre. Data for monoglutamyl folate are inconsistent. Ristow et al. (1982) found little or no binding to purified dietary fibre, using equilibrium dialysis. Matrix effects by lima beans, cabbage and yeast indicate that bioavailability depends on the dietary level. Conjugase inhibitors have been reported in pulses, yeast and cabbage (Butterworth et al. 1974).

EXCRETION

Faecal folate levels are sometimes higher than intakes, presumably reflecting folate biosynthesis by the microflora of the lower gastrointestinal tract. Bile contains high levels of folate, owing to enterohepatic circulation. Most folate in bile is reabsorbed, but c. 100 µg is lost daily through incomplete absorption (Brody et al. 1984). In animals, bile duct cannulation for only 6 h leads to a 60–70% fall in serum folate (Steinberg et al. 1979).

CONCLUSIONS AND FUTURE RESEARCH PRIORITIES

More information is needed about the requirements of people with elevated demands, e.g. pregnant and lactating women, the elderly, slimmers with a low energy intake, and people with gastrointestinal disorders or using drugs such as diphenylhydantoin, salicylazosulphapyridine. Ethanol may also cause malabsorption of folate (Halstead, 1990).

Human data are limited to two studies performed twenty years ago in which urinary excretion after presaturation was measured, after single food items and in groups of less than ten subjects. These showed wide individual variations and hence low reliability, with an overall absorption from mixed diets of about 70% (Tamura & Stokstad, 1973; Babu & Srikantia, 1976).

Further research is needed, with more sensitive and specific methods. Stable isotopes have enabled availability studies in man over physiologically relevant ranges (Gregory et al. 1991a). Improvements in analysis of different forms of folates, absorption kinetic studies and in vitro absorption techniques for dietary interactions and effects of food processing are now needed.

VITAMIN B_{12}

Vitamin B_{12} comprises a number of active metabolites, which differ only in the β-ligand on the central cobalt. This ligand can be CH_3 (methylcobalamin) or adenosyl (adenosylcobalamin) in the human body or alternatively OH (hydroxocobalamin) or CN (cyanocobalamin) in pharmaceuticals owing to their greater stability. The latter are most stable at pH 4–4.5. Cobalamins can be destroyed by heavy metals or by strong oxidizing or reducing agents (like ascorbate). They are sensitive to light, but relatively stable to heat.

VITAMIN B_{12} IN FOOD

Only microorganisms are able to synthesize cobalamin and therefore higher plants contain B_{12} only if processed microbiologically or contaminated. Food from animal origin (especially liver and kidney) contains B_{12}. Other rich sources are fish, milk and eggs. Some foods or microorganisms contain other corrinoids, with varying biological activity, some of which may have antivitamin activity. Paradoxically, some compounds with growth factor activity for *Lactobacillus leichmannii* are described as having B_{12} activity even if they have none for man (Herbert, 1988). Non-vegetarian humans typically have intakes in the region of 1–10 µg/d, usually considerably greater than the requirement, and therefore accumulate a reserve (typically 1–2 mg) of B_{12} in the liver (Heinrich & Gabbe, 1990). Smokers may develop a secondary deficiency associated with conversion of B_{12} to the cyano form (Linnell et al. 1968; Shaw et al. 1987).

ABSORPTION

There are two separate mechanisms for the absorption of vitamin B_{12}. In the active physiological one ingested B_{12} is freed from proteins by gastric acid and enzymes, and then bound to haptocorrin (also called R-protein). The B_{12}–haptocorrin complex is destroyed by trypsin, and B_{12} is transferred to intrinsic factor (IF, synthesized by gastric parietal cells), to form the B_{12}–IF complex. Ileal mucosa cell receptors bind the B_{12}–IF complex, take up the cobalamin and set the IF free. Ca^{2+} and a pH of 6 or higher are necessary to release cobalamin. This active mechanism depends on IF, and on the function of the exocrine pancreas and the ileal receptors. In a normal subject, the number of receptors limits absorption. Absorption is affected by gastrectomy, lack of intrinsic factor or trypsin, or intestinal infections. Atrophic gastritis, especially in elderly people, can lead to B_{12} deficiency.

The contribution of active and passive mechanisms to total absorption depends on the quantity of vitamin B_{12} (Heinrich & Wolfsteller, 1966). IF can transport up to 1·5 μg B_{12} at a time; passive diffusion plays a role for higher doses. B_{12} is not absorbed in the colon (Herbert, 1988) and gut bacteria typically produce non-cobalamin B_{12} analogues.

RETENTION

There is a difference in retention between CN-Cbl and OH-Cbl. Most pharmacological preparations contain CN-Cbl, owing to its greater stability. The conversion of CN-Cbl into the natural forms in the human body takes several days. OH-cobalamin is also used in pharmacological preparations and is better retained by the human body. Two phases can be differentiated in the elimination of parenterally administered OH-Cbl: during the first 5–7 h, serum concentrations decline rapidly (distribution phase), while during the elimination phase (half time of elimination 21–29 h) serum concentrations decline more slowly (Loew et al. 1988). Differences between CN-Cbl and OH-Cbl might be due to a different rate of protein binding, leading to gradual accumulation of OH-Cbl (Loew, 1991). Plasma levels of CN-Cbl after parenteral doses decline with a half life of 7 h.

Vitamin B_{12} is very effectively reabsorbed in the ileum (hepato-biliary recirculation); this is interpreted by some authors as a regulatory mechanism to exclude analogues, which are reabsorbed less effectively (Kanazawa & Herbert, 1982; Herzlich & Herbert, 1984). The hepato-biliary recirculation is a major reason why strict vegetarians become B_{12} deficient only after several years of very low dietary intake of B_{12}. Excretion through the kidneys (0–0·25 μg/d) is lower than excretion in bile.

CONCLUSION

Further studies are needed on the factors which determine the release of protein bound B_{12} in food, and on the diversity of forms of available B_{12} in foods and contamination sources.

BIOTIN

In humans, biotin (vitamin B_8 or H) acts as cofactor for at least four carboxylases, which are key enzymes of intermediary metabolism, namely pyruvate carboxylase (EC 6.4.1.1), propionyl-CoA carboxylase (EC 6.4.1.3), methylcrotonyl-CoA carboxylase (EC 6.4.1.4) and acetyl-CoA carboxylase (EC 6.4.1.2) (Wood & Barden 1977). Biotin ($C_{10}H_{16}O_3N_2S$)

comprises an imidazolidone ring fused with a tetrahydrothiopene ring bearing an aliphatic chain with five asymmetric carbon atoms. There are several known derivatives (oxybiotin, biotin sulphoxide, biocytin, biotin sulphone, biotinol), but of these only biocytin (ϵ-N-biotinyl-L-lysine) is found in nature and has vitamin activity. The lysine bond is a link to protein (Marquet, 1977; Cowan, 1984). Avidin, a glycoprotein component of egg white (hen, duck, goose, turkey), has a great affinity for biotin, with which it combines stoichiometrically. This complex, which resists the digestive enzymes, is unavailable for intestinal absorption, unlike biocytin (Bonjour, 1977; Gravel et al. 1980).

ABSORPTION AND EXCRETION

Biotin metabolism in mammals involves gastrointestinal absorption, transport and catabolism. Two major sources for man are the diet and the biotin producing flora in the intestinal tract distal to the caecum. The intake from all sources is about 150–300 μg/d. The pancreatic enzyme, biotinidase, probably splits the biotin–lysine bond, releasing the vitamin in the intestinal lumen. The mechanisms of intestinal transport are not completely known. Transport occurs in the proximal small intestine and jejunum, coupled with Na$^+$ and against the concentration gradient (Spencer & Brody, 1964; Berger et al. 1972; Said et al. 1987). Urinary levels are approximately 160 nmol/24 h or 70 nmol/l. A blood plasma, or serum level of about 1500 pmol/l suggests an adequate supply of biotin in man (Bonjour, 1984).

DIETARY SOURCES

Biotin is widely present in meats (chicken, pork, beef, lamb), vegetables (cauliflowers, mushrooms, carrots, tomatoes, spinach, beans and peas) and fruits (apples, oranges), and it is found in milk (human and bovine), in cheese, eggs and marine fish. The human requirement for biotin is not known, and recommended dietary allowances for it have not been formulated. A 'safe and adequate intake' has been proposed by the Food and Nutrition Board of the U.S. National Academy of Sciences (Food and Nutrition Board, 1989): for adults 0·1–0·4 μmol (30–100 μg) per day, based on typical intakes in Western society. No depletion–repletion studies are available, to define human requirements. There are no detailed studies which indicate the contribution from bacterial synthesis in humans.

REQUIREMENTS AND NUTRITIONAL STATUS

Knowledge about biotin requirements is limited to the range of usual intakes by human populations in which deficiency is not observed. Clearcut deficiency has only been recorded in people eating large quantities of egg white. For this reason, intake recommendations can only reflect the biotin content of human diets.

Plasma (or serum) levels and urinary biotin excretion are used to evaluate biotin nutritional status in humans. Biotin deficiency has been demonstrated in certain infants with inborn metabolic errors, but cases of human deficiency are rare. Deficiency may occur in (a) individuals receiving long term parenteral nutrition without biotin (Velazquez et al. 1990); (b) prolonged feeding of undenatured egg white, rich in avidin (Bonjour, 1977; Gravel et al. 1980; Sweetman et al. 1981); or (c) individuals affected by genetic biotinidase deficiency (Cowan et al. 1979; Munnich et al. 1981; Thoene & Wolf, 1983; Wolf et al. 1983). Clinical manifestations of induced biotin deficiency are mild and non-specific (non-pruritic dermatitis, seborrhoea, fatigue, muscular pain, anorexia, nausea, anaemia, hypercholesterolaemia).

CONCLUSION

Biotin deficiency is rare in humans. Studies of genetic multiple carboxylase deficiencies have permitted research on its metabolic functions. Absorption and catabolism in humans requires further research.

PANTOTHENIC ACID

Pantothenic acid or vitamin B_5 is an essential dietary component for humans, composed of β-alanine and the dihydroxyacid, pantoic acid. It occurs in both the bound and free form in food, and is ubiquitously distributed. Meat, avocado, broccoli, bran and molasses are excellent sources (Machlin, 1984). It is stable at high temperatures at pH 5–7 (Pike & Brown, 1975). The Food and Nutrition Board (1989) of the National Research Council concluded that there was insufficient evidence to set a recommended dietary allowance for this vitamin, and proposed a 'safe and adequate intake' of 4–7 mg/d in adults, and 2·0–3·0 mg/d in infants, based on usual intakes, but no depletion–repletion studies are available.

RELEASE AND ABSORPTION

Bioavailability of food pantothenic acid for man ranges from 40 to 61%, with a mean of 50% (Tarr *et al.* 1981). Studies *in vivo* (Reibel *et al.* 1981) and *in vitro* (Sugarman & Munro, 1980) have identified one mechanism of transmembrane transport. In the adipocyte of rat the transport of ^{14}C-pantothenate is fast, with a high affinity (half saturation at a concentration of 10^{-5} M), saturable, specific, and energy and temperature dependent. A specific intestinal transport mechanism, unsaturated at physiological concentration, was therefore proposed (Munnich *et al.* 1990).

EXCRETION

No degradation products of pantothenic acid are known (Sugarman & Munro, 1980); 1–7 mg/d of intact pantothenic acid are excreted in the urine. Mean excretion increased from 2·3 to 3·9 mg/d as intake increased from 5 to 10 mg (Foss, 1981). Excretion of 9·7 and 36·2 mg was reported at 17 and 117 mg/d intakes respectively (Lin, 1981; Kies *et al.* 1982).

NUTRITIONAL STATUS AND INDUCED DEFICIENCY

Deficiency of pantothenic acid occurs only in severe malnutrition (Machlin, 1984); however, deficiency symptoms have been produced by administering a metabolic antagonist, ω-methylpantothenic acid (Bean & Hodges, 1954; Lubin *et al.* 1956). The subjects developed a burning sensation in the feet, vomiting, depression, fatigue, insomnia, tenderness in the heels, and muscular weakness. A change in glucose tolerance was also reported, associated with increased sensitivity to insulin and a decrease in antibody production (Hodges *et al.* 1958). Other antagonists are pantoyltaurine and phenyl-pantothenate (Machlin, 1984). Whole blood, serum levels and urine pantothenic acid excretion are used to evaluate pantothenic acid status in humans.

CONCLUSION

The physiological roles of pantothenic acid and coenzyme A are known, but there is a lack of knowledge about intestinal absorption, plasma transport and catabolism. No genetic diseases related to pantothenic acid metabolism are known, probably because the diseases would be incompatible with life. Further research is needed on the metabolic pathways of pantothenic acid and coenzyme A.

VITAMIN C

Vitamin C exists in three generally recognized forms in living tissues: ascorbate, the non-ionizable, oxidized form, dehydroascorbate, and the intermediate oxidation state, ascorbate free radical (also known as semidehydroascorbate or monodehydroascorbate). The stability of the three species diminishes in that order. Ascorbate free radical can only be detected transiently, e.g. by electron spin resonance, but the other two forms exist as food components, and their bioavailability has been studied in some detail. The availability of vitamin C is closely bound up with its stability to oxidation, since it is the least stable of all the vitamins in foods, and its biological activity and retention within the body are also dependent on the linkage with various redox cycles *in vivo*. A dietary requirement for vitamin C is encountered only in higher primates and in a few other mammals, birds and insects. Other species normally synthesize it from hexose sugars.

FOOD SOURCES

It is well established that the most important food sources of vitamin C are fresh fruit and vegetables, especially soft and citrus fruits, and the growing tips of plants; potatoes can also be a significant source because of the relatively large amount commonly eaten. Ungerminated seeds and the foods made from them contain very little vitamin C. Vitamin C is easily destroyed by exposure to light and oxygen, by prolonged storage, by alkaline cooking conditions, or by exposure to redox transition metals such as iron and copper, unless these are complexed, for instance, within proteins. Ascorbate oxidases, released from vegetables during cutting or wilting, help to destroy vitamin C. Both ascorbate and dehydroascorbate are found in food; the reduced form predominates in fresh food and in the living animal. Analytical methods for vitamin C in food or tissue extracts may or may not include dehydroascorbate or dioxogulonic acid. Of these two compounds, only the first is biologically active. Ascorbate or its analogue D-isoascorbate (erythrobate) are frequently added to manufactured foods, not only to increase their vitamin C content but also to protect other oxygen sensitive molecules and thus to prevent discolouration, nitrosamine formation etc. Ascorbate is used in bread making, as a flour improver, but none of it survives in the finished loaf.

INTESTINAL ABSORPTION

In man and guineapigs, both of which have an absolute requirement for exogenous vitamin C in their diet, there is an ouabain sensitive sodium dependent saturable active transport system for ascorbic acid at the brush border of the duodenum and upper ileum, and another sodium independent transfer process at the basolateral membrane (Stevenson, 1974; Toggenburger *et al.* 1979; Hornig, 1981; Patterson *et al.* 1982; Bianchi *et al.* 1986; Rose, 1988). In isolated intestinal preparations from guineapigs, active concentration of ascorbic acid against an electrochemical gradient has been demonstrated (Bianchi *et al.* 1986). Active transport seems absent in rats, rabbits or hamsters, species which do not require a dietary source (Toggenburger *et al.* 1979).

Dehydroascorbate is absorbed by a carrier mediated passive mechanism, both in the intestinal and in the buccal mucosa (Bianchi *et al.* 1986; Rose, 1988). In the guineapig it is reduced to ascorbate before reaching the basolateral membrane (Bianchi *et al.* 1986).

Since dehydroascorbate is unstable, this may account for poor agreement between studies of its transport.

At low levels of intake vitamin C is very efficiently absorbed and retained by humans. Divided doses are absorbed more efficiently than a single bolus, and at intakes up to c. 100–180 mg/d, 80–95% is absorbed (Hornig, 1981; Kallner et al. 1977). At higher intakes, the absorption mechanism becomes overloaded, so that of a single 1·5 g dose only 50% is absorbed, of 6 g 25% is absorbed and of 12 g 16% is absorbed (Rivers, 1987). At high intakes, and provided the subject is initially near saturation, the absorption efficiency can be estimated by the proportion of the dose that is recovered in the urine within 24 h. Peak excretion is usually at around 4 h after dosing. The unabsorbed fraction is virtually all destroyed, presumably by bacterial processes, in the lower bowel. Clearly, although gram doses are not very efficiently absorbed, yet the total amount that can be absorbed increases steadily across a wide range of increasing intakes up to a maximum of c. 1·2 g at high single doses (Hornig et al. 1980; Yung et al. 1982; Melethil et al. 1986; Rivers, 1987). Since the amount absorbed also determines the concentration achieved in the tissues, it is possible to increase tissue ascorbate levels progressively, by increasing the intakes to high levels. However, the greatest tissue sensitivity to intake variation occurs over a low intake range, 0–50 mg/d.

TURNOVER AND EXCRETION AND FACTORS WHICH AFFECT AVAILABILITY

Turnover amounts to c. 3%/d of the ascorbate body pool (Baker et al. 1969). Saturation kinetics with vitamin C at the brush border of the ileum implies that large doses (1 g or more) cannot be completely absorbed. Enhanced absorption can, however, be achieved either by dividing the dose or by using a slow release preparation (Yung et al. 1982). Some investigators have claimed that certain food components such as the bioflavonoids can enhance the availability of food ascorbate over that observed with the pure vitamin, possibly by protecting it from oxidation (Vinson & Bose, 1988). Since the optical isomer of natural ascorbate, D-iso-ascorbate (= D-erythrobate), is absorbed less efficiently, it is clear that intestinal absorption of L-ascorbate is a stereospecific process (Hornig, 1975). Aspirin apparently inhibits the absorption of ascorbic acid (Basu, 1982), although the precise mechanism is not fully understood.

TRANSPORT AT OTHER SITES

Most organs concentrate ascorbate from plasma indicating active transport against a concentration gradient. At some sites—white blood cells, the adrenal medulla, brain, connective tissue cells, placenta, kidney, intestine, and the ciliary epithelium of the eye—ascorbate transport predominates (Rose, 1988; Choi & Rose, 1989; Bergsten et al. 1990). Other tissues appear to transport dehydroascorbate, reducing it to ascorbate within the cell. Ascorbate homeostasis and its failure during conditions such as infection, diabetes and atrophic gastritis require further study (Schorah, 1992).

CONTRASTS BETWEEN GROUPS OF PEOPLE

Differences in requirements for ascorbate exist between the sexes (Blanchard, 1991), between smokers and non-smokers (Murata, 1991) and possibly between elderly and young adults (Davies et al. 1984; Blanchard et al. 1990). Diabetics may have disturbed ascorbate economy, although the mechanism is uncertain (Schorah, 1992). There is no clear consensus on whether these differences are attributable to differences in the efficiency of ascorbate absorption; differences in turnover may be important.

CONCLUSION

Vitamin C is absorbed by active transport in man. This transport process is highly efficient at low to moderate intakes, but becomes progressively less efficient at high intakes of single doses. Active transport at other sites, including the kidney, ensures retention of appropriate levels of the vitamin and its characteristic distribution between tissues. Further studies are needed on the factors which determine efficiency of absorption from food, and turnover of vitamin C.

REFERENCES

Abad, A. R. & Gregory, J. F. (1987). Determination of folate bioavailability with a rat bioassay. *Journal of Nutrition* **117**, 866–873.

Akiyama, T., Selhub, J. & Rosenberg, I. H. (1982). FMN phosphatase and FAD pyrophosphatase in rat intestinal brush borders: role in intestinal absorption of dietary riboflavin. *Journal of Nutrition* **112**, 263–268.

Allison, P. M., Mummah-Schendel, L. L., Kindberg, C. G., Harms, C. S., Bang, N. U. & Suttie, J. W. (1987). Effects of a vitamin K-deficient diet and antibiotics in normal human volunteers. *Journal of Laboratory and Clinical Medicine* **110**, 180–188.

Aw, T. Y., Jones, D. P. & McCormick, D. B. (1983). Uptake of riboflavin by isolated rat liver cells. *Journal of Nutrition* **113**, 1249–1254.

Babu, S. & Srikantia, S. G. (1976). Availability of folates from some foods. *American Journal of Clinical Nutrition* **29**, 376–379.

Baessler, K-H., Gruehn, E., Loew, D. & Pietrzik, K. (1992). *Vitamin-Lexikon*, pp. 154–173. Stuttgart: Gustav Fischer Verlag.

Baker, E. M., Hodges, R. E., Hood, J., Sauberlich, H. E. & March, S. C. (1969). Metabolism of ascorbic-1-^{14}C acid in experimental human scurvy. *American Journal of Clinical Nutrition* **22**, 549–558.

Baker, H. & Frank, O. (1976). Absorption, utilization and clinical effects of allithiamins compared to water-soluble vitamins. *Journal of Nutritional Science and Vitaminology* **22**, 63–66.

Barua, A. B., Batres, R. O. & Olson, J. A. (1989). Characterization of retinyl β-glucuronide in human blood. *American Journal of Clinical Nutrition* **50**, 370–374.

Basilico, V., Ferrari, G., Rindi, G. & D'Andrea, G. (1979). Thiamine intestinal transport and phosphorylation: a study in vitro of potential inhibitors of small intestinal thiamine-pyrophosphokinase using a crude enzymatic preparation. *Archives Internationales de Physiologie et de Biochimie* **87**, 981–995.

Basu, T. K. (1982). Vitamin C-aspirin interactions. In *Vitamin C: New Clinical Applications in Immunology, Lipid Metabolism and Cancer* (International Journal of Vitamin and Nutrition Research, Suppl. 23) 83–90 [A. Hanck, editor].

Batchelor, A. J. & Compston, J. E. (1983). Reduced plasma half-life of radio-labelled 25-hydroxyvitamin D_3 in subjects receiving a high-fibre diet. *British Journal of Nutrition* **49**, 213–216.

Bauernfeind, J. (1980). Tocopherols in foods. In *Vitamin E: A Comprehensive Treatise*, pp. 99–167. [L. J. Machlin, editor]. New York: Deckker.

Bean, W. B. & Hodges, R. E. (1954). Pantothenic acid deficiency induced in human subjects. *Proceedings of the Society for Experimental Biology and Medicine* **86**, 693–698.

Bechgaard, H. & Jespersen, S. (1977). Gastrointestinal absorption of niacin in humans. *Journal of Pharmaceutical Science* **66**, 871–872.

Bender, D. A. & Bender, A. E. (1986). Niacin and tryptophan metabolism: the biochemical basis of niacin requirements and recommendations. *Nutrition Abstracts and Reviews A* **56**, 695–719.

Bendich, A. & Langseth, L. (1989). Safety of vitamin A. *American Journal of Clinical Nutrition* **49**, 358–371.

Berger, E., Long, E. & Semenza, G. (1972). The sodium activation of biotin absorption in hamster small intestine in vitro. *Biochimica et Biophysica Acta* **255**, 873–887.

Bergsten, P., Amitai, G., Kehrl, J., Dhariwal, K. R., Klein, H. G. & Levine, M. (1990). Millimolar concentrations of ascorbic acid in purified human mononuclear leukocytes. Depletion and reaccumulation. *Journal of Biological Chemistry* **265**, 2584–2587.

Bianchi, J., Wilson, F. A. & Rose, R. C. (1986). Dehydroascorbic acid and ascorbic acid transport systems in the guinea pig ileum. *American Journal of Physiology* **250**, G461–468.

Bitsch, R. & Schramm, W. (1992). Free and bound vitamin B_6 derivatives in plant foods. In *Chemical Reactions in Foods. II. (FECS Event No. 174)*, pp. 285–290.

Bjørneboe, A., Bjørneboe, G-E. A., Bodd, E., Hagen, B. F., Kveseth, N. & Drevon, C. A. (1986). Transport and distribution of α-tocopherol in lymph, serum and liver cells in rats. *Biochimica et Biophysica Acta* **889**, 310–315.

Bjørneboe, A., Bjørneboe, G-E. A. & Drevon, C. A. (1990). Absorption, transport and distribution of vitamin E. *Journal of Nutrition* **120**, 233–242.

Bjornsson, T. D., Meffin, P. J., Swezey, S. E. & Blaschke, T. F. (1979). Effects of clofibrate and warfarin alone and in combination on the disposition of vitamin K_1. *Journal of Pharmacology & Experimental Therapeutics* **210**, 322–326.

Bjornsson, T. D., Meffin, P. J., Swezey, S. E. & Blaschke, T. F. (1980). Disposition and turnover of vitamin K in man. In *Vitamin K Metabolism and Vitamin K-dependent Proteins*, pp. 328–332 [J. W. Suttie, editor]. Baltimore, Md: University Park.

Blanchard, J. (1991). Effects of gender on vitamin C pharmacokinetics in man. *Journal of the American College of Nutrition* **10**, 453–459.

Blanchard, J., Conrad, K. A., Mead, R. A. & Garry, P. J. (1990). Vitamin C disposition in young and elderly men. *American Journal of Clinical Nutrition* **51**, 837–845.

Blomstrand, R. & Forsgren, L. (1968). Vitamin K_1-^{3}H in man. Its intestinal absorption and transport in the thoracic duct lymph. *Internationale Zeitschift für Vitaminforschung* **38**, 45–64.

Boland, R. L. (1986). Plants as a source of vitamin D_3 metabolites. *Nutrition Reviews* **44**, 1–8.

Bonjour, J. P. (1977). Biotin in man's nutrition and therapy—a review. *International Journal for Vitamin and Nutrition Research* **47**, 107–118.

Bonjour, J. P. (1984). Biotin. In *Handbook of Vitamins, Nutritional, Biochemical and Clinical Aspects*, pp. 403–435 [L. J. Machlin, editor]. New York: Dekker.

Bowman, B. B., McCormick, D. B. & Rosenberg, I. H. (1989). Epithelial transport of water-soluble vitamins. *Annual Review of Nutrition* **9**, 187–199.

Brody, T., Shane, B. & Stokstad, E. L. R. (1984). Folic acid. In *Handbook of Vitamins, Nutritional, Biochemical and Clinical Aspects*, pp. 460–491 [L. J. Machlin, editor]. New York: Dekker.

Brown, E. D., Micozzi, M. S., Craft, N. E., Bieri, J. G., Beecher, G., Edwards, B. K., Rose, A., Taylor, P. R. & Smith, J. C. (1989). Plasma carotenoids in normal men after a single ingestion of vegetables or purified β-carotene. *American Journal of Clinical Nutrition* **49**, 1258–1265.

Burton, G. W. (1989). Antioxidant action of carotenoids. *Journal of Nutrition* **119**, 109–111.

Burton, G. W., Ingold, K. U., Foster, D. O., Cheng, S. C., Webb, A., Hughes, L. & Lusztyk, E. (1988). Comparison of free α-tocopherol and α-tocopheryl acetate as sources of vitamin E in rats and humans. *Lipids* **23**, 834–840.

Butterworth, C. E., Newman, A. J. & Krumdieck, C. L. (1974). Tropical sprue: a consideration of possible etiologic mechanism with emphasis on pteroyl polyglutamate metabolism. *Transactions of the American Clinical and Climatological Association* **86**, 11–22.

Canton, M. C. & Cremin, F. M. (1990). The effect of dietary zinc depletion and repletion on rats: Zn concentration in various tissues and activity of pancreatic γ-glutamyl hydrolase (EC 3.4.22.12) as indices of Zn status. *British Journal of Nutrition* **64**, 201–209.

Carter, E. G. A. & Carpenter, K. J. (1982). The bioavailability for humans of bound niacin from wheat bran. *American Journal of Clinical Nutrition* **36**, 855–861.

Choi, J.-L. & Rose, R. C. (1989). Transport and metabolism of ascorbic acid in human placenta. *American Journal of Physiology* **257**, C110–113.

Clements, M. R., Davies, M., Fraser, D. R., Lumb, G. A., Mawer, E. B. & Adams, P. H. (1987). Metabolic inactivation of vitamin D is enhanced in primary hyperparathyroidism. *Clinical Science* **73**, 659–664.

Clifford, A., Jones, A. D. & Bills, N. D. (1990). Bioavailability of folates in selected foods incorporated into amino acid-based diets fed to rats. *Journal of Nutrition* **120**, 1640–1647.

Combs, G. F. (1992). Thiamin. In *The Vitamins: Fundamental Aspects in Nutrition and Health*, pp. 253–268. San Diego, CA: Academic Press.

Coursin, D. B. (1954). Convulsive seizures in infants with pyridoxine-deficient diet. *Journal of the American Medical Association* **154**, 406–408.

Cowan, M. J. (1984). Biotin responsive metabolic disorders in early childhood. In *Recent Vitamin Research*, pp. 2–23 [M. H. Briggs, editor]. Ohio: CRC Press.

Cowan, M. J., Wara, D. W., Packman, S., Amman, A. J., Yoshino, M., Sweetman, L. & Nyhan, W. (1979). Multiple biotin-dependent carboxylase deficiencies associated with defects in T-cell and B-cell immunity. *Lancet* **ii**, 115–118.

Craft, N. E. & Soares, J. H. (1992). Relative solubility, stability, and absorptivity of lutein and β-carotene in organic solvents. *Journal of Agricultural and Food Chemistry* **40**, 431–434.

Dagnelie, P. C., Vergote, F. J. V. R. A., van Staveren, W. A., van den Berg, H., Dingjan, P. G. & Hautvast, J. G. A. J. (1990). High prevalence of rickets in infants on macrobiotic diets. *American Journal of Clinical Nutrition* **51**, 202–208.

Dancis, J., Lehanka, J. & Levitz, M. (1985). Transfer of riboflavin by the perfused human placenta. *Pediatric Research* **19**, 1143–1146.

Daniel, H., Wille, U. & Rehner, G. (1983). In vitro kinetics of the intestinal transport of riboflavin in rats. *Journal of Nutrition* **113**, 636–643.

Davies, H. E. F., Davies, J. E. W., Hughes, R. E. & Jones, E. (1984). Studies on the absorption of L-xyloascorbic acid (vitamin C) in young and elderly subjects. *Human Nutrition: Clinical Nutrition* **38C**, 463–471.

Davies, M. K., Gregory, M. E. & Henry, K. M. (1959). The effect of heat on the vitamin B$_6$ of milk. II. A comparison of biological and microbiological tests of evaporated milk. *Journal of Dairy Research* **26**, 215–220.

Davis, R. E. & Icke, G. C. (1983). Clinical chemistry of thiamin. *Advances in Clinical Chemistry* **23**, 93–140.

DeLuca, H. F. (1988). The vitamin D story: a collaborative effort of basic science and clinical medicine. *FASEB Journal* **2**, 224–236.

Deutsche Gesellschaft für Ernährung (1988). *Erganzungsband zum Ernährungsbericht 1988*. Frankfurt: DGE.

Drevon, C. A. (1991). Absorption, transport and metabolism of vitamin E. *Free Radical Research Communications* **14**, 229–246.

Dutta, P., Pinto, J. & Rivlin, R. (1985). Antimalarial effects of riboflavin deficiency. *Lancet* **ii**, 1040–1043.

Evans, W. B. & Wollaeger, E. E. (1966). Incidence and severity of nutritional deficiency states in chronic exocrine pancreatic insufficiency: comparison with non-tropical sprue. *American Journal of Digestive Diseases* **11**, 594–606.

FAO/WHO. (1967). Requirements of vitamin A, thiamine, riboflavine and niacin, pp. 271–276. Rome: FAO (*FAO Nutrition Meetings Report Series* no. 41; *WHO Technical Report Series* no. 362).

FAO/WHO (1988). Requirements of vitamin A, iron, folate and vitamin B$_{12}$, pp. 17–18, 29–30. Rome: FAO (*FAO Food and Nutrition Series* no. 23).

Ferland, G., Sadowski, J. A. & O'Brien, M. E. (1993). Dietary induced subclinical vitamin K deficiency in normal human subjects. *Journal of Clinical Investigation* **91**, 1761–1768.

Food and Nutrition Board (1989). Recommended Dietary Allowances, 10th Edn, pp. 169–173. Washington, DC: National Academy of Sciences.

Ford, J. E., Salter, D. N. & Scott, K. J. (1969). The folate-binding protein in milk. *Journal of Dairy Research* **36**, 435–446.

Foss, Z. M. S. (1981). Thesis, University of Nebraska, Lincoln, Nebraska, USA.

Fraser, D. R. (1983). The physiological economy of vitamin D. *Lancet* **i**, 969–972.

Friedrich, W. (1987). *Handbuch der Vitamine*, pp. 305–308. Munich: Urban & Schwarzenberg.

Fukui, E., Kurohara, H., Kageyu, A., Kurosaki, Y., Nakayama, T. & Kimura, T. (1989). Enhancing effect of medium-chain tryglycerides on intestinal absorption of d-α-tocopherol acetate from lecithin-dispersed preparations in the rat. *Journal of Pharmacobio-Dynamics* **12**, 80–86.

Gallo-Torres, H. E. (1980). Absorption, transport and metabolism. In *Vitamin E: a Comprehensive Treatise*, pp. 170–267 [L. J. Machlin, editor]. New York: Dekker.

Gascon-Barré, M. (1986). Is there any physiological significance to the enterohepatic circulation of vitamin D sterols. *Journal of the American College of Nutrition* **5**, 317–324.

Ghitis, J. (1967). The folate binding in milk. *American Journal of Clinical Nutrition* **20**, 1–4.

Gilbert, J. A. & Gregory, J. F. (1992). Pyridoxine-5'-β-D-glucoside affects the metabolic utilization of pyridoxine in rats. *Journal of Nutrition* **122**, 1029–1035.

Goodman, D. S., Blomstrand, R., Werner, B., Huang, H. S. & Shiratori, T. (1966). The intestinal absorption and metabolism of vitamin A and β-carotene in man. *Journal of Clinical Investigation* **45**, 1615–1623.

Goodman, D. S. & Huang, H. S. (1965). Biosynthesis of vitamin A with rat intestinal enzymes. *Science* **149**, 879–880.

Gravel, R. A., Lam, K. F., Mahuran, D. & Kronis, A. (1980). Purification of human liver propionyl-CoA carboxylase by carbon tetrachloride extraction and monomeric avidin affinity chromatography. *Archives of Biochemistry and Biophysics* **201**, 669–673.

Greer, F. R., Mummah-Schendel, L. L., Marshall, S. & Suttie, J. W. (1988). Vitamin K$_1$ (phylloquinone) and vitamin K$_2$ (menaquinone) status in newborns during the first week of life. *Pediatrics* **81**, 137–140.

Gregory, J. F. (1980). Effect of ε-pyridoxyllysine bound to dietary protein on the vitamin B$_6$ status of rats. *Journal of Nutrition* **110**, 995–1005.

Gregory, J. F., Bhandari, S. D., Bailey, L. B., Toth, J. P., Baumgartner, T. G. & Cerda, J. J. (1991a). Relative bioavailability of deuterium-labelled monoglutamyl and hexaglutamyl folates in human subjects. *American Journal of Clinical Nutrition* **53**, 736–740.

Gregory, J. F. & Ink, S. L. (1987). Identification and quantification of pyridoxine-β-glucoside as a major form of vitamin B$_6$ in plant-derived foods. *Journal of Agricultural and Food Chemistry* **35**, 76–82.

Gregory, J. F. & Kirk, J. R. (1977). Interaction of pyridoxal and pyridoxal phosphate with peptides in a model food system during thermal processing. *Journal of Food Science* **42**, 1554–1557, 1561.

Gregory, J. F., Trumbo, P. R., Bailey, L. B., Toth, J. P., Baumgartner, T. G. & Cerda, J. J. (1991b). Bioavailability of pyridoxine-5-β-D-glucoside determined in humans by stable-isotope methods. *Journal of Nutrition* **121**, 177–186.

Gronowska-Senger, A. & Wolf, G. (1970). Effect of dietary protein on the enzyme from rat and human intestine which converts β-carotene to retinal. *Journal of Nutrition* **100**, 300–308.

Gubler, C. J. (1991). Thiamin. In *Handbook of Vitamins, Nutritional, Biochemical and Clinical Aspects*, pp. 233–281 [L. J. Machlin, editor]. New York: Dekker.

Guillaumont, M., Sann, L., Leclerq, M., Dostalova, L., Vignal, B. & Frederich, A. (1993). Changes in hepatic vitamin K$_1$ levels after prophylactic administration to the newborn. *Journal of Pediatric Gastroenterology and Nutrition* **16**, 10–14.

Halstead, C. (1990). Intestinal absorption of dietary folates. In *Folic Acid Metabolism in Health and Disease*, pp. 23–45 [M. F. Picciano, E. L. R. Stokstad & J. F. Gregory, editors] New York: Wiley-Liss.

Hansen, C., Leklem, J., Hardin, K. & Miller, L. (1991). Vitamin B_6 status of women with constant protein intake and four levels of vitamin B_6. *FASEB Journal* **5**, 556A.

Hathcock, J. N. (1985). Quantitative evaluation of vitamin safety. *Pharmacy Times* May, 104–113.

Heinrich, H. C. & Gabbe, E. E. (1990). Experimental basis of oral and parenteral therapy with cyano- or aquacobalamin. *Biomedicine and Physiology of Vitamin B_{12}*. London: Children's Medical Charity.

Heinrich, H. C. & Wolfsteller, E. (1966). [High dosage oral vitamin B_{12} therapy. Experimental basis and practical application.] *MedizinischeKlinik* **61**, 756–763.

Heldenberg, D., Tenenbaum, G. & Weisman, Y. (1992). Effect of iron on serum 25-hydroxyvitamin D and 24,25-dihydroxyvitamin D concentrations. *American Journal of Clinical Nutrition* **56**, 533–536.

Henderson, C. T., Mobarhan, S., Bowen, P., Stacewicz-Sapuntzakis, M., Langenberg, P., Kiani, R., Lucchesi, D. & Sugerman, S. (1989). Normal serum response to oral beta-carotene in humans. *Journal of the American College of Nutrition* **8**, 625–635.

Henderson, L. M. (1983). Niacin. *Annual Review of Nutrition* **3**, 289–307.

Herbert, V. (1988). Vitamin B_{12}: plant sources, requirements, and assay. *American Journal of Clinical Nutrition* **48**, 852–858.

Herzlich, B. & Herbert, V. (1984). The role of the pancreas in cobalamin (vitamin B_{12}) absorption. *American Journal of Gastroenterology* **79**, 489–493.

Heymann, W. (1936). Absorption of carotene. *American Journal of Diseases of Children* **51**, 273–283.

Hodges, R. E., Ohlson, M. A. & Bean, W. B. (1958). Pantothenic acid deficiency in man. *Journal of Clinical Investigation* **37**, 1642–1657.

Hollander, D. E. (1973). Vitamin K_1 absorption by everted intestinal sacs of the rat. *American Journal of Physiology* **225**, 360–364.

Hollander, D. E. (1981). Intestinal absorption of vitamins A, E, D, and K. *Journal of Laboratory and Clinical Medicine* **97**, 449–462.

Hollander, D. E. & Dadufalza, V. (1989). Lymphatic and portal absorption of vitamin E in aging rats. *Digestive Diseases and Sciences* **34**, 768–772.

Hollander, D. E., Rim, E. & Ruble, P. E. (1977). Vitamin K_2 colonic and ileal in vivo absorption: bile, fatty acids, and pH effects on transport. *American Journal of Physiology* **233**, E124–129.

Hollander, D. E. & Ruble, P. E. (1978). β-Carotene intestinal absorption: bile, fatty acid, pH, and flow rate effects on transport. *American Journal of Physiology* **235**, E686–691.

Hollis, B. W., Roos, B. A., Draper, H. H. & Lambert, P. W. (1981a). Vitamin D and its metabolites in human and bovine milk. *Journal of Nutrition* **111**, 1240–1248.

Hollis, B. W., Roos, B. A., Draper, H. H. & Lambert, P. W. (1981b). Occurrence of vitamin D sulfate in human milk whey. Vitamin D and its metabolites in human and bovine milk. *Journal of Nutrition* **111**, 384–390.

Holmberg, I., Aksnes, L., Berlin, T., Lindback, B., Zemgals, J. & Lindeke, B. (1990). Absorption of a pharmacological dose of vitamin D_3 from two different lipid vehicles in man: comparison of peanut oil and a medium chain triglyceride. *Biopharmaceutics and Drug Deposition* **11**, 807–815.

Hoppner, K. & Lampi, B. (1986). Bioavailability of food folacin as determined by rat liver bioassay. *Nutrition Reports International* **34**, 489–494.

Hornig, D. (1975). Distribution of ascorbic acid, metabolites and analogues in man and animals. *Annals of the New York Academy of Sciences* **258**, 103–118.

Hornig, D. (1981). Metabolism and requirements of ascorbic acid in man. *South African Medical Journal* **60**, 818–823.

Hornig, D., Vuilleumier, J-P. & Hartmann, D. (1980). Absorption of large, single, oral intakes of ascorbic acid. *International Journal for Vitamin and Nutrition Research* **50**, 309–314.

Horst, R. L., Goff, J. P. & Reinhardt, T. A. (1990). Advancing age results in reduction of intestinal and bone 1,25-dihydroxyvitamin D receptor. *Endocrinology* **126**, 1053–1057.

Horwitt, M. K., Harper, A. E. & Henderson, L. M. (1981). Niacin-tryptophan relationships for evaluating niacin equivalents. *American Journal of Clinical Nutrition* **34**, 423–427.

Howard, L., Wagner, C. & Schenker, S. (1974). Malabsorption of thiamin in folate-deficient rats. *Journal of Nutrition* **104**, 1024–1032.

Hoyumpa, A. M. (1982). Characterization of normal intestinal thiamin transport in animals and man. *Annals of the New York Academy of Sciences* **378**, 337–343.

Hoyumpa, A. M., Middleton, H. M., Wilson, F. A. & Schenker, S. (1975). Thiamin transport across the rat intestine. I. Normal characteristics. *Gastroenterology* **68**, 1218–1227.

Hoyumpa, A. M., Strickland, R., Sheehan, J. J., Yarborough, G. & Nichols, S. (1982). Dual system of intestinal thiamine transport in humans. *Journal of Laboratory and Clinical Medicine* **99**, 701–708.

Hume, E. M. & Krebs, H. A. (1949). *Vitamin A Requirements of Human Adults* (Special Report Series no. 264). London: Medical Research Council.

Ingold, K. U., Burton, G. W., Foster, D. O. & Hughes, L. (1990). Further studies of a new vitamin E analogue more active than α-tocopherol in the rat curative myopathy bioassay. *FEBS Letters* **267**, 63–65.

Ingold, K. U., Burton, G. W., Foster, D. O., Hughes, L., Lindsay, D. A. & Webb, A. (1987). Biokinetics of and discrimination between dietary *RRR*- and *SRR*-α-tocopherols in the male rat. *Lipids* **22**, 163–172.

James, D. R., Owen, G., Campbell, I. A. & Goodchild, M. C. (1992). Vitamin A absorption in cystic fibrosis: risk of hypervitaminosis A. *Gut* **33**, 707–710.

Jensen, C. D., Howes, T. W., Spiller, G. A., Pattison, T. S., Whittam, J. H. & Scala, J. (1987). Observations on the effects of ingesting cis- and trans-beta-carotene isomers on human serum concentrations. *Nutrition Reports International* **35**, 413–442.

Johnson, E. J. & Russell, R. M. (1992). Distribution of orally administered β-carotene among lipoproteins in healthy men. *American Journal of Clinical Nutrition* **56**, 128–135.

Jolly, D. W., Craig, C. & Nelson, T. E. (1977). Estrogen and prothrombin synthesis; effect of estrogen on absorption of vitamin K. *American Journal of Physiology* **232**, H12–17.

Jusko, W. J. & Levy, G. (1975). Absorption, protein binding and elimination of riboflavin. In *Riboflavin*, pp. 99–152 [R. S. Rivlin, editor]. New York: Plenum Press.

Kabir, H., Leklem, J. E. & Miller, L. T. (1983). Relationship of the glycosylated vitamin B_6 content of foods to vitamin B_6 bioavailability in humans. *Nutrition Reports International* **28**, 709–716.

Kahlon, T. S., Chow, F. I., Hoefer, J. L. & Betschart, A. A. (1986). Bioavailability of vitamins A and E as influenced by wheat bran and bran particle size. *Cereal Chemistry* **63**, 490–493.

Kallner, A., Hartmann, D. & Hornig, D. (1977). On the absorption of ascorbic acid in man. *International Journal for Vitamin and Nutrition Research* **47**, 383–388.

Kanazawa, S. & Herbert, V. (1982). Vitamin B_{12} analogue content in human red cells, liver and brain. *Clinical Research* **30**, 504A.

Kaplan, L. A., Lau, J. M. & Stein, E. A. (1990). Carotenoid composition, concentrations and relationships in various human organs. *Clinical Physiology and Biochemistry* **8**, 1–10.

Kasparek, S. (1980). Chemistry of tocopherols and tocotrienols. In *Vitamin E: A Comprehensive Treatise*, pp. 7–65 [L. J. Machlin, editor]. New York: Dekker.

Kato, S. (1959). The effect of nicotine on thiamin levels in the chick. *Folia Pharmacologica Japonica* **55**, 5–12.

Keagy, P. M., Shane, B. & Oace, S. M. (1988). Folate bioavailability in humans: effects of wheat bran and beans. *American Journal of Clinical Nutrition* **47**, 80–88.

Kern, F., Birkner, H. J. & Ostrower, V. S. (1978). Binding of bile acids by dietary fiber. *American Journal of Clinical Nutrition* **31**, S175–179.

Kies, C., Wishart, C., McGee, M., Foss, Z., Yong, L. C., Quillian, J. & Fox, H. M. (1982). Pantothenic acid levels in urine, blood serum, and whole blood of adult humans fed graded levels of pantothenic acid. *Federation Proceedings* **41**, 276A.

Kimura, M. & Itokawa, Y. (1977). Effects of calcium and magnesium deficiency on thiamine distribution in rat brain and liver. *Journal of Neurochemistry* **28**, 389–393.

Lakshmi, A. V. & Bamji, M. S. (1979). Metabolism of [2-^{14}C]pyridoxine in riboflavin deficiency. *Biochemical Medicine* **22**, 274–281.

Lala, V. R. & Reddy, V. (1970). Absorption of β-carotene from green leafy vegetables in undernourished children. *American Journal of Clinical Nutrition* **23**, 110–113.

Lane, P. A. & Hathaway, W. E. (1985). Vitamin K in infancy. *Journal of Pediatrics* **106**, 351–359.

Lawson, D. E. M. (1980). Metabolism of vitamin D. In *Vitamin D: Molecular Biology and Clinical Nutrition* [A. W. Norman, editor]. New York: Dekker.

Lin, M. M. S. (1981). Thesis. University of Nebraska, Lincoln, Nebraska, USA.

Linnell, J. C., Smith, A. D. M., Smith, C. L., Wilson, J. & Matthews, D. M. (1968). Effects of smoking on metabolism and excretion of vitamin B_{12}. *British Medical Journal* **ii**, 215–216.

Loew, D. (1991). [Pharmacokinetics of the cobalamins: cyano-, hydroxy-, methylcobalamin.] In *Pharmakologie und klinische Anwendung hochdosierter B-Vitamine* [N. Reitbrock, editor], Darmstadt.

Loew, D. *et al.* (1988). [Pharmacokinetics of hydroxycobalamin and folic acid.] *Vitaminspur* **3**, 168–172.

Lubin, R., Daum, K. A. & Bean, W. B. (1956). Studies of pantothenic acid metabolism. *American Journal of Clinical Nutrition* **4**, 420–430.

McCormick, D. B. (1972). The fate of riboflavin in the mammal. *Nutrition Reviews* **30**, 75–79.

McCormick, D. B. (1989). Two interconnected B vitamins: riboflavin and pyridoxine. *Physiological Reviews* **69**, 1170–1198.

Machlin, L. J. (1984). Pantothenic acid. In *Handbook of Vitamins, Nutritional, Biochemical and Clincial Aspects*, pp. 437–457 [L. J. Machlin, editor]. New York: Dekker.

McLaren, D. S. & Zekian, B. (1971). Failure of enzymic cleavage of β-carotene. *American Journal of Diseases of Children* **121**, 278–280.

Maiani, G., Mobarhan, S., Ceccanti, M., Ranaldi, L., Gettner, S., Bowen, P., Friedman, H., de Lorenzo, A. & Ferro-Luzzi, A. (1989). Beta-carotene serum response in young and elderly females. *European Journal of Clinical Nutrition* **43**, 749–761.

Maislos, M., Silver, J. & Fainaru, M. (1981). Intestinal absorption of vitamin D sterols: differential absorption into lymph and portal blood in the rat. *Gastroenterology* **80**, 1528–1534.

Marquet, A. (1977). New aspects of the chemistry of biotin and of some analogs. *Pure and Applied Chemistry* **49**, 183–196.

Mason, J. B., Gibson, N. & Kodicek, E. (1973). The chemical nature of the bound nicotinic acid of wheat bran: studies of nicotinic acid-containing macromolecules. *British Journal of Nutrition* **30**, 297–311.

Matsuda, I., Endo, F. & Motohara, K. (1991). Vitamin K deficiency in infancy. *World Review of Nutrition and Dietetics* **64**, 85–108.

Matsuura, A., Iwashima, A. & Nose, Y. (1975). Purification of thiamine-binding protein from *Escherichia coli* by affinity chromatography. *Biochemical and Biophysical Research Communications* **51**, 241–246.

Matthews, D. M. (1974). Absorption of water-soluble vitamins. In *Biomembranes: Intestinal Absorption*, vol. 4B, pp. 847–915 [D. H. Smyth, editor]. New York: Plenum Press.

Mawer, E. B., Backhouse, J., Holman, C. A., Lumb, G. A. & Stanbury, S. W. (1972). The distribution and storage of vitamin D and its metabolites in human tissues. *Clinical Science* **43**, 413–431.

Mayersohn, M., Feldman, S. & Gibaldi, M. (1969). Bile salt enhancement of riboflavin and flavin mononucleotide absorption in man. *Journal of Nutrition* **98**, 288–296.

Melethil, S., Mason, W. D. & Chang, C.-J. (1986). Dose-dependent absorption and excretion of vitamin C in humans. *International Journal of Pharmaceutics* **31**, 83–90.

Merrill, A. H. & Henderson, J. M. (1987). Diseases associated with defects in vitamin B_6 metabolism or utilization. *Annual Review of Nutrition* **7**, 137–156.

Micozzi, M. S., Brown, E. D., Edwards, B. K., Bieri, J. G., Taylor, P. R., Khachik, F., Beecher, G. R. & Smith, J. C. (1992). Plasma carotenoid response to chronic intake of selected foods and β-carotene supplements in men. *American Journal of Clinical Nutrition* **55**, 1120–1125.

Mills, C. A., Cottingham, E. & Taylor, E. (1976). The effect of advancing age on dietary thiamine requirements. *Archives of Biochemistry* **9**, 221–227.

Monk, B. E., (1982). Metabolic carotenaemia. *British Journal of Dermatology* **106**, 485–487.

Mrochek, J. E., Jolley, R. L., Young, D. S. & Turner, W. J. (1976). Metabolic response of humans to ingestion of nicotinic acid and nicotinamide. *Clinical Chemistry* **22**, 1821–1827.

Munnich, A., Ogier, H. & Saudubray, J-M. (1990). [Pantothenic acid.] In *Le Vitamine: Aspetti Metabolici, Genetici, Nutrizionali e Terapeutici*, pp. 185–201. Rome: Masson.

Munnich, A., Saudubray, J. M., Carré, G., Coudé, F. X., Ogier, H., Charpentier, C. & Frézal, J. (1981). Defective biotin absorption in multiple carboxylase deficiency. *Lancet* **ii**, 263.

Murata, A. (1991). Smoking and vitamin C. *World Review of Nutrition and Dietetics* **64**, 31–57.

Murata, K. (1965). Review of Japanese literature on beriberi and thiamin, p. 220. Tokyo: Vitamin B1 Research Committee of Japan.

Murty, C. V. R. & Adiga, P. R. (1982). Pregnancy suppression by active immunization against gestation-specific riboflavin carrier protein. *Science* **216**, 191–193.

National Research Council (1989). *Recommended Dietary Allowances*, 10th Edn, pp. 137–142. Washington, DC: National Academy Press.

Need, G. N., Morris, H. A., Horowitz, M. Y. & Nordin, B. E. C. (1993). Effects of skin thickness, age, body fat, and sunlight on serum 25-hydroxyvitamin D. *American Journal of Clinical Nutrition* **58**, 882–885.

Nishino, K. & Itokawa, Y. (1977). Thiamin malabsorption in vitamin B_6 or vitamin B_{12} deficient rats. *Journal of Nutrition* **107**, 775–782.

Nose, Y. & Iwashima, A. (1965). Intestinal absorption of thiamin propyl disulfide. *Journal of Vitaminology* **11**, 165–170.

Olson, J. A. (1987). Recommended Dietary Intakes (RDI) of vitamin K in humans. *American Journal of Clinical Nutrition* **45**, 687–692.

Olson, J. A. & Hayaishi, O. (1965). The enzymatic cleavage of β-carotene into vitamin A by soluble enzymes of rat liver and intestine. *Proceedings of the National Academy of Sciences, USA* **54**, 1364–1370.

Omaye, S. T. & Chow, F. I. (1984). Effect of hard red spring wheat bran on the bioavailability of lipid-soluble vitamins and growth of rats fed for 56 days. *Journal of Food Science* **49**, 504–506.

Parker, R. S., Viereck, S. M., Spielman, A. B., Brenna, J. T. & Goodman, K. J. (1992). Metabolism and biokinetics of ^{13}C-β-carotene in humans following a small oral dose. *FASEB Journal* **6**, 1645A.

Patterson, L. T., Nahrwold, D. L. & Rose, R. C. (1982). Ascorbic acid uptake in guinea pig intestinal mucosa. *Life Sciences* **31**, 2783–2791.

Pearson, W. N. (1967). Blood and urinary vitamin levels as potential indices of body stores. *American Journal of Clinical Nutrition* **20**, 514–525.

Pietrzik, K. (1993). Problems of folate bioavailability. In *Nutritional, Chemical and Food Processing Implications of Nutrient Availability. Proceedings, Part 2, 'Bioavailability 93'*, pp. 289–299 [U. Schlemmer, editor].

Pike, R. L. & Brown, M. L. (1975). Water-soluble vitamins. *Nutrition: an Integrated Approach*, 2nd edn, pp. 80–139. New York: Wiley.

Pinto, J., Huang, Y. P. & Rivlin, R. S. (1981). Inhibition of riboflavin metabolism in rat tissues by chlorpromazine, imipramine and amitriptyline. *Journal of Clinical Investigation* **67**, 1500–1506.

Prentice, A. M. & Bates, C. J. (1980). Refection in rats fed on a sucrose-based riboflavin-deficient diet. *British Journal of Nutrition* **43**, 171–177.

Prince, M. R., Frisoli, J. K., Goetschkes, M. M., Stringham, J. M. & LaMuraglia, G. M. (1991). Rapid serum carotene loading with high-dose β-carotene: clinical implications. *Journal of Cardiovascular Pharmacology* **17**, 343–347.

Rafsky, H. A., Newman, B. & Jolliffe, N. (1947). The relationship of gastric acidity to thiamine excretion in the aged. *Journal of Laboratory and Clinical Medicine* **32**, 118–123.

Reibel, D. K., Wyse, B. W., Berkich, D. A., Palko, W. M. & Neely, J. R. (1981). Effects of diabetes and fasting on pantothenic acid metabolism in rats. *American Journal of Physiology* **240**, E597–601.

Reisenauer, A. M., Krumdieck, C. L. & Halstead, C. H. (1977). Folate conjugase: two separate activities in human jejunum. *Science* **198**, 196–197.
Reynolds, R. D. (1988). Bioavailability of vitamin B_6 from plant foods. *American Journal of Clinical Nutrition* **48**, 863–867.
Rhode, B. M., Cooper, B. A. & Farmer, F. A. (1983). Effect of orange juice, folic acid, and oral contraceptives on serum folate in women taking a folate-restricted diet. *Journal of the American College of Nutrition* **2**, 221–230.
Rietz, P., Gloor, U. & Wiss, O. (1970). [Menaquinones from human liver and sewage sludge.] *Internationale Zeitschrift für Vitaminforschung* **40**, 351–362.
Ristow, K. A., Gregory, J. F. & Damron, B. L. (1982). Effect of dietary fiber on the bioavailability of folic acid monoglutamate. *Journal of Nutrition* **112**, 750–758.
Rivers, J. M. (1987). Safety of high-level vitamin C ingestion. *Annals of the New York Academy of Sciences* **498**, 445–453.
Rock, C. L. & Swendseid, M. E. (1992). Plasma β-carotene response in humans after meals supplemented with dietary pectin. *American Journal of Clinical Nutrition* **55**, 96–99.
Rose, R. C. (1988). Transport of ascorbic acid and other water-soluble vitamins. *Biochimica et Biophysica Acta* **947**, 335–366.
Russell, R. M., Dahr, G. J., Dutta, S. K. & Rosenberg, I. H. (1979). Influence of intraluminal pH on folate absorption. Studies in control subjects and in patients with pancreatic insufficiency. *Journal of Laboratory and Clinical Medicine* **93**, 428–436.
Sadoogh-Abasian, F. & Evered, D. F. (1980). Absorption of nicotinic acid and nicotinamide from rat small intestine in vitro. *Biochimica et Biophysica Acta* **598**, 385–391.
Sadowski, J., Bacon, D., Hood, S., Davidson, K., Gauter, C., Haroon, Y. & Shepard, D. (1988). The application of methods used for the evaluation of vitamin K nutritional status in human and animal studies. In *Current Advances in Vitamin K Research*, pp. 453–463 [J. W. Suttie, editor]. Amsterdam: Elsevier Science Publishing Co.
Said, H. M. & Hollander, D. (1985). Does aging affect the intestinal transport of riboflavin? *Life Sciences* **36**, 69–73.
Said, H. M., Hollander, D. & Duong, Y. (1985). A dual, concentration-dependent transport system for riboflavin in rat intestine *in vitro*. *Nutrition Research* **5**, 1269–1279.
Said, H. M., Horne, D. W. & Wagner, C. (1986). Effect of human milk folate binding protein on folate intestinal transport. *Archives of Biochemistry and Biophysics* **251**, 114–120.
Said, H. M., Ong, D. E. & Shingleton, J. L. (1989). Intestinal uptake of retinol: enhancement by bovine milk β-lactoglobulin. *American Journal of Clinical Nutrition* **49**, 690–694.
Said, H. M., Redha, R. & Nylander, W. (1987). A carrier-mediated, Na+ gradient-dependent transport for biotin in human intestinal brush-border membrane vesicles. *American Journal of Physiology* **253**, G631–636.
Salter, D. N. & Blakeborough, P. (1988). Influence of goat's milk folate-binding protein on transport of 5-methyltetrahydrofolate in neonatal goat small intestinal brush border membrane vesicles. *British Journal of Nutrition* **59**, 497–507.
Sampson, D. A. & Chung, S. (1991). Dietary protein quality and vitamin B_6 nutritional status in lactating rats. *FASEB Journal* **5**, 557A.
Satyanarayana, U. & Narasinga Rao, B. S. (1980). Dietary tryptophan level and the enzymes of tryptophan NAD pathway. *British Journal of Nutrition* **43**, 107–113.
Sauberlich, H. E., Hodges, R. E., Wallace, D. L., Kolder, H., Canham, J. E., Hood, J., Raica, N. & Lowry, L. K. (1974). Vitamin A metabolism and requirements in the human studied with the use of labeled retinol. *Vitamins and Hormones* **32**, 251–275.
Saupe, J., Shearer, M. J. & Kohlmeier, M. (1994). Phylloquinone (vitamin K_1) transport and its' influence on gamma-carboxyglutamic (Gla)-residues of osteocalcin in patients on maintenance hemodialysis. *American Journal of Clinical Nutrition* (In the press.)
Schaller, K. & Höller, H. (1974). Thiamine absorption in the rat. 1. Intestinal permeability and active transport of thiamine; passage and cleavage of thiamine pyrophosphate *in vitro*. *International Journal for Vitamin and Nutrition Research* **44**, 443–451.
Schaus, E. E., de Lumen, B. O., Chow, F. I., Reyes, P. & Omaye, S. T. (1985). Bioavailability of vitamin E in rats fed graded levels of pectin. *Journal of Nutrition* **115**, 263–270.
Schorah, C. J. (1992). The transport of vitamin C and effects of disease. *Proceedings of the Nutrition Society* **51**, 189–198.
Schubiger, G., Tonz, O., Gruter, J. & Shearer, M. J. (1993). Vitamin K_1 concentration in breast-fed neonates after oral or intramuscular administration of a single dose of a new mixed micellar preparation of phylloquinone. *Journal of Pediatric Gastroenterology and Nutrition* **16**, 435–439.
Selhub, J., Powell, G. M. & Rosenberg, I. H. (1984). Intestinal transport of 5-methyltetrahydrofolate. *American Journal of Physiology* **246**, G515–520.
Shaw, S., Meyers, S., Colman, N., Jayatilleke, E. & Herbert, V. (1987). The ileum is the major site of absorption of vitamin B_{12}. *Federation Proceedings* **46**, 1004A.
Shearer, M. J. (1990). Vitamin K and vitamin K-dependent proteins. *British Journal of Haematology* **75**, 156–162.
Shearer, M. J., Barkhan, P. & Webster, G. R. (1970). Absorption and excretion of an oral dose of tritiated vitamin K_1 in man. *British Journal of Haematology* **18**, 297–308.

Shearer, M. J., McBurney, A. & Barkhan, P. (1974). Studies on the absorption and metabolism of phylloquinone (vitamin K₁) in man. *Vitamins and Hormones* **32**, 513–542.

Shearer, M. J., Rahim, S., Barkhan, P. & Stimmer, L. (1982). Plasma vitamin K₁ in mothers and their newborn babies. *Lancet* **ii**, 460–463.

Shinzawa, T., Mura, T., Tsunei, M. & Shiraki, K. (1989). Vitamin K absorption capacity and its association with vitamin K deficiency. *American Journal of Diseases of Children* **143**, 686–689.

Simpson, K. L. & Chichester, C. O. (1981). Metabolism and nutritional significance of carotenoids. *Annual Review of Nutrition* **1**, 351–374.

Sitrin, M. D. & Bengoa, J. M. (1987). Intestinal absorption of cholecalciferol and 25-hydroxycholecalciferol in chronic cholestatic liver disease. *American Journal of Clinical Nutrition* **46**, 1011–1015.

Somogyi, J. C. (1971). On antithiamine factors of fern. *Journal of Vitaminology* **17**, 165–174.

Spector, R. & Boose, B. (1979). Active transport of riboflavin by the isolated choroid plexus *in vitro*. *Journal of Biological Chemistry* **254**, 10286–10289.

Spector, R. & Boose, B. (1982). Riboflavin transport by rabbit kidney slices: characterization and relation to cyclic organic acid transport. *Journal of Pharmacology and Experimental Therapeutics* **221**, 394–398.

Spencer, R. P. & Brody, K. R. (1964). Biotin transport by small intestine of rat, hamster, and other species. *American Journal of Physiology* **206**, 653–657.

Stahl, W., Schwarz, W., Sundquist, A. R. & Sies, H. (1992). *cis-trans* Isomers of lycopene and β-carotene in human serum and tissues. *Archives of Biochemistry and Biophysics* **294**, 173–177.

Stamp, T. C. B. (1975). Factors in human vitamin D nutrition and in the production and cure of classical rickets. *Proceedings of the Nutrition Society* **34**, 119–130.

Steinberg, S. E., Campbell, C. L. & Hillman, R. S. (1979). Kinetics of the normal folate enterohepatic cycle. *Journal of Clinical Investigation* **64**, 83–88.

Stenflo, J., Fernlund, P., Egan, W. & Roepstorff, P. (1974). Vitamin K dependent modifications of glutamic acid residues in prothrombin. *Proceedings of the National Academy of Sciences, USA* **71**, 2730–2733.

Stevenson, N. R. (1974). Active transport of L-ascorbic acid in the human ileum. *Gastroenterology* **67**, 952–956.

Strauss, L. H. & Scheer, P. (1939). [Effect of nicotine on vitamin C metabolism.] *Zeitschrift für Vitaminforschung* **9**, 39–48.

Sugarman, B. & Munro, H. N. (1980). [¹⁴C]pantothenate accumulation by isolated adipocytes from adult rats of different ages. *Journal of Nutrition* **110**, 2297–2301.

Suttie, J. W. (1984). Vitamin K. In *Handbook of Vitamins, Nutritional, Biochemical and Clinical Aspects*, pp. 147–198 [L. J. Machlin, editor]. New York: Dekker.

Suttie, J. W., Mummah-Schendel, L. L., Shah, D. V., Lyle, B. J. & Greger, J. L. (1988). Vitamin K deficiency from dietary vitamin K restriction in humans. *American Journal of Clinical Nutrition* **47**, 475–480.

Sweetman, L., Surh, L., Baker, H., Peterson, R. M. & Nyhan, W. L. (1981). Clinical and metabolic abnormalities in a boy with dietary deficiency of biotin. *Pediatrics* **68**, 553–558.

Tadera, K., Arima, M., Yoshino, S., Yagi, F. & Kobayashi, A. (1986). Conversion of pyridoxine into 6-hydroxypyridoxine by food components, especially ascorbic acid. *Journal of Nutritional Science and Vitaminology* **32**, 267–277.

Tadera, K., Kaneko, T. & Yagi, F. (1988). Isolation and structural elucidation of three new pyridoxine-glycosides in rice bran. *Journal of Nutritional Science and Vitaminology* **34**, 167–177.

Tamura, T. & Stokstad, E. L. R. (1973). The availability of food folate in man. *British Journal of Haematology* **25**, 513–532.

Tanaka, Y., Frank, H. & DeLuca, H. F. (1973). Biological activity of 1,25-dihydroxyvitamin D₃ in the rat. *Endocrinology* **92**, 417–422.

Tarr, J. B., Tamura, T. & Stokstad, E. L. R. (1981). Availability of vitamin B₆ and pantothenate in an average American diet in man. *American Journal of Clinical Nutrition* **34**, 1328–1337.

Thoene, J. & Wolf, B. (1983). Biotinidase deficiency in juvenile multiple carboxylase deficiency. *Lancet* **ii**, 398.

Toggenburger, G., Landolt, M. & Semenza, G. (1979). Na⁺-dependent electroneutral L-ascorbate transport across brush border membrane vesicles from human small intestine. Inhibition by D-erythorbate. *FEBS Letters* **108**, 473–476.

Traber, M. G., Burton, G. W., Ingold, K. U. & Kayden, H. J. (1990). *RRR*- and *SRR*-α-tocopherols are secreted without discrimination in human chylomicrons, but *RRR*-α-tocopherol is preferentially secreted in very low density lipoproteins. *Journal of Lipid Research* **31**, 675–685.

Traber, M. G., Ingold, K. U., Burton, G. W. & Kayden, H. J. (1988). Absorption and transport of deuterium-substituted 2*R*,4′*R*,8′*R*-α-tocopherol in human lipoproteins. *Lipids* **23**, 791–797.

Traber, M. G. & Kayden, H. J. (1989). Preferential incorporation of α-tocopherol vs γ-tocopherol in human lipoproteins. *American Journal of Clinical Nutrition* **49**, 517–526.

Traber, M. G., Kayden, H. J., Balmer Green, J. & Green, M. H. (1986). Absorption of water-miscible forms of vitamin E in a patient with cholestasis and in thoracic duct-cannulated rats. *American Journal of Clinical Nutrition* **44**, 914–923.

Trumbo, P. R. (1991). Influence of dietary protein on vitamin B₆ bioavailability and secretion in milk. *FASEB Journal* **5**, 557A.

van den Berg, H. (1991). *Modern Lifestyles, Lower Energy Intake and Micronutrient Status*, pp. 21–36. Berlin: Springer-Verlag.

van den Berg, H. (1993). General aspects of bioavailability of vitamins. In: Proceedings Bioavailability '93; Nutritional, chemical and food processing in implications of nutrient availability. *Berichte der Bundesforschunganstalt für Ernährung*, pp. 267–278 [U. Schlemmer, editor].

van Vliet, T., van Schaik, F. & van den Berg, H. (1992). [Beta-carotene metabolism: the enzymatic cleavage to retinal.] *Voeding* **53**, 186–190.

Vaziri, N. D., Said, H. M., Hollander, D., Barbari, A., Patel, N., Dang, D. & Kariger, R. (1985). Impaired intestinal absorption of riboflavin in experimental uremia. *Nephron* **41**, 26–29.

Velázquez, A., Zamudio, S., Báez, A., Murguia-Corral, R., Rangel-Peniche, B. & Carrasco, A. (1990). Indicators of biotin status: a study of patients on prolonged total parenteral nutrition. *European Journal of Clinical Nutrition* **44**, 11–16.

Verdon, C. P. & Blumberg, J. B. (1988). Influence of dietary vitamin E on the intermembrane transfer of α-tocopherol as mediated by an α-tocopherol binding protein. *Proceedings of the Society for Experimental Biology and Medicine* **189**, 52–60.

Vieth, R. (1990). The mechanisms of vitamin D toxicity. *Bone and Mineral* **11**, 267–272.

Villard, L. & Bates, C. J. (1986). Carotene dioxygenase (EC 1.13.11.21) activity in rat intestine: effects of vitamin A deficiency and of pregnancy. *British Journal of Nutrition* **56**, 115–122.

Vinson, J. A. & Bose, P. (1988). Comparative bioavailability to humans of ascorbic acid alone or in a citrus extract. *American Journal of Clinical Nutrition* **48**, 601–604.

von Kries, R., Greer, F. R. & Suttie, J. W. (1993). Assessment of vitamin K status of the newborn infant. *Journal of Pediatric Gastroenterology and Nutrition* **16**, 231–238.

Wagner, C. (1985). Folate-binding proteins. *Nutrition Reviews* **43**, 293–299.

Wang, X-D., Tang, G-W., Fox, J. G., Krinsky, N. I. & Russell, R. M. (1991). Enzymatic conversion of β-carotene into β-apo-carotenals and retinoids by human, monkey, ferret, and rat tissues. *Archives of Biochemistry and Biophysics* **285**, 8–16.

Warkany, J. (1975). Riboflavin deficiency and congenital malformations. In *Riboflavin*, pp. 279–302 [R. S. Rivlin, editor]. New York: Plenum Press.

Webb, A. R., Kline, L. & Holick, M. F. (1988). Influence of season and latitude on the cutaneous synthesis of vitamin D_3: exposure to winter sunlight in Boston and Edmonton will not promote vitamin D_3 synthesis in human skin. *Journal of Clinical Endocrinology and Metabolism* **67**, 373–378.

Weber, F. (1981). Absorption mechanism for fat-soluble vitamins and the effect of other food constituents. In *Nutrition in Health and Disease and International Development; Proceedings of the XII International Congress of Nutrition*, pp. 119–135. New York: Liss.

White, H. B. & Merrill, A. H. (1988). Riboflavin-binding proteins. *Annual Review of Nutrition* **8**, 279–299.

Whyte, M. P., Haddad, J. G., Walters, D. D. & Stamp, T. C. B. (1979). Vitamin D bioavailability: serum 25-hydroxyvitamin D levels in man after oral, subcutaneous, intramuscular, and intravenous vitamin D administration. *Journal of Clinical Endocrinology and Metabolism* **48**, 906–911.

Wolf, B., Grier, R. E., Allen, R. J., Goodman, S. I. & Kien, C. L. (1983). Biotinidase deficiency: the enzymatic defect in late-onset multiple carboxylase deficiency. *Clinica Chimica Acta* **131**, 273–281.

Wolf, H. (1971). Hormonal alteration of efficiency of conversion of tryptophan to urinary metabolites of niacin in man. *American Journal of Clinical Nutrition* **24**, 792–799.

Wood, H. G. & Barden, R. E. (1977). Biotin enzymes. *Annual Review of Biochemistry* **46**, 385–413.

Wu Leung, W. T., Busson, F. & Jardin, C. (1968). *Food Composition Table for Use in Africa*. Bethesda, MD: US Department of Health, Education and Welfare, and Rome: FAO.

Yang, Y-M., Simon, N., Maertens, P., Brigham, S. & Liu, P. (1989). Maternal-fetal transport of vitamin K_1 and its effects on coagulation in premature infants. *Journal of Pediatrics* **115**, 1009–1013.

Yung, S., Mayersohn, M. & Robinson, J. B. (1982). Ascorbic acid absorption in humans: a comparison among several dosage forms. *Journal of Pharmaceutical Science* **71**, 282–285.

BIOLOGY OF ZINC AND BIOLOGICAL VALUE OF DIETARY ORGANIC ZINC COMPLEXES AND CHELATES

JOHANNES W. G. M. SWINKELS[1,2,3], ERVIN T. KORNEGAY[1] AND MARTIN W. A. VERSTEGEN[4]

[1] Department of Animal Science, Virginia Polytechnic Institute and State University, Blacksburg, VA 24061, USA
[2] Current address: Research Institute for Pig Husbandry, P.O. Box 83, 5240 AB Rosmalen, The Netherlands
[3] The author was a Pratt Fellow and this research was supported by the John Lee Pratt Animal Nutrition Foundation
[4] Department of Animal Nutrition, Agricultural University of Wageningen, Haagsteeg 4, 6708 PM Wageningen, The Netherlands

CONTENTS

INTRODUCTION	130
BIOLOGY OF ZINC	130
ZINC IN PORCINE TISSUES AND FLUIDS	130
INTRACELLULAR DISTRIBUTION OF ZINC	130
BIOLOGICAL FUNCTIONS OF INTRACELLULAR ZINC	131
ZINC HOMEOSTASIS	133
ASSESSMENT OF ZINC STATUS	133
BIOLOGICAL ADAPTATIONS DURING ZINC DEFICIENCY	135
ABSORPTION OF ZINC	136
ROLE OF INTESTINAL METALLOTHIONEIN IN ZINC ABSORPTION	136
SITE(S) OF ZINC ABSORPTION	137
MINERALS INTERACTING WITH ZINC DURING ZINC ABSORPTION	139
INTRINSIC FACTORS AFFECTING ZINC ABSORPTION	140
BIOLOGICAL VALUE OF DIETARY COMPLEXES AND CHELATES OF ZINC	140
METAL COMPLEXES AND CHELATES	140
CONSIDERATIONS WHEN DETERMINING BIOAVAILABILITY OF ZINC	141
AVAILABILITY OF ZINC COMPLEXED WITH PICOLINIC ACID	142
AVAILABILITY OF ZINC COMPLEXED WITH METHIONINE OR OTHER AMINO ACIDS	142
CONCLUSIONS	145
REFERENCES	145

INTRODUCTION

Growth retardation and an abnormal hair coat, induced in rats by feeding a purified diet low in Zn (1·6 p.p.m. Zn), were the first clinical signs associated with dietary Zn deficiency (Todd *et al.* 1934). Two decades later, a direct relationship between Zn and growth was reported for swine (Tucker & Salmon, 1955). In addition to impaired growth in the Zn deficient pigs, Tucker & Salmon (1955) also observed dermatosis that was previously called 'parakeratosis' by Kernkamp & Ferrin in 1953. Parakeratosis is the classic characteristic associated with severe Zn deficiency in swine (National Research Council, 1979).

Since the recognition of Zn as an essential nutrient, many researchers have studied the role of Zn in biology and nutrition. In the first section of this paper the biology of Zn is reviewed. This is followed by an overview of our current understanding of possible modes of action of Zn absorption. In the last part the biological value of dietary organic Zn complexes and Zn chelates is discussed.

BIOLOGY OF ZINC

ZINC IN PORCINE TISSUES AND FLUIDS

The concentration of Zn in the whole body of the pig, expressed on a fat free basis, is 25 mg/kg (Spray & Widdowson, 1950). This is within the 20–30 mg/kg range reported for fat free bodies of rats, cats, man (Spray & Widdowson, 1950), sheep (Grace, 1983) and dairy cows (Miller *et al.* 1974). In Table 1, concentrations of Zn in tissues, fluids, bone and integuments of the pig are presented. These concentrations were similar to those found in body compartments of rats, sheep, cows, monkeys and men (Hambidge *et al.* 1986; Jackson, 1989).

In pigs, the highest concentration of Zn is found in hair. However, relative to the Zn content of the whole body the total amount of Zn in hair is small. The largest pool of Zn, approximately 60%, is found in skeletal muscle tissue because of its bulk and fairly high Zn concentration (Jackson, 1989). The concentration of Zn varies with the type of skeletal muscle, being highest in red and lowest in white skeletal muscle (Cassens *et al.* 1967). The remainder of the Zn pool of the body is primarily located in bone and organs. Body fluids contain only a small proportion of total body Zn (Hambidge *et al.* 1986).

The total Zn content of the body on a weight basis remains fairly constant from birth to maturity (Spray & Widdowson, 1950). The relative proportion of body Zn found in the liver gradually increases from birth to weaning. After weaning, the liver Zn concentration rapidly decreases to approximately the level of Zn present at birth (Spray & Widdowson, 1950).

INTRACELLULAR DISTRIBUTION OF ZINC

In mouse liver cells, the largest proportion of intracellular Zn was found in the light fraction containing organelles other than mitochondria and nuclei (Bartholomew *et al.* 1959). Conversely, in porcine muscle cells the largest concentration of Zn was found in the heavy fraction containing myofibrils and nuclei (Cassens *et al.* 1967). Moreover, it was found that the level of Zn in the heavy fraction of red skeletal muscle was almost four times as high as in white skeletal muscle. The Zn level in the light fraction was equally low for both the red and white muscle types (Cassens *et al.* 1967). Thus, the intracellular distribution of Zn varies among tissues.

It is much harder to determine the nature of intracellular Zn than its location. Most biochemical techniques currently available involve destruction of the cell allowing Zn to

Table 1. *Concentration of Zn in body fluids and tissues of fast growing pigs*[a]

Item	Concentration
Blood	
Plasma (μg/l)	740
Serum (μg/l)	600
Erythrocytes (μg/g packed cells)	7·7
Leucocytes[b]	21·5
Tissues[b]	
Bone	113
Brain	70
Heart	96
Kidney	141
Liver	151
Red muscle	137
Mixed muscle	89
White muscle	67
Pancreas	161
Spleen	107
Integuments[b]	
Hair	201
Skin	28

[a] Data compiled from Hoekstra *et al.* (1956, 1967), Cassens *et al.* (1967), Miller *et al.* (1968), Crofton *et al.* (1983) and Zhou *et al.* (1994).
[b] Data of leucocytes, tissues and integuments are expressed as p.p.m. on a DM basis.

exchange ligands prior to analysis of Zn binding compounds (Jackson, 1989). Nevertheless, it has become apparent that Zn is part of many cellular metalloenzymes (Galdes & Vallee, 1983) and that Zn binds readily to the thiolate ligands present in the cellular protein metallothionein (Vasak & Kägi, 1983).

The distribution and nature of intracellular Zn is depicted in Fig. 1. Zinc is found in many cell compartments and is largely bound to cellular proteins (Williams, 1984). However, this does not exclude the existence of substantial amounts of free intracellular Zn or intracellular Zn complexed with one or more free amino acids (Jackson, 1989).

BIOLOGICAL FUNCTIONS OF INTRACELLULAR ZINC

To date, it has been suggested that Zn is involved in the following biological processes: (1) catalysis, (2) structural arrangement of protein, and (3) regulation of cellular events (Williams, 1989). For each of the processes, Zn exerts its biological activity almost entirely as part of complex molecules.

The catalytic function of Zn is clearly demonstrated in the enzyme carbonic anhydrase (Galdes & Vallee, 1983). The only physiological reaction known to be catalysed by carbonic anhydrase is the reversible hydration of carbon dioxide ($H_2O + CO_2 = H^+ + HCO_3^-$). The Zn ion of carbonic anhydrase is thought to participate in the first step of the catalytic reaction. Basically, Zn functions as an electron acceptor, or Lewis acid, and binds to the H_2O molecule. Due to the neutral imidazole ligands of the enzyme, the complex $Zn(H_2O)$ attains maximum acidity making ionization of H_2O to OH^- possible at pH 7 (Williams, 1989). The reactivity of the nucleophilic OH^- group is sufficient for carbonic anhydrase to attack the electrophilic CO_2 molecule. As a result, the end product HCO_3^- is formed (Galdes & Vallee, 1983). Zinc is thought to behave in a similar way in other Zn metalloenzymes which contain Zn at the active site.

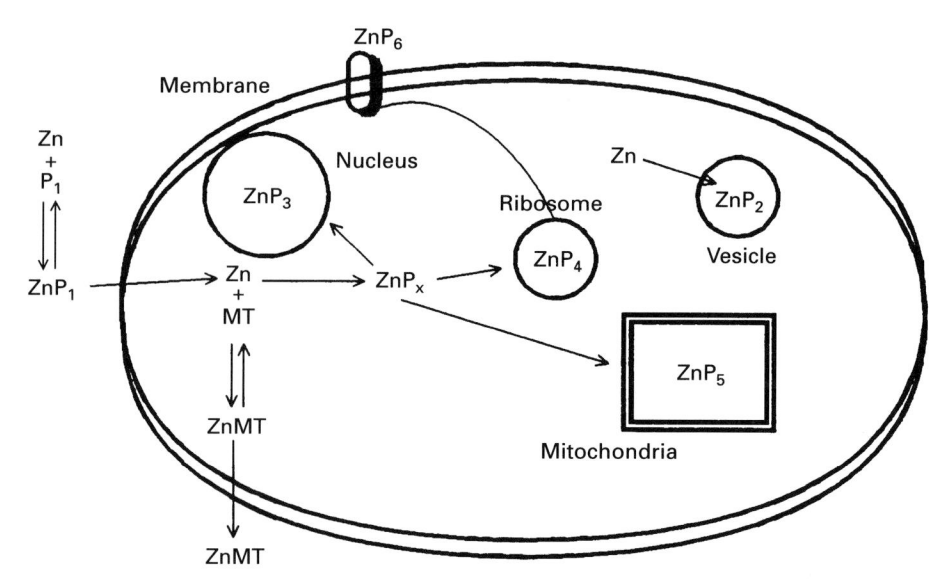

Fig. 1. Outline of the distribution of intracellular Zn (Williams, 1984). Zinc is carried to the cell by an extracellular protein carrier (ZnP$_1$). After being transported into the cell, free Zn can bind to intracellular metallothionein (MT) or other intracellular proteins (ZnP$_x$). Metallothionein binds Zn (ZnMT) and functions either as an intracellular storage protein or as an extracellular protein carrier. Zinc can also be transferred outside the cell as a constituent of enzymes or hormones (ZnP$_2$). Intracellularly, Zn is transported to different cellular compartments in which it carries out important biological functions. Nuclear (ZnP$_3$), ribosomal (ZnP$_4$) and mitochondrial (ZnP$_5$) proteins may be involved in polymerization, catalysis or protection. Zinc-containing enzymes or transporter proteins (ZnP$_6$), produced at the ribosomes, can be incorporated into the cell membrane.

Zinc may play a structural role in enzymes whenever it is located in a site not critical for catalysis. Furthermore, Zn is considered of critical importance in maintaining the structure of metalloproteins such as insulin and growth hormone. In this sense, Zn can be viewed as a replacement for a disulphide bond, a common feature in many proteins that provides stability by interlinking polypeptide chains (Stryer, 1988).

One disadvantage of the disulphide bond is that in a reducing environment the sulphur containing amino acids of the protein chain can be protonated. Protonation causes a breakup of the disulphide bridge so that protein conformation is lost. A second disadvantage of the disulphide bridge is that it allows little motion about itself and therefore restricts protein conformation. In contrast to the disulphide bridge, Zn cannot be reduced and puts very little stereochemical demand on the protein (Williams, 1984). Thus, Zn provides enzymes and other Zn-containing proteins with conformational stability at various pH without causing much steric hindrance.

A possible cooperative role of Zn and enzymes has been recognized in regulation of metabolic processes and synthesis (Jackson, 1989). Moreover, evidence has been presented that Zn is involved in gene expression of metallothionein (Seguin & Hamer, 1987). Recently, Cousins & Lee-Ambrose (1992) used rats to investigate the interactions of dietary Zn intake, nuclear Zn uptake and metallothionein gene expression. They reported that increases in dietary Zn were proportional to nuclear uptake of ingested Zn as well as to the level of metallothionein gene expression in the kidney, liver, intestine, spleen and heart. Using heparin-Sepharose chromatography and South-Western blotting, several Zn-binding protein fractions were isolated. One of the isolated protein fractions was able to bind an oligonucleotide in addition to Zn. This oligonucleotide was a DNA fragment of a transcription factor of the metallothionein gene (Cousins & Lee-Ambrose, 1992). These results support the earlier findings of Seguin & Hamer (1987) that Zn is involved in regulation of metallothionein gene expression and suggest that this occurs in several tissues.

It is not always easy to distinguish among the catalytic, structural and regulatory functions of Zn. A good illustration is provided by Zn present in RNA and DNA polymerases. The nature of the function of Zn in these enzymes may be catalytic by binding substrate, primer or template. Alternatively, Zn may be involved in maintaining conformation and not be part of the active site of the enzymes. A third possibility is that Zn acts in a regulatory manner by supplying specificity to proteins involved in gene replication and transcription (Wu & Wu, 1983).

ZINC HOMEOSTASIS

Homeostasis can be considered effective when the animal is able to maintain optimum health and function (Aggett, 1991). Initially, the animal is able to maintain Zn homeostasis by varying the rates of Zn absorption and excretion (Fig. 2). In the short term the animal can further adjust its Zn status. At low dietary Zn intakes, redistribution of Zn occurs to those Zn pools that are important in metabolism. When dietary Zn intakes are high, Zn is sequestered in several body tissues such as liver and bone. Conditions of long term deprivation or excess of dietary Zn lead to inadequacy of processes involved in maintaining Zn homeostasis (Fig. 2).

The mechanisms which enable the animal to maintain Zn homeostasis are not exactly understood (Aggett, 1991). This lack of understanding is an important factor contributing to the problems encountered in determining Zn status.

ASSESSMENT OF ZINC STATUS

Levels of plasma or serum Zn and activities of Zn-containing metalloenzymes are frequently measured in studies with man (Prasad et al. 1971). When animals are used, these easily obtainable indicators of Zn status are often supplemented with measurements of Zn content and Zn metalloenzyme activities in various tissues (Giugliano & Millward, 1984). It is questionable, though, whether these indicators are sensitive enough to provide the accuracy required for a reliable interpretation of Zn status.

Plasma or serum contains only a small proportion of the whole body Zn content (Jackson, 1989). Moreover, Zn levels in plasma or serum respond directly to increases in dietary Zn intake and also to minor catabolic processes occurring in skeletal muscle and other tissues with large Zn stores.

There are clear effects of dietary Zn intake on tissue Zn levels, but they are not uniform among tissues. Feeding diets low in Zn leads to reductions in Zn concentrations of the pancreas, liver, kidney, heart, intestine, skin, hair and bones of the Zn depleted pigs compared with pigs pair-fed diets with National Research Council (1988) recommended levels of Zn (Hoekstra et al. 1956, 1967; Miller et al. 1968; Crofton et al. 1983; Dørup & Clausen, 1991). Zinc concentrations in skeletal muscle tissue, however, are not affected by dietary Zn levels (Crofton et al. 1983; Dørup & Clausen, 1991). The large Zn pool found in skeletal muscle appears to be important for the biological functioning of the animal.

The relationship between dietary Zn intake and tissue Zn levels was studied by Cousins & Lee-Ambrose (1992). Following an overnight fast, diets containing different levels of Zn were administered via a stomach tube to Zn adequate rats. Two hours after feeding, the largest portion of ^{65}Zn was found in the small intestinal tissue, followed by liver, bone

Excess Zn

EXCRETION AND TISSUE SEQUESTRATION

DECREASED ABSORPTION AND INCREASED INTESTINAL SECRETION

Zinc Homeostasis

INCREASED ABSORPTION AND DECREASED INTESTINAL SECRETION

TISSUE REDISTRIBUTION TO PROTECT VITAL Zn POOLS

Zn Deficiency

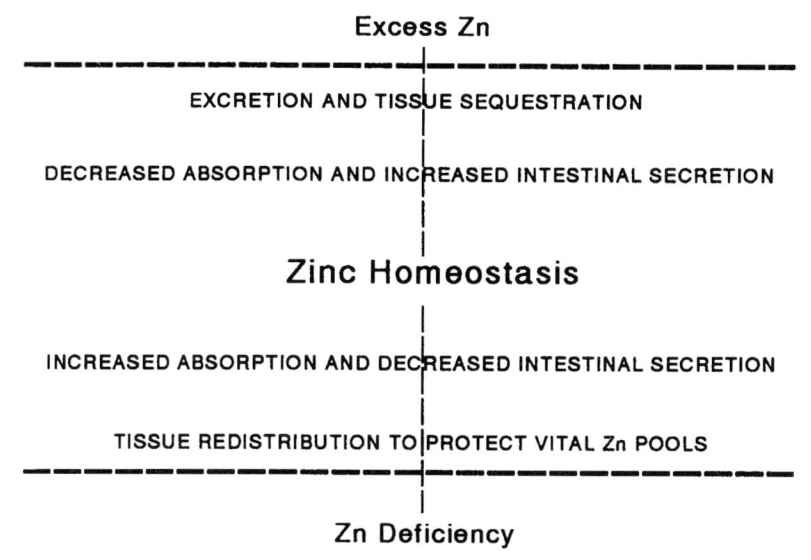

Fig. 2. Adaptive processes used to maintain Zn homeostasis (Aggett, 1991). During short periods of low or high dietary Zn intake, animals can initially maintain Zn homeostasis by adjusting intestinal Zn absorption and secretion. Other mechanisms that are enacted when intestinal mechanisms become insufficient are redistribution of Zn to tissues containing functional Zn pools and sequestration of Zn in tissues containing exchangeable Zn pools. During prolonged periods of low or high dietary Zn intake, animals cannot maintain Zn homeostasis and show signs of Zn deficiency and Zn toxicity respectively.

marrow, bone, skin, kidney, serum, thymus and skeletal muscle (Cousins & Lee-Ambrose, 1992).

It may be argued that pools which contain Zn needed for metabolic functions, the functional Zn pools, are located in those tissues and body fluids least affected by dietary Zn intake. Tissues that respond strongly to variations in dietary Zn intake may contain Zn pools less important for metabolism, the exchangeable Zn pools. The Zn content of many tissues and body fluids increases or decreases as dietary Zn intake changes. This suggests that there are numerous small sized exchangeable Zn pools.

In clinical nutrition, two approaches have been proposed to identify the early onset of Zn deficiency. The first involves prediction of exchangeable Zn pool sizes. This could be done by kinetic modelling studies in which either radioisotopes or the more widely used stable isotopes of Zn are used. Recently, Fairweather-Tait *et al.* (1993) demonstrated the feasibility of using an intravenous infusion of ^{70}Zn stable isotope and measuring plasma kinetics to estimate body pool sizes in man. They were able to distinguish between small rapidly exchanging and large slowly exchanging Zn pools containing less than 10 mg and approximately 350 mg of Zn respectively. Before this method can be used for assessment of Zn status, the relative sensitivity of body Zn pools to dietary and (or) physiological change must be established (Fairweather-Tait *et al.* 1993).

An alternative approach for identifying the exchangeable Zn pools is to diagnose factors involved in Zn redistribution between exchangeable and functional Zn pools. Some progress has been made in identifying a key factor involved in Zn redistribution, namely the Zn metalloprotein metallothionein (Golden, 1989). For accurately assessing Zn status in man, King (1990) proposed the use of a combination of plasma levels of Zn and metallothionein. Low plasma Zn and low metallothionein levels would indicate depletion of exchangeable Zn pools as a result of inadequate Zn intake. On the other hand, low plasma Zn in combination with high plasma metallothionein levels could be interpreted as tissue redistribution of Zn from exchangeable to functional Zn pools. Although plasma Zn is low in both events, the latter condition is not necessarily caused by low dietary Zn intake (King, 1990). A balanced diet is not available to every human and most of the time there is a lack of information on dietary intake. Consequently, cases of Zn deficiency are frequently reported (Prasad, 1988), especially if Zn demands are increased due to growth or pregnancy (Yasodhara *et al.* 1991). The development of diagnostic tools which allow easy and accurate assessment of Zn status is essential. Without these indicators, it remains impossible to diagnose cases of Zn deficiency at an early stage.

In animal nutrition, optimum levels of Zn have been defined for the different domestic species. The US National Research Council (1979) and the UK Agricultural Research Council (1981) give recommended intakes, and diets are generally formulated accordingly. Therefore, Zn deficiency is rarely observed in modern livestock production and thus the assessment of Zn status is not common practice.

BIOLOGICAL ADAPTATIONS DURING ZINC DEFICIENCY

Experimentally depleted animals have been used to study the biological effects of Zn deficiency. Early investigations showed that Zn depletion reduces levels of Zn and Zn metalloenzymes in many tissues and body fluids (Hoekstra *et al.* 1956, 1967; Prasad *et al.* 1971). More recently, it was found that Zn deficiency causes increased osmotic fragility of red blood cell membranes (Johanning *et al.* 1990) and depression in both the humoral and cellular immune responses (Gupta *et al.* 1985; Verma *et al.* 1988; Spears *et al.* 1991).

Of particular interest are the studies investigating the relationships between dietary Zn intake and growth. Growth retardation observed at low intakes of Zn can only partly be accounted for by overall depression of feed intake. This was demonstrated by Miller *et al.* (1968). In their study, performance was determined for three groups of pigs receiving different amounts of dietary Zn. Pigs fed a Zn deficient diet had lower gains and poorer feed conversion efficiencies than both pair-fed controls and control pigs with *ad lib.* access to feed. Feed conversion efficiency was similar in the two control groups (Miller *et al.* 1968).

Growth retardation of Zn deficient animals has been associated with reductions in levels of blood insulin-like growth factor I (Cossack, 1986; Dørup *et al.* 1991), and insulin (Giugliano & Millward, 1987; Dørup *et al.* 1991; Droke *et al.* 1993). Moreover, decreases in serum mitogenic activity and depressions of total pituitary RNA levels and growth hormone mRNA expression were found in pigs fed low dietary levels of Zn (Swinkels *et al.* 1994*c*). The observed decreases in both pituitary growth hormone mRNA (Swinkels *et al.* 1994*c*) and blood insulin-like growth factor I levels (Cossack, 1986; Dørup *et al.* 1991; Droke *et al.* 1993) could not be linked to reduced serum growth hormone levels (Dørup *et al.* 1991). After injection of a growth hormone-releasing factor analogue, Droke *et al.* (1993) even observed an increase in serum growth hormone levels in Zn deficient lambs. Thus, more research is warranted to determine whether synthesis of growth hormone is affected by Zn deficiency and to identify those growth mechanisms in which Zn is of critical importance.

In fast growing animals, Zn deficiency primarily affects protein metabolism. Reduced protein accretion has been found to occur in skeletal muscle, heart, thymus (Giugliano & Millward, 1987; Dørup & Clausen, 1991), and small intestinal tissues (Southon *et al.* 1986) of Zn deficient animals compared with Zn adequate controls. A direct association between Zn metabolism and protein metabolism in tissues like the small intestine may be present.

The magnitude of the effects of Zn deficiency on metabolism is an indication of the importance of Zn as a nutrient. The changes in growth, gross anatomy and histology are directly related to inadequate supply of intracellular Zn. Consequently, levels and/or activities of Zn metalloenzymes, Zn metalloproteins, and Zn transcription factors are reduced thereby impairing metabolism.

ABSORPTION OF ZINC

In general, absorption refers to one of the components in nutrient balance studies. More precisely, apparent absorption can be defined as the fraction of the dietary intake that does not appear in the faecal secretions. True absorption corrects the apparent absorption for endogenous losses occurring with intestinal secretions and mucosal sloughing which are not reabsorbed (O'Dell, 1984).

The process of Zn absorption can be physiologically divided into two separate events: firstly, uptake of Zn from the lumen into the cell, and secondly Zn transport from the cell into the circulatory system. In a review, Cousins (1989) has summarized current knowledge on mechanisms suspected to be involved in Zn uptake and transport (Fig. 3).

Uptake or cellular entry of Zn appears to occur by means of active transport and facilitated diffusion, both saturable processes (Davies, 1980; Menard & Cousins, 1983; Blakeborough & Salter, 1987). A small portion of Zn uptake and transport may be non-saturable, occurring through simple diffusion (Steel & Cousins, 1985) and paracellular movement of Zn, i.e. solvent drag (Bronner, 1987). The saturable uptake of Zn may involve binding of Zn by low molecular weight ligands which are present within the intestinal lumen. The Zn ligand complex either enters the cell intact or donates Zn to a membrane bound receptor. Subsequently, the receptor releases Zn intracellularly.

The capacity of the small intestine for Zn transport (V_{max}) depends on the body Zn status. Using isolated intestinal brush border membrane vesicles, Menard & Cousins (1983) found that Zn transport in rats fed adequate levels of Zn was twice as low as in Zn depleted rats. The affinity for Zn (K_m) was not affected by previous dietary Zn intakes. Thus, the increase in transport rate at low dietary Zn intakes is due only to an increase in number of receptors for free Zn or Zn bound to a low molecular weight ligand.

Evidence presented thus far suggests that Zn absorption is initiated largely by saturable uptake of Zn from the intestinal lumen into the cell. Using basolateral membrane vesicles of rat intestine, Oestreicher & Cousins (1989) studied transport of Zn out of the cell into the vascular system. Uptake of Zn in basolateral membrane vesicles was saturable and not affected by dietary Zn intake (Oestreicher & Cousins, 1989). A saturable Zn uptake indicates that a carrier mediated mechanism exists for Zn to enter the blood circulation. The lack of effect of dietary Zn intake on vesicular Zn uptake further suggests that Zn absorption is not regulated at the basolateral membrane (Oestreicher & Cousins, 1989).

ROLE OF INTESTINAL METALLOTHIONEIN IN ZINC ABSORPTION

Metallothionein is a protein of low molecular weight, about 6500 D, with a high metal-binding capacity, 7–10 atoms/mole (Bremner, 1983). For a thorough review of the role of metallothionein in Zn metabolism the reader is referred to Bremner (1983), Cousins (1985), Richards (1989) and Bremner & Beattie (1990).

A regulatory role of intestinal metallothionein in Zn absorption was first proposed by Richards & Cousins (1975). Intestinal metallothionein is synthesized in proportion to dietary Zn intake (Cousins & Lee-Ambrose, 1992). Metallothionein reduces Zn absorption by sequestering Zn within the enterocyte due to its higher affinity for Zn compared with other identified intestinal proteins (Starcher et al. 1980; Menard et al. 1981).

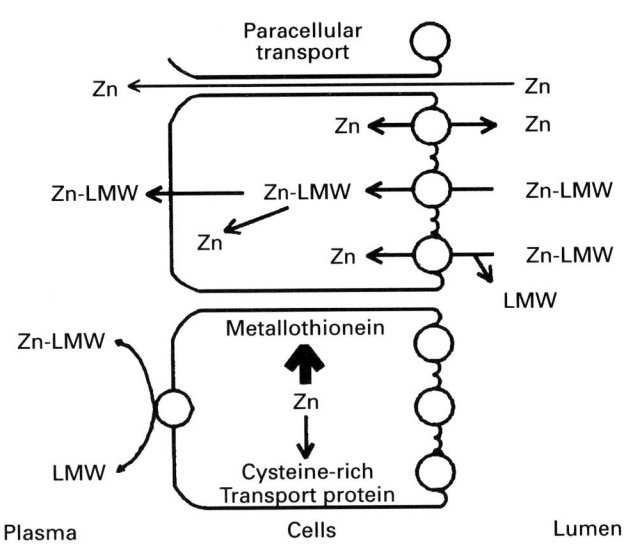

Fig. 3. A model for zinc absorption (Cousins, 1989; Hempe & Cousins, 1992). Zinc absorption can be physiologically divided into two processes: uptake of Zn from the GI lumen into the enterocyte (top) and transport of Zn from the enterocyte into the circulatory system (bottom). Within the GI lumen Zn may be presented for uptake into the enterocyte as free Zn or as Zn bound to a low molecular weight (LMW) ligand. The uptake of free Zn or Zn-LMW may involve carrier mediated and non-mediated mechanisms (top). Within the enterocyte, Zn transport may involve a cysteine-rich transcellular transport protein. Metallothionein competes for Zn with the transcellular transport protein and, therefore, may play a regulatory role in Zn absorption. Export of Zn from the enterocyte into the circulatory system may involve active mechanisms (bottom). A small portion of Zn uptake and transport may occur through simple diffusion and paracellular transport of free Zn (top).

As shown in Fig. 3, Hempe & Cousins (1992) included an interaction of a cysteine-rich intestinal protein, identified by Hempe & Cousins in 1989, with metallothionein in the Zn absorption model proposed by Cousins (1989). The cysteine-rich protein may enhance Zn absorption by transporting Zn transcellularly, from the intestinal brush border to the basolateral membrane. Intestinal metallothionein competitively inhibits binding of Zn to the transport protein and thereby regulates Zn absorption (Hempe & Cousins, 1992).

The regulatory role of intestinal metallothionein in Zn absorption has been partly challenged by Flanagan et al. (1983) and Coppen & Davies (1987). In these studies, criticism was focused on the proportional response of intestinal metallothionein synthesis to dietary Zn levels. Flanagan et al. (1983) observed only a transitory effect of dietary Zn level on intestinal metallothionein synthesis. They concluded that intestinal metallothionein synthesis was most likely induced by the nutritional stress associated with Zn deficiency and not by dietary Zn level (Flanagan et al. 1983). Moreover, Coppen & Davies (1987) found that dietary Zn intake did induce intestinal metallothionein synthesis in rats, but only at Zn levels of 5 to 80 mg per kg diet. At higher levels of Zn, 80 to 160 mg per kg diet, no further induction of intestinal metallothionein was observed (Coppen & Davies, 1987).

SITE(S) OF ZINC ABSORPTION

Many researchers have investigated the capacity for Zn absorption of various sites of the gut using a variety of methods. Some have reported that net Zn absorption in rats occurs primarily in the small intestine (Underwood, 1977) with negligible Zn absorption occurring

Table 2. *Apparent absorption coefficients of Zn determined within seven gut segments of depleted pigs fed an isolated soya protein semipurified Zn depletion diet supplemented with 15 and 45 p.p.m. Zn as ZnSO$_4$*[a]

gut segment[bc]	15 p.p.m. ZnSO$_4$	45 p.p.m. ZnSO$_4$	SEM
	%	%	
Stomach	−20.4	−16.1	6.6
Small intestine			
Proximal	−24.3	−46.9	18.7
Medial	5.4	0.7	14.1
Distal	25.5	17.6	5.8
Large intestine			
Caecum	18.0	18.4	4.7
Colon			
Proximal	20.3	15.7	4.9
Distal	16.7	17.4	6.2

[a] Apparent absorption coefficients were determined using the indirect indicator (0.25% Cr$_2$O$_3$) method. Each depleted mean represents five pigs that had been depleted for a 32-d period using an isolated soyabean semipurified diet containing 17 p.p.m. Zn. Pigs were killed 0, 3, 6, 12 and 24 d (one pig per d) after the start of a 24 d Zn repletion period. Digesta were collected exactly 2.5 h after feeding the last meal using a total digesta collection procedure. The gut was divided into seven segments: stomach, three small intestinal segments of equal length, caecum and two colonic segments of equal length.
[b] Linear, quadratic and cubic increases from stomach to distal colon segment ($P < 0.05$).
[c] Zn level by gut segment interaction ($P < 0.01$).

in other segments of the gut (Underwood, 1977; Davies, 1980). Others, however, did observe substantial Zn absorption in the large intestine of rats (Wapnir et al. 1985; Seal & Mathers, 1989), pigs (Partridge, 1978), sheep (Grace, 1975) and cattle (Bertoni et al. 1976). Furthermore, absorption of Zn anterior to the small intestine was observed in chickens (Miller & Jensen, 1966) and dairy cattle (Miller & Cragle, 1965).

Duodenal and ileal segments of the small intestine have been suggested as primary sites for Zn absorption. Infusion of ^{65}Zn into a ligated duodenal loop led to the highest ^{65}Zn recovery in blood, liver, kidneys and heart (Van Campen & Mitchell, 1965) or the whole body (Davies, 1980). With the use of an *in vivo* intestinal perfusion technique, it was found that the ileum had the highest capacity for Zn absorption (Antonson et al. 1979). Recently, Swinkels et al. (1994b) determined the site of apparent Zn absorption using a total digesta collection procedure and the indirect indicator method. The pigs used in this study had been depleted of Zn for a 5-week period by feeding an isolated soya protein semipurified diet. The diet contained 17 p.p.m. Zn and 3% cellulose. Following the depletion period, pigs were repleted by feeding the same diet supplemented with either 15 or 45 p.p.m. Zn as ZnSO$_4$. As shown in Table 2, jejunal and ileal segments of the small intestine were identified as the main sites of Zn absorption. The large intestine did not seem to contribute to the overall apparent Zn absorption (Swinkels et al. 1994b).

Partridge (1978) and Seal & Mathers (1989) examined the role of different amounts and sources of dietary fibre or non-starch polysaccharides on capacity and site of apparent Zn absorption. Using re-entrant cannulas in Zn adequate pigs given a casein semipurified diet containing about 50 p.p.m. Zn as ZnCO$_3$ and 3% cellulose, Partridge (1978) found that the large intestine was the primary site of Zn absorption. However, the gut sites anterior to the terminal ileum became the more important sites of apparent Zn absorption when 9% instead of 3% cellulose was included in the diet. The inclusion of a high level of cellulose in the diet also decreased overall apparent Zn absorption (Partridge, 1978). Feeding different non-starch polysaccharide sources to rats, Seal & Mathers (1989) found similar rates of Zn absorption from everted gut sacs of duodenal, ileal and colonic segments. Analysis of the everted gut sacs showed that the absorbed Zn was largely accumulated in all intestinal tissues, particularly in the duodenal segment, and that only a small amount of Zn was transferred across the serosal surface. The source of non-starch polysaccharides in the diet appeared to affect the capacity of the large intestine for Zn absorption. Rats previously fed diets containing high levels of pectin showed higher rates of Zn transfer by colonic tissues than rats previously fed diets with or without non-starch polysaccharides from wheat bran (Seal & Mathers, 1989).

From the above mentioned studies it appears that all segments of the gut have the capacity to absorb Zn. As stated by Seal & Mathers (1989), some of the differences observed among the studies may relate to experimental technique, animal species and dietary composition. With regard to the diet, it appears that the capacity of the large intestine to absorb Zn is expressed when a fibre or non-starch polysaccharide source is included in the diet (Grace, 1975; Bertoni et al. 1976; Partridge, 1978; Seal & Mathers, 1989). However, inclusion of high amounts of dietary fibre or non-starch polysaccharide (Partridge, 1978) or a high body need for Zn in the experimental animals (Swinkels et al. 1994b) appeared to increase the relative contribution of the small intestine to overall Zn absorption.

MINERALS INTERACTING WITH ZINC DURING ABSORPTION

Transport of Zn from the intestinal lumen into the enterocyte can be impeded by other minerals. Iron and Cd have been shown to inhibit Zn uptake from an open-ended duodenal loop in Fe deficient mice (Hamilton et al. 1978). An interaction between Zn and Fe was also observed from jejunal segments of Fe adequate rats. Addition of Zn to the perfusate reduced absorption of Fe by 34% (El-Shobaki & Srour, 1989). Substantial inhibition exerted by Fe on Zn absorption in men was reported by Solomons & Jacob (1981). The inhibition became more apparent with increasing ratios of dietary Fe to Zn. In a subsequent study, it was shown that there is a competition between Fe and Zn at intraluminal and intracellular sites (Solomons et al. 1983).

Copper uptake by intestinal brush border membrane vesicles of rats given either high levels of Zn, adequate Cu and Zn, or Cu deficient diets was studied by Fischer & L'Abbe (1985). They observed the highest Cu uptake by vesicles of rats given high levels of Zn (Fischer & L'Abbe, 1985). In pigs given a high dietary level of Cu, an increase in plasma Cu and in liver Cu and Zn contents was observed together with a concurrent decrease in plasma and liver Fe (Shurson et al. 1990). A tissue specific association between Zn and Cu was found by Swinkels et al. (1994a). In their Zn depletion–repletion study, they examined the bioavailability of Zn from different Zn sources by determining the Zn contents in liver, kidney, pancreas, brain and gut tissues. In the kidney, Cu was depleted and replaced with Zn, whereas kidney Fe levels and both Cu and Fe levels in other tissues were not affected by the tissue Zn status (Swinkels et al. 1994a).

Zinc retention was not different in pigs fed different levels of Ca (Morgan et al. 1969). In rats, a negative effect of Zn on Ca uptake by intestinal brush border membrane vesicles was only observed at high Zn to Ca ratios (Roth-Bassell & Clydesdale, 1991).

It has become clear that the presence of Cu and Fe in the chyme can affect Zn absorption. It is not clear whether both minerals actually interfere with cellular uptake of Zn or whether Cu and Fe interact with Zn to form non-absorbable complexes within the lumen of the gut. Interference with cellular uptake may occur when the minerals compete either for common transporters in the cell membrane or for common cytosolic proteins that are involved in intracellular transport of Zn. The interaction between Cu and Zn may result from competition for binding with metallothionein. Metallothionein has a high affinity for both Cu and Zn (Cousins, 1985), but thermodynamically binding with Cu is preferred to binding with Zn (Williams, 1984).

INTRINSIC FACTORS AFFECTING ZINC ABSORPTION

Zinc absorption is affected by dietary and endogenous factors. In rat jejunal segments, an excess of unhydrolysed glucose polymers and slowly absorbed sugars reduces Zn absorption (Wapnir et al. 1989). Absorption of Zn is also affected by the protein source (Miller & Jensen, 1966; O'Dell et al. 1972) as well as the protein concentration (Hunt & Larson, 1990; Hunt & Johnson, 1992).

The negative association of protein with Zn absorption may be due to other dietary components that contaminate the protein source. An allegedly higher Zn absorption from diets containing an animal protein source compared with diets based on cereal protein proved to be due to the presence of phytate in the cereal protein sources (Harmuth-Hoene & Meuser, 1987). Phytate and also fibre are two well known intrinsic factors which have been shown to affect the absorption of Zn negatively (Davies & Reid, 1979; Simons et al. 1990).

Endogenous factors may also interfere with Zn absorption from the gut lumen. A decrease in the plasma Zn level was observed by Sturniolo et al. (1991) after selective inhibition of gastric acid secretion in man. Their explanation was that a more alkaline environment in the stomach induces the formation of insoluble Zn compounds which cannot be absorbed further down the gut (Sturniolo et al. 1991).

As shown in the above mentioned studies both dietary and endogenous factors may affect the absorption of Zn to some extent but, unless dietary Zn intake is too low or too high for a prolonged period, the animal will be able to maintain Zn homeostasis.

BIOLOGICAL VALUE OF DIETARY COMPLEXES AND CHELATES OF ZINC

METAL COMPLEXES AND CHELATES

Metal complexes are compounds of a central metal atom together with ligands which contain at least one ligand atom with a free electron pair. Proteins and carbohydrates including their derivatives, lipids, and many synthetic compounds which contain an O, S or N atom may function as ligand (Kratzer & Vohra, 1986). The number of ligands that bind the metal atom usually exceeds the number expected from valency considerations (Smith, 1990). Binding of the ligand to the metal occurs through donation of the free electron pair of the ligand atom to the metal atom which acts as an electron acceptor (Fig. 4). This type of bond is referred to as a coordinate or dative bond. Coordinate bonds are mostly formed between the transitional elements and the electronegative atoms oxygen, sulphur and nitrogen (Kratzer & Vohra, 1986).

A metal chelate is a special form of a metal complex. A metal complex is considered a metal chelate when instead of one ligand atom, two or more atoms of the ligand donate their electron pairs to the metal in the formation of coordinate bonds. The chemical ring structure formed between the ligand and the metal resembles a pincer-like claw, for which the Greek word is 'Chēlē' (Fig. 4). Formation of metal complexes and chelates are both reversible processes. A continuous exchange of ligands occurs with a change in intraluminal or intracellular conditions. The free metal prefers those ligands with which it can form the

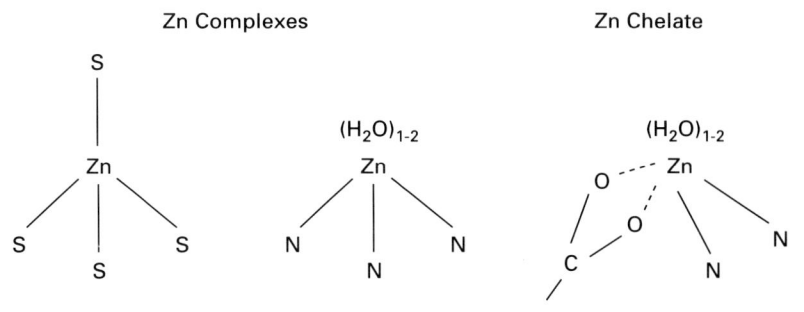

Fig. 4. Complexes and chelates of zinc as found at some known Zn sites of Zn-containing proteins (Williams, 1984). In Zn complexes only one atom of each participating ligand donates its free electron pair to Zn. this occurs at the structure sites of metallothionein (shown in part) and alcohol dehydrogenase (left figure) and at the active site of carbonic anhydrase (centre figure). In Zn chelates two or more atoms of each participating ligand donate their free electron pairs to Zn. This occurs at the active site of carboxypeptidases (right figure).

chemically most stable complexes or chelates under the given conditions (Kratzer & Vohra, 1986).

CONSIDERATIONS WHEN DETERMINING BIOAVAILABILITY OF ZINC

The term 'bioavailability' is generally used to describe the properties of absorption and utilization of nutrients (O'Dell, 1984). Only absorbed nutrients which can participate in the biological processes in the animal are considered utilizable, and thus bioavailable. Free Zn or Zn bound to low molecular weight ligands could have been absorbed and be present in body fluids and tissues, but it may not have been utilized. In determining the Zn availability of mineral sources, therefore, Zn metalloenzyme activities and metallothionein levels in body fluids and tissues are probably more reliable indicators of Zn availability than blood or tissue Zn concentrations.

A second important consideration in Zn availability studies is that animals maintain homeostasis by secreting part of the excess body Zn into the gut lumen. To prevent this mechanism from occurring, the animals should be depleted of Zn prior to the study. In a study conducted by Hallmans *et al.* (1987), bioavailability of Zn from one test food was estimated using two groups of rats with different needs for Zn to maintain Zn homeostasis. To increase the need for Zn in rats of the treatment group, anabolic processes that require Zn were stimulated by an intraperitoneal injection of a solution containing amino acids. The controls were injected with physiological saline. Results of the study showed that rats receiving the amino acid solution had a 40% increase in Zn absorption from the test food (Hallmans *et al.* 1987).

In addition to the use of animals with a high body need for Zn, the level of Zn used to replete the animals with Zn has to be carefully selected. If the levels of Zn are too low, the appetite of the depleted animals may remain depressed. On the other hand, if levels of Zn are too high, repletion of the depleted tissues with Zn may occur very rapidly. In both cases, the availability of Zn from the experimental diets cannot be determined. Thus, a dose response experiment has to be conducted to determine an optimum level of Zn. With an optimum Zn level, the change in Zn status of the depleted animal can be monitored over time as an indication of Zn availability in the experimental diets. Effectiveness and duration of Zn repletion can be determined by monitoring performance, repetitive measurement of serum or plasma Zn or metallothionein levels, the activities of metalloenzymes and serum mitogenic activity (Miller *et al.* 1968; Prasad *et al.* 1971; Swinkels *et al.* 1994*a*). Moreover, tissue levels of Zn and Zn metalloenzyme activities can be determined at one or more times during Zn repletion. Each of these measurements or several of them together may be referred to as a Zn bioassay (Wedekind & Baker, 1990).

A final consideration in mineral availability studies is that the mineral source used as the control may influence the outcome of the comparison. For example, based on tibia Zn, Wedekind & Baker (1990) estimated a 61% availability of Zn from ZnO relative to $ZnSO_4$. Both inorganic Zn sources are frequently used as control treatments in Zn bioassays.

AVAILABILITY OF ZINC COMPLEXED WITH PICOLINIC ACID

Picolinic acid (pyridine 2-carboxylic acid), a minor metabolite of tryptophan, was one of the first organic ligands studied for its possible promoting effect on Zn availability. The interest in picolinic acid originated from the finding that human milk provides more available Zn than cows' milk. In this research, Evans & Johnson (1979) characterized picolinic acid as a strong Zn binding ligand in human milk.

In subsequent studies, Evans & Johnson (1980*a–c*) showed that absorption of Zn was increased and growth rate stimulated in rats given diets supplemented with picolinic acid. In the later studies of Roth & Kirchgessner (1985) and Hill *et al.* (1986) these findings were not confirmed. Addition of picolinic acid to the diet improved neither the body gains, serum Zn levels, and Zn contents in the testes, femur and whole body of rats (Roth & Kirchgessner, 1985) nor bone Zn concentrations in pigs (Hill *et al.* 1986).

Using everted sacs of rat duodenum and ileum, Seal & Heaton (1983) studied the uptake of Zn with a variety of ligands including 2-picolinic and 4-picolinic acid. Salient features of this study are presented in Fig. 5. Adding 2-picolinic acid to the mixture improved the uptake of Zn in both everted duodenal and ileal sacs when compared with inorganic sulphate. Of the organic ligands tested, sulphate proved to be the most effective in enhancing Zn uptake. Compared to sulphate, the amino acids histidine and cysteine did improve Zn uptake from the ileal but not from the duodenal sac (Seal & Heaton, 1983). As part of the same study, the most promising ligands were included in the diets of intact rats housed in metabolism cages. The ligand 2-picolinic acid did improve apparent Zn absorption as expected from the results of the *in vitro* study. Zinc retention, however, was not improved primarily because of increased urinary Zn excretion (Seal & Heaton, 1983). These findings suggest that Zn was so tightly bound to 2-picolinic acid that it could not be utilized after being absorbed in complexed form, and therefore was excreted *via* the kidneys (Seal & Heaton, 1985).

AVAILABILITY OF ZINC COMPLEXED WITH METHIONINE OR OTHER AMINO ACIDS

Amino acids are frequently used as dietary ligands for synthesizing Zn complexes or chelates. Of all Zn amino acid complexes studied Zn methionine has received by far the most attention.

An improvement in Zn availability from the complex Zn methionine compared with $ZnSO_4$ was observed after measuring levels of Zn in the tibia of chicks (Wedekind *et al.* 1992). In other studies, Zn availability from Zn methionine, determined by measuring performance and serum Zn levels, was not different from an inorganic Zn salt in pigs (Kornegay & Thomas, 1975; Hill *et al.* 1986) or in heifers (Spears, 1989). Although apparent Zn absorption from Zn methionine was not different from ZnO, Spears (1989) did observe an increase in Zn retention of lambs fed Zn methionine. The increase in Zn

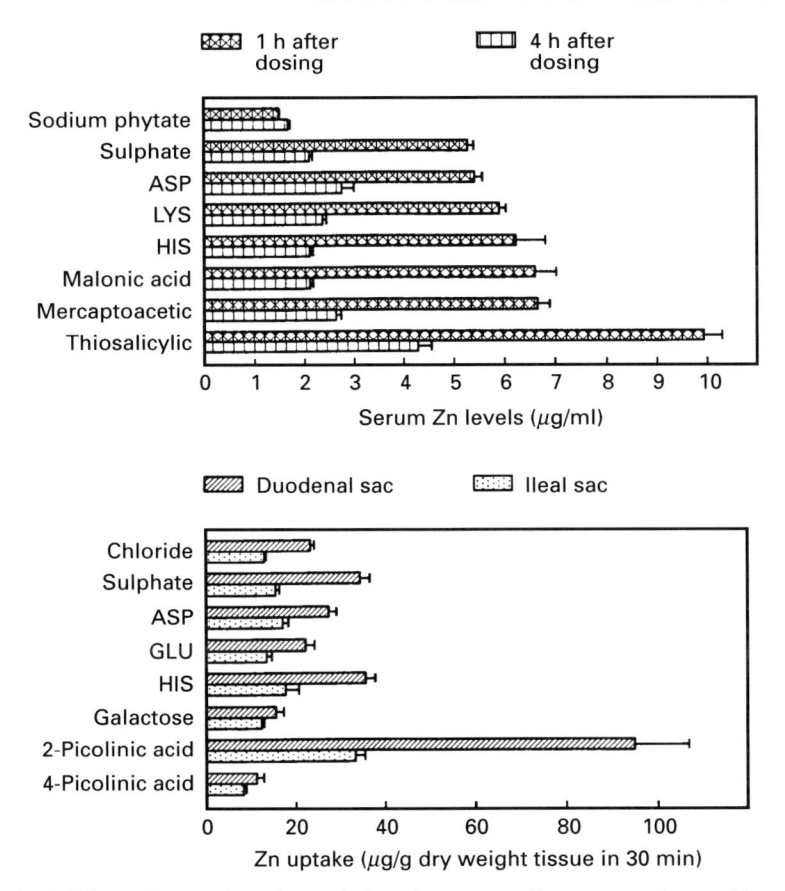

Fig. 5. Effect of inorganic and organic ligands on serum Zn concentrations and intestinal Zn uptake in rats. Serum Zn levels (top) were measured in rats 1 and 4 h after oral dosage of ZnSO$_4$ or mixtures of ZnCl$_2$ and different organic ligands containing 10 p.p.m. Zn (Giroux & Prakash, 1977). Intestinal Zn uptake (bottom) was measured using duodenal and ileal everted gut sacs after incubating with a buffer containing 0·0003 M-Zn as ZnCl$_2$, ZnSO$_4$ or mixtures of ZnCl$_2$ and organic ligands (Seal & Heaton, 1983). Organic ligands containing negatively charged carboxyl or thiol groups appeared to be the most effective in stimulating serum Zn concentrations and intestinal Zn uptake.

retention was caused by a slight reduction in urinary Zn excretion. A higher Zn retention with no difference in apparent Zn absorption (Spears, 1989) suggests that the complex Zn methionine provided more utilizable Zn than ZnO. In a 2-year study with beef cows and calves, Spears & Kegley (1991) observed a slight improvement in over-all performance with the use of Zn methionine and Mn methionine compared with ZnO and MnO. Addition of Zn methionine instead of ZnO to chick diets containing already adequate amounts of Zn slightly increased the content of Zn in the pancreas (Pimentel et al. 1991). Growth and concentrations of Zn in tibiotarsus and liver were not affected by the Zn source (Pimentel et al. 1991). Carcass quality of steers was improved with a Zn methionine supplement to the diet instead of ZnO (Greene et al. 1988). The better carcass quality was not associated with improved performance.

The influence of a variety of amino acids and their chemical homologues on Zn uptake from perfused jejunal, ileal and colonic segments of rats was studied by Wapnir & Stiel (1986). In the small intestine, perfusion with tryptophan, histidine, cysteine and proline achieved a higher Zn uptake compared with results after perfusion with their respective homologues tryptophol, imidazole, *N*-acetyl-L-cysteine and pyroglutamate. It appeared that both mediated and non-mediated transport mechanisms were involved in Zn uptake of the small intestine when the perfusate contained one of the amino acids. When the perfusate contained one of the amino acid homologues only non-mediated transport mechanisms appeared to be activated. In the colon, uptake of Zn was increased only with imidazole, the homologue of histidine, which may be explained by the high structural affinity of imidazole for Zn (Wapnir & Stiel, 1986). In humans, a 25% increase in serum Zn level was observed after ingestion of Zn as a Zn histidine complex compared with $ZnSO_4$ (Schölmerich *et al.* 1987). The apparent higher absorption of Zn histidine, however, was associated with increased urinary Zn excretion. Performance of grower pigs was not improved by supplementing Zn adequate diets with 1% histidine or 289 p.p.m. EDTA (Dahmer *et al.* 1972; Owen *et al.* 1973). However, inclusion of histidine appeared to alleviate skin lesions of the Zn deficient pigs used by Dahmer *et al.* (1972). This may indicate that dietary histidine did improve Zn availability. Alternatively, histidine may have stimulated the healing process independent of Zn. After measuring several serum and tissue variables in pigs that had been depleted of Zn, Swinkels *et al.* (1994*a*) reported similar availabilities of Zn from an amino acid chelate and $ZnSO_4$. However, the apparent Zn absorption coefficients were not consistent with the serum and tissue Zn measurements (Swinkels *et al.* 1994*b*).

Uptake, mucosal retention and absorption of Zn from $ZnCl_2$, $ZnCl_2$ with methionine, Zn complexes with methionine and $ZnCl_2$ with EDTA were studied by Hempe & Cousins (1989) with the aid of ligated rat duodenal loops. After 60 min incubation, Zn uptake and absorption were lowest for Zn methionine and the Zn EDTA mixture. It was suggested that the low Zn absorption was associated with reduced binding of Zn to an unidentified low molecular weight protein present in the mucosa (Hempe & Cousins, 1989). Later on, the protein was assigned a possible role in the transcellular transport of Zn as shown in Fig. 3. Uptake of Zn, determined with everted duodenal sacs of pigs, was not different among $ZnSO_4$, Zn methionine and Zn lysine (Hill *et al.* 1987).

The hypothesis that organic ligands are actively involved in Zn absorption was examined by Giroux & Prakash (1977). Salient features of the results of this study are shown in Fig. 5. In their study, Giroux & Prakash (1977) gave different ligand and $ZnSO_4$ mixtures (10 p.p.m. Zn) by stomach tube after a 24 h fast. One and 4 h after force feeding the rats, they determined Zn absorption by measuring serum Zn levels. As shown in Fig. 5, ligands containing thiol and carboxylic acid groups, such that formation of five- or six-membered Zn chelates could occur, proved to be the most effective in increasing levels of serum Zn 1 h after feeding. A 1:1 mixture of phytate and $ZnSO_4$ reduced serum Zn levels about 3·5 times compared with the control $ZnSO_4$. Of the amino acids tested, a 2:1 mixture of glycine and $ZnSO_4$ was most effective. Increases in serum Zn levels observed with the amino acids lysine, histidine and cysteine were only slightly lower. Serum Zn levels, measured 4 h after feeding, returned to the $ZnSO_4$ control value for most amino acid and $ZnSO_4$ mixtures. In contrast to the amino acid ligands, changes in levels of serum Zn observed with phytate and several ligands containing thiol or carboxylic acid groups 1 h after feeding were still maintained 4 h after feeding (Giroux & Prakash, 1977). Also, in the study of Seal & Heaton (1983), a ligand containing a carboxylic acid group, 2-picolinic acid, improved the uptake of Zn when compared with inorganic sulphate (Fig. 5).

It is difficult to deduce a general mode of action for absorption of organic Zn complexes or chelates from the studies of Giroux & Prakash (1977) and Seal & Heaton (1983). In both studies it appears, however, that absorption of Zn may be enhanced in the presence of organic ligands containing highly negatively charged atom groups that form negatively charged Zn complexes or chelates. This observation is supported by the findings of Tacnet *et al.* (1990) in a study using intestinal membrane vesicles of rats. They also found a considerable increase in vesicular Zn uptake when $ZnSO_4$ was substituted for $Zn(SCN)_4^{2-}$, a highly negatively charged anion.

CONCLUSIONS

Studies on the mechanisms underlying Zn absorption have suggested that Zn may be absorbed as part of an intact complex or chelate formed between Zn and one or more organic ligands. To date, this hypothesis has been examined primarily using Zn complexed to either picolinate or methionine. Although both organic Zn forms have been shown to affect Zn absorption, a consistent improvement in availability of Zn from these sources was not found. More basic research focusing on conditions within the gut lumen and on mechanisms underlying Zn absorption is needed to elucidate which characteristics of a dietary Zn form are essential in determining its biological value. Within the gut lumen, dietary Zn complexes or chelates should be stable enough to withstand luminal conditions in all gut segments prior to the site of absorption. In order to be utilized, the stability of the dietary organic Zn form should be low enough to allow for the release or donation of Zn either during or after Zn absorption. Previous research has shown that dietary ligands may affect both absorption and utilization of Zn. A complete understanding of the mechanisms underlying these processes, however, is necessary to determine the specific importance of ligands. When these mechanisms are elucidated, it should be possible to develop dietary Zn complexes or chelates with a high biological value more effectively.

REFERENCES

Aggett, P. J. (1991). The assessment of zinc status: a personal view. *Proceedings of the Nutrition Society* **50**, 9–17.
Agricultural Research Council. (1981). *The Nutrient Requirements of Pigs.* Farnham Royal: Commonwealth Agricultural Bureaux.
Antonson, D. L., Barak, A. J. & Vanderhoof, J. A. (1979). Determination of the site of zinc absorption in rat small intestine. *Journal of Nutrition* **109**, 142–147.
Bartholomew, M. E., Tupper, R. & Wormall, A. (1959). Incorporation of [65]Zn in the sub-cellular fractions of the liver and spontaneously occurring mammary tumours of mice after the injection of zinc-glycine containing [65]Zn. *Biochemical Journal* **73**, 256–261.
Bertoni, G., Watson, M. J., Savage, G. P. & Armstrong, D. G. (1976). [The movements of minerals in the digestive tract of dry and lactating Jersey cows. 2. Net movements of Cu, Fe, Mn and Zn.] *Zootecnica e Nutrizione Animale* **2**, 185–191.
Blakeborough, P. & Salter, D. N. (1987). The intestinal transport of zinc studied using brush-border-membrane vesicles from the piglet. *British Journal of Nutrition* **57**, 45–55.
Bremner, I. (1983). The roles of metallothionein in the metabolism of copper and zinc. *Annual Report of Studies in Animal Nutrition and Allied Sciences, Rowett Research Institute* **39**, 13–28.
Bremner, I. & Beattie, J. H. (1990). Metallothionein and the trace minerals. *Annual Review of Nutrition* **10**, 63–83.
Bronner, F. (1987). Intestinal calcium absorption: mechanisms and applications. *Journal of Nutrition* **117**, 1347–1352.
Cassens, R. G., Hoekstra, W. G., Faltin, E. C. & Briskey, E. J. (1967). Zinc content and subcellular distribution in red vs. white porcine skeletal muscle. *American Journal of Physiology* **212**, 688–692.
Coppen, D. E. & Davies, N. T. (1987). Studies on the effects of dietary zinc dose on [65]Zn absorption in vivo and on the effects of Zn status on [65]Zn absorption and body loss in young rats. *British Journal of Nutrition* **57**, 35–44.
Cossack, Z. T. (1986). Somatomedin-C and zinc status in rats as affected by Zn, protein and food intake. *British Journal of Nutrition* **56**, 163–169.
Cousins, R. J. (1985). Absorption, transport, and hepatic metabolism of copper and zinc: special reference to metallothionein and ceruloplasmin. *Physiological Reviews* **65**, 238–309.
Cousins, R. J. (1989). Theoretical and practical aspects of zinc uptake and absorption. In *Mineral Absorption in the Monogastric GI Tract*, pp. 3–12 [F. R. Dintzis and J. A. Laszlo, editors]. New York, NY: Plenum Press.
Cousins, R. J. & Lee-Ambrose, L. M. (1992). Nuclear zinc uptake and interactions and metallothionein gene expression are influenced by dietary zinc in rats. *Journal of Nutrition* **122**, 56–64.
Crofton, R. W., Clapham, M., Humphries, W. R., Aggett, P. J. & Mills, C. F. (1983). Leucocyte and tissue zinc concentrations in the growing pig. *Proceedings of the Nutrition Society* **42**, 128A (Abstr.).

Dahmer, E. J., Coleman, B. W., Grummer, R. H. & Hoekstra, W. G. (1972). Alleviation of parakeratosis in zinc deficient swine by high levels of dietary histidine. *Journal of Animal Science* **35**, 1181–1189.

Davies, N. T. (1980). Studies on the absorption of zinc by rat intestine. *British Journal of Nutrition* **43**, 189–203.

Davies, N. T. & Reid, H. (1979). An evaluation of the phytate, zinc, copper, iron and manganese contents of, and Zn availability from, soya-based textured-vegetable-protein meat-substitutes or meat-extenders. *British Journal of Nutrition* **41**, 579–589.

Dørup, I. & Clausen, T. (1991). Effects of magnesium and zinc deficiencies on growth and protein synthesis in skeletal muscle and the heart. *British Journal of Nutrition* **66**, 493–504.

Dørup, I., Flyvbjerg, A., Everts, M. E. & Clausen, T. (1991). Role of insulin-like growth factor-1 and growth hormone in growth inhibition induced by magnesium and zinc deficiencies. *British Journal of Nutrition* **66**, 505–521.

Droke, E. A., Spears, J. W., Armstrong, J. D., Kegley, E. B. & Simpson, R. B. (1993). Dietary zinc affects serum concentrations of insulin and insulin-like growth factor I in growing lambs. *Journal of Nutrition* **123**, 13–19.

El-Shobaki, F. A. & Srour, M. G. (1989). The influence of ascorbic acid and lactose on the interaction of iron with each of cobalt and zinc during intestinal absorption. *Zeitschrift für Ernährungswissenschaft* **28**, 310–315.

Evans, G. W. & Johnson, E. C. (1980a). Zinc absorption in rats fed a low-protein diet and a low-protein diet supplemented with tryptophan or picolinic acid. *Journal of Nutrition* **110**, 1076–1080.

Evans, G. W. & Johnson, E. C. (1980b). Zinc concentration of liver and kidneys from rat pups nursing dams fed supplemental zinc dipicolinate or zinc acetate. *Journal of Nutrition* **110**, 2121–2124.

Evans, G. W. & Johnson, E. C. (1980c). Growth stimulating effect of picolinic acid added to rat diets. *Proceedings of the Society for Experimental Biology and Medicine* **165**, 457–461.

Evans, G. W. & Johnson, P. E. (1979). Purification and characterization of a zinc-binding ligand in human milk. *Federation Proceedings* **38**, 703.

Fairweather-Tait, S. J., Jackson, M. J., Fox, T. E., Wharf, S. G., Eagles, J. & Croghan, P. C. (1993). The measurement of exchangeable pools of zinc using the stable isotope ^{70}Zn. *British Journal of Nutrition* **70**, 221–234.

Fischer, P. W. F. & L'Abbe, M. R. (1985). Copper transport by intestinal brush border membrane vesicles from rats fed high zinc or copper deficient diets. *Nutrition Research* **5**, 759–767.

Flanagan, P. R., Haist, J. & Valberg, L. S. (1983). Zinc absorption, intraluminal zinc and intestinal metallothionein levels in zinc-deficient and zinc-repleted rodents. *Journal of Nutrition* **113**, 962–972.

Galdes, A. & Vallee, B. L. (1983). Categories of zinc metalloenzymes. In *Zinc and its Role in Biology and Nutrition (Metal Ions in Biological Systems* vol. 15), pp. 1–54 [H. Sigel and A. Sigel, editors]. New York: Marcel Dekker, Inc.

Giroux, E. & Prakash, N. J. (1977). Influence of zinc-ligand mixtures on serum zinc levels in rats. *Journal of Pharmaceutical Sciences* **66**, 391–395.

Giugliano, R. & Millward, D. J. (1984). Growth and zinc homeostasis in the severely Zn-deficient rat. *British Journal of Nutrition* **52**, 545–560.

Giugliano, R. & Millward, D. J. (1987). The effects of severe zinc deficiency on protein turnover in muscle and thymus. *British Journal of Nutrition* **57**, 139–155.

Golden, B. E. (1989). Zinc in cell division and tissue growth: physiological aspects. In *Zinc in Human Biology*, pp. 119–128 [C. F. Mills, editor]. London: Springer-Verlag.

Grace, N. D. (1975). Studies on the flow of zinc, cobalt, copper and manganese along the digestive tract of sheep given fresh perennial ryegrass, or white or red clover. *British Journal of Nutrition* **34**, 73–82.

Grace, N. D. (1983). Amounts and distribution of mineral elements associated with fleece-free empty body weight gains in the grazing sheep. *New Zealand Journal of Agricultural Research* **26**, 59–70.

Greene, L. W., Lunt, D. K., Byers, F. M., Chirase, N. K., Richmond, C. E., Knutson, R. E. & Schelling, G. T. (1988). Performance and carcass quality of steers supplemented with zinc oxide or zinc methionine. *Journal of Animal Science* **66**, 1818–1823.

Gupta, R. P., Verma, P. C. & Gupta, R. K. P. (1985). Experimental zinc deficiency in guinea-pigs: clinical signs and some haematological studies. *British Journal of Nutrition* **54**, 421–428.

Hallmans, G., Nilsson, U., Sjöstrom, R., Wetter, L. & Wing, K. (1987). The importance of the body's need for zinc in determining Zn availability in food: a principle demonstrated in the rat. *British Journal of Nutrition* **58**, 59–64.

Hambidge, K. M., Casey, C. E. & Krebs, N. F. (1986). Zinc. In *Trace Elements in Human and Animal Nutrition*, 5th Edn, vol. 2, pp. 1–137 [W. Mertz, editor]. London: Academic Press.

Hamilton, D. L., Bellamy, J. E. C., Valberg, J. D. & Valberg, L. S. (1978). Zinc, cadmium, and iron interactions during intestinal absorption in iron-deficient mice. *Canadian Journal of Physiology and Pharmacology* **56**, 384–389.

Harmuth-Hoene, A. E. & Meuser, F. (1987). [Biological availability of zinc in whole-grain cereal products with various phytate contents.] *Zeitschrift für Ernährungswissenschaft* **26**, 250–267.

Hempe, J. M. & Cousins, R. J. (1989). Effect of EDTA and zinc-methionine complex on zinc absorption by rat intestine. *Journal of Nutrition* **119**, 1179–1187.

Hempe, J. M. & Cousins, R. J. (1992). Cysteine-rich intestinal protein and intestinal metallothionein: an inverse relationship as a conceptual model for zinc absorption in rats. *Journal of Nutrition* **122**, 89–95.

Hill, D. A., Peo, E. R. & Lewis, A. J. (1987). Influence of picolinic acid on the uptake of ^{65}Zn-amino acid complexes by the everted rat gut. *Journal of Animal Science* **65**, 173–178.

Hill, D. A., Peo, E. R., Lewis, A. J. & Crenshaw, J. D. (1986). Zinc-amino acid complexes for swine. *Journal of Animal Science* **63**, 121–130.

Hoekstra, W. G., Faltin, E. C., Lin, C. W., Roberts, H. F. & Grummer, R. H. (1967). Zinc deficiency in reproducing gilts fed a diet high in calcium and its effect on tissue zinc and blood serum alkaline phosphatase. *Journal of Animal Science* **26**, 1348–1357.

Hoekstra, W. G., Lewis, P. K., Phillips, P. H. & Grummer, R. H. (1956). The relationship of parakeratosis, supplemental calcium and zinc to the zinc content of certain body components of swine. *Journal of Animal Science* **15**, 752–764.

Hunt, J. R. & Johnson, L. K. (1992). Dietary protein, as egg albumen: effects on bone composition, zinc bioavailability and zinc requirements of rats, assessed by a modified broken-line model. *Journal of Nutrition* **122**, 161–169.

Hunt, J. R. & Larson, B. J. (1990). Meal protein and zinc levels interact to influence zinc retention by the rat. *Nutrition Research* **10**, 697–705.

Jackson, M. J. (1989). Physiology of zinc: general aspects. In *Zinc in Human Biology*, pp. 1–14 [C. F. Mills, editor]. London: Springer-Verlag.

Johanning, G. L., Browning, J. D., Bobilya, D. J., Veum, T. L. & O'Dell, B. L. (1990). Effect of zinc deficiency and food restriction in the pig on erythrocyte fragility and plasma membrane composition. *Nutrition Research* **10**, 1463–1471.

Kernkamp, H. C. H. & Ferrin, E. F. (1953). Parakeratosis in swine. *Journal of American Veterinary Medical Association* **123**, 217–220.

King, J. C. (1990). Assessment of zinc status. *Journal of Nutrition* **120**, 1474–1479.

Kornegay, E. T. & Thomas, H. R. (1975). Zinc-proteinate supplement studied. *Hog Farm Management* (August), 50–51.

Kratzer, F. H. & Vohra, P. (1986). *Chelates in Nutrition*. Boca Raton, FL: CRC Press, Inc.

Menard, M. P. & Cousins, R. J. (1983). Zinc transport by brush border membrane vesicles from rat intestine. *Journal of Nutrition* **113**, 1434–1442.

Menard, M. P., McCormick, C. C. & Cousins, R. J. (1981). Regulation of intestinal metallothionein biosynthesis in rats by dietary zinc. *Journal of Nutrition* **111**, 1353–1361.

Miller, E. R., Luecke, R. W., Ullrey, D. E., Baltzer, B. V., Bradley, B. L. & Hoefer, J. A. (1968). Biochemical, skeletal and allometric changes due to zinc deficiency in the baby pig. *Journal of Nutrition* **95**, 278–286.

Miller, J. K. & Cragle, R. G. (1965). Gastrointestinal sites of absorption and endogenous secretion of zinc in dairy cattle. *Journal of Dairy Science* **48**, 370–373.

Miller, J. K. & Jensen, L. S. (1966). Effect of dietary protein source on zinc absorption and excretion along the alimentary tracts of chicks. *Poultry Science* **45**, 1051–1053.

Miller, W. J., Neathery, M. W., Gentry, R. P., Blackmon, D. M. & Stake, P. E. (1974). Adaptations in zinc metabolism by lactating cows fed a low-zinc practical-type diet. In *Trace Element Metabolism in Animals-2*, pp. 550–552 [W. G. Hoekstra, J. W. Suttie, H. E. Ganther and W. Mertz, editors]. Baltimore, MD: University Park Press.

Morgan, D. P., Young, E. P., Earle, I. P., Davey, R. J. & Stevenson, J. W. (1969). Effects of dietary calcium and zinc on calcium, phosphorus and zinc retention in swine. *Journal of Animal Science* **29**, 900–905.

National Research Council, Subcommittee on Zinc. (1979). *Zinc*. Baltimore, MD: University Park Press.

National Research Council. (1988). *Nutrient Requirements of Swine*, 10th ed. Washington, DC: National Academy Press.

O'Dell, B. L. (1984). Bioavailability of trace elements. *Nutrition Reviews* **42**, 301–308.

O'Dell, B. L., Burpo, C. E. & Savage, J. E. (1972). Evaluation of zinc availability in foodstuffs of plant and animal origin. *Journal of Nutrition* **102**, 653–660.

Oestreicher, P. & Cousins, R. J. (1989). Zinc uptake by basolateral membrane vesicles from rat small intestine. *Journal of Nutrition* **119**, 639–646.

Owen, A. A., Peo, E. R., Cunningham, P. J. & Moser, B. D. (1973). Effect of EDTA on utilization of dietary zinc by G-F swine. *Journal of Animal Science* **37**, 470–478.

Pallauf, J., Höhler, D. & Rimbach, G. (1992). [Effect of microbial phytase supplementation to a maize-soya diet on the apparent absorption of Mg, Fe, Cu, Mn and Zn and parameters of Zn status in piglets]. *Journal of Animal Physiology and Nutrition* **68**, 1–9.

Partridge, I. G. (1978). Studies on digestion and absorption in the intestines of growing pigs. 4. Effects of dietary cellulose and sodium levels on mineral absorption. *British Journal of Nutrition* **39**, 539–545.

Pimentel, J. L., Cook, M. E. & Greger, J. L. (1991). Bioavailability of zinc-methionine for chicks. *Poultry Science* **70**, 1637–1639.

Prasad, A. S. (1988). Zinc in growth and development and spectrum of human zinc deficiency. *Journal of the American College of Nutrition* **7**, 377–384.

Prasad, A. S., Oberleas, D., Miller, E. R. & Luecke, R. W. (1971). Biochemical effects of zinc deficiency: changes in activities of zinc-dependent enzymes and ribonucleic acid and deoxyribonucleic acid content of tissues. *Journal of Laboratory and Clinical Medicine* **77**, 144–152.

Richards, M. P. (1989). Recent developments in trace element metabolism and function: role of metallothionein in copper and zinc metabolism. *Journal of Nutrition* **119**, 1062–1070.

Richards, M. P. & Cousins, R. J. (1975). Mammalian zinc homeostasis: requirement for RNA and metallothionein synthesis. *Biochemical and Biophysical Research Communications* **64**, 1215–1223.

Roth, H. P. & Kirchgessner, M. (1985). Utilization of zinc from picolinic or citric acid complexes in relation to dietary protein source in rats. *Journal of Nutrition* **115**, 1641–1649.

Roth-Bassell, H. A. & Clydesdale, F. M. (1991). The influence of zinc, magnesium, and iron on calcium uptake in brush border membrane vesicles. *Journal of the American College of Nutrition* **10**, 44–49.

Schölmerich, J., Freudemann, A., Köttgen, E., Wietholz, H., Steiert, B., Löhle, B., Häussinger, D. & Gerok, W. (1987). Bioavailability of zinc from zinc-histidine complexes. I. Comparison with zinc sulfate in healthy men. *American Journal of Clinical Nutrition* **45**, 1480–1486.

Seal, C. J. & Heaton, F. W. (1983). Chemical factors affecting the intestinal absorption of zinc in vitro and in vivo. *British Journal of Nutrition* **50**, 317–324.

Seal, C. J. & Heaton, F. W. (1985). Effect of dietary picolinic acid on the metabolism of exogenous and endogenous zinc in the rat. *Journal of Nutrition* **115**, 986–993.

Seal, C. J. & Mathers, J. C. (1989). Intestinal zinc transfer by everted gut sacs from rats given diets containing different amounts and types of dietary fibre. *British Journal of Nutrition* **62**, 151–163.

Seguin, C. & Hamer, D. H. (1987). Regulation in vitro of metallothionein gene binding factors. *Science* **235**, 1383–1387.

Shurson, G. C., Ku, P. K., Waxler, G. L., Yokoyama, M. T. & Miller, E. R. (1990). Physiological relationships between microbiological status and dietary copper levels in the pig. *Journal of Animal Science* **68**, 1061–1071.

Simons, P. C. M., Versteegh, H. A. J., Jongbloed, A. W., Kemme, P. A., Slump, P., Bos, K. D., Wolters, M. G. E., Beudeker, R. F. & Verschoor, G. J. (1990). Improvement of phosphorus availability by microbial phytase in broilers and pigs. *British Journal of Nutrition* **64**, 525–540.

Smith, D. W. (1990). *Inorganic Substances.* Cambridge: University Press.

Solomons, N. W. & Jacob, R. A. (1981). Studies on the bioavailability of zinc in humans: effects of heme and nonheme iron on the absorption of zinc. *American Journal of Clinical Nutrition* **34**, 475–482.

Solomons, N. W., Pineda, O., Viteri, F. & Sandstead, H. H. (1983). Studies on the bioavailability of zinc in humans: mechanism of the intestinal interaction of nonheme iron and zinc. *Journal of Nutrition* **113**, 337–349.

Southon, S., Gee, J. M., Bayliss, C. E., Wyatt, G. M., Horn, N. & Johnson, I. T. (1986). Intestinal microflora, morphology and enzyme activity in zinc-deficient and Zn-supplemented rats. *British Journal of Nutrition* **55**, 603–611.

Spears, J. W. (1989). Zinc methionine for ruminants: relative bioavailability of zinc in lambs and effects of growth and performance of growing heifers. *Journal of Animal Science* **67**, 835–843.

Spears, J. W. & Kegley, E. B. (1991). Effect of zinc and manganese methionine on performance of beef cows and calves. *Journal of Animal Science* **69** Suppl. 1, 59.

Spears, J. W., Kegley, E. B. & Ward, J. D. (1991). Bioavailability of organic, inorganic trace minerals explored. *Feedstuffs* (November), 12–20.

Spray, C. M. & Widdowson, E. M. (1950). The effect of growth and development on the composition of mammals. *British Journal of Nutrition* **4**, 332–352.

Starcher, B. C., Glauber, J. G. & Madaras, J. G. (1980). Zinc absorption and its relationship to intestinal metallothionein. *Journal of Nutrition* **110**, 1391–1397.

Steel, L. & Cousins, R. J. (1985). Kinetics of zinc absorption by luminally and vascularly perfused rat intestine. *American Journal of Physiology* **248**, G46–G53.

Stryer, L. (1988). *Biochemistry*, 3rd ed. New York, NY: W. H. Freeman and Co.

Sturniolo, G. C., Montino, C., Rossetto, L., Martin, A., D'Inca, R., D'Odorico, A. & Naccarato, R. (1991). Inhibition of gastric acid secretion reduces zinc absorption in man. *Journal of American College of Nutrition* **10**, 372–375.

Swinkels, J. W. G. M., Kornegay, E. T., Zhou, W., Lindemann, M. D., Webb, K. E. & Verstegen, M. W. A. (1994a). Effectiveness of a zinc amino acid chelate and zinc sulfate in repleting serum and soft tissue zinc pools when fed to zinc depleted pigs. *Journal of Animal Science*, submitted.

Swinkels, J. W. G. M., Kornegay, E. T., Zhou, W., Lindemann, M. D., Webb, K. E. & Verstegen, M. W. A. (1994b). In vivo assessment of rate and site of apparent zinc, copper and iron absorption as affected by Zn source using Zn depleted pigs. *Journal of Animal Science*, submitted.

Swinkels, J. W. G. M., Kornegay, E. T., Zhou, W., Wong, E. A., Lindemann, M. D. & Verstegen, M. W. A. (1994c). Serum mitogenic activity, total pituitary RNA and growth hormone mRNA concentrations of experimentally zinc depleted pigs. *Journal of Nutrition* (In press).

Tacnet, F., Watkins, D. W. & Ripoche, P. (1990). Studies of zinc transport into brush-border membrane vesicles isolated from pig small intestine. *Biochimica et Biophysica Acta* **1024**, 323–330.

Todd, W. R., Elvehjem, C. A. & Hart, E. B. (1934). Zinc in the nutrition of the rat. *American Journal of Physiology* **107**, 146–156.

Tucker, H. F. & Salmon, W. D. (1955). Parakeratosis or zinc deficiency disease in the pig. *Proceedings of the Society for Experimental Biology and Medicine* **88**, 613–616.

Underwood, E. J. (1977). *Trace Elements in Human and Animal Nutrition*, 4th ed. London: Academic Press.

Van Campen, D. R. & Mitchell, E. A. (1965). Absorption of Cu^{64}, Zn^{65}, Mo^{99}, and Fe^{59} from ligated segments of the rat gastrointestinal tract. *Journal of Nutrition* **86**, 120–124.

Vasak, M. & Kägi, J. H. R. (1983). Spectroscopic properties of metallothionein. In *Zinc and its Role in Biology and Nutrition* (*Metal Ions in Biological Systems* vol. 15), pp. 213–273 [H. Sigel and A. Sigel, editors]. New York, NY: Marcel Dekker, Inc.

Verma, P. C., Gupta, R. P., Sadana, J. R. & Gupta, R. K. P. (1988). Effect of experimental zinc deficiency and repletion on some immunological variables in guinea-pigs. *British Journal of Nutrition* **59**, 149–154.

Wapnir, R. A., Garcia-Aranda, J. A., Mevorach, D. E. K. & Lifshitz, F. (1985). Differential absorption of zinc and low-molecular-weight ligands in the rat gut in protein-energy malnutrition. *Journal of Nutrition* **115**, 900–908.

Wapnir, R. A. & Stiel, L. (1986). Zinc intestinal absorption in rats: specificity of amino acids as ligands. *Journal of Nutrition* **116**, 2171–2179.

Wapnir, R. A., Stiel, L. & Lee, S.-Y. (1989). Zinc intestinal absorption: effect of carbohydrates. *Nutrition Research* **9**, 1277–1284.

Wedekind, K. J. & Baker, D. H. (1990). Zinc bioavailability in feed-grade sources of zinc. *Journal of Animal Science* **68**, 684–689.

Wedekind, K. J., Hortin, A. E. & Baker, D. H. (1992). Methodology for assessing zinc bioavailability: efficacy estimates for zinc-methionine, zinc sulfate and zinc oxide. *Journal of Animal Science* **70**, 178–187.

Williams, R. J. P. (1984). Zinc: what is its role in biology? *Endeavour* **8**, 65–70.

Williams, R. J. P. (1989). An introduction to the biochemistry of zinc. In *Zinc in Human Biology*, pp. 15–31 [C. F. Mills, editor]. London: Springer-Verlag.

Wu, F. Y.-H. & Wu, C.-W. (1983). The role of zinc in DNA and RNA polymerases. In *Zinc and its Role in Biology and Nutrition* (*Metal Ions in Biological Systems* vol. 15), pp. 157–192 [H. Sigel and A. Sigel, editors]. New York, NY: Marcel Dekker, Inc.

Yasodhara, P., Ramaraju, L. A. & Raman, L. (1991). Trace minerals in pregnancy. 1. Copper and zinc. *Nutrition Research* **11**, 15–21.

Zhou, W., Kornegay, E. T., Lindemann, M. D., Swinkels, J. W. G. M., Welten, M. K. & Wong, E. A. (1994). Stimulation of growth by intravenous injection of copper in weanling pigs. *Journal of Animal Science* (In press).

ZINC NUTRITION IN DEVELOPING COUNTRIES

ROSALIND S. GIBSON

Division of Applied Human Nutrition, University of Guelph, Guelph, Ontario, N1G 2W1, Canada

CONTENTS

INTRODUCTION	151
AETIOLOGY OF ZINC DEFICIENCY IN DEVELOPING COUNTRIES	153
DIETARY FACTORS: LOW INTAKE AND POOR BIOAVAILABILITY OF DIETARY ZINC	153
EXCESSIVE LOSSES	154
HIGH PHYSIOLOGICAL REQUIREMENTS	154
ZINC INTAKES IN RELATION TO ESTIMATED REQUIREMENTS	155
LABORATORY ASSESSMENT OF ZINC STATUS	157
BIOCHEMICAL INDICES OF ZINC STATUS	157
PHYSIOLOGICAL FUNCTIONAL INDICES OF ZINC STATUS	160
ZINC DEFICIENCY THROUGHOUT THE LIFE CYCLE	160
INFANCY AND CHILDHOOD	160
PREGNANCY	164
LACTATION	166
NUTRITION INTERVENTION STRATEGIES TO PREVENT ZINC DEFICIENCY IN DEVELOPING COUNTRIES	167
REFERENCES	168

INTRODUCTION

Recently the United Nations has urged that priority should be given to developing programmes in less industrialized countries to prevent deficiencies of iodine, vitamin A, and Fe (United Nations, 1991). Nutritional Fe deficiency is associated with plant based diets which contain high levels of dietary fibre and phytate, components known to inhibit non-haem Fe absorption, and low levels of flesh foods, rich sources of readily available haem iron (Monsen, 1988). Such plant based diets will also induce Zn deficiency. The consequences of Zn deficiency on human health in developing countries, however, have not yet been recognized. This is unfortunate because even mild Zn deficiency may contribute to pregnancy complications, low birth weight, impaired immune competence, maternal and infant mortality and morbidity, and growth failure in infancy and childhood (Swanson & King, 1987; Hambidge, 1989; National Academy of Sciences, 1991; United Nations, 1991). Hence Zn deficiency may have far reaching consequences on maternal, infant, and child health in many developing countries.

Table 1. *Zinc†, phytic acid† and [phytate]:[zinc] molar ratios of some foods and composite dishes consumed in Ghana and Malawi*

Food, and scientific name or recipe	Zn	Phy	Phy:Zn	% H$_2$O
Cereals				
Maize flour, 95% extraction (*Zea mays* L.)	2·2	792	36	10
Maize flour, 65% extraction	0·9	211	23	10
Maize bran	3·7	1089	29	10
Maize dough	1·4	n.a.	n.a.	50
Sorghum flour (*Sorghum bicolor* (L.) Moench)	1·4	446	32	10
Rice (*Oryza sativa*)	1·6	n.a.	n.a.	10
Legumes				
Ground nuts, boiled (*Arachis hypogaea* L.)	1·4	505	35	49
Ground nuts, flour	2·8	1297	45	8
Pigeon peas, fresh (*Cajanus cajan* (L.) Millsp.)	0·9	255	27	63
Pigeon peas, dry	2·2	727	33	8
Kidney beans, fresh (*Phaseolus vulgaris* L.)	1·5	557	36	52
Cowpeas, boiled (*Vigna unguiculata* (L.) Walp.)	1·0	349	37	68
Lima beans, fresh (*Phaseolus lunatus* L.)	1·5	238	16	66
Bengal beans, fresh (*Stizolobium aterrimum* Piper & Tracey)	1·0	166	17	68
Vegetables (boiled)				
Pumpkin leaf (*Cucurbita maxima* Duch. ex Lam.)	0·7	34	5	89
Chinese cabbage (*Brassica chinensis* L.)	0·7	5	1	94
Okra leaf (*Hibiscus esculentus* (L.))	1·8	97	5	79
Okra (*Hibiscus esculentus* (L.))	0·5	13	3	91
Cassava leaf (*Manihot esculenta* Crantz)	1·2	42	3	78
Cocoyam leaves (*Xanthosoma* sp. Schott.)	0·6	19	3	88
Amaranth leaves (*Amaranth* sp. L.)	0·3	n.a.	n.a.	93
Roots and plantain (boiled)				
Sweet potato (*Ipomoea batatas* L.)	0·2	10	5	70
Yam (*Dioscorea* sp. L.)	0·3	50	13	68
Cocoyam (*Xanthosoma* sp.)	0·5	37	7	60
Cassava (*Manihot* sp.)	0·3	54	18	65
Cassava dough, fermented	0·4	48	12	51
Gari: dry fermented cassava, not boiled	0·7	51	4	12
Plantain, ripe (*Musa paradisiaca* L.)	0·2	0	0	73
Plantain, unripe (*Musa paradisiaca* L.)	0·2	1	1	65
Water yam (*Dioscorea alata* L.)	0·2	26	16	72
Fruits				
Avocado pear (*Persea americana* Mill.)	0·3	11	3	78
Banana (*Musa paradisiaca* L.)	0·2	22	9	72
Mango, raw (*Mangifera indica* L.)	0·1	25	23	82
Composite dishes – home-prepared snacks				
Chitumbuwa (mixture of water, maize flour and pounded bananas formed into a round cake and fried in oil)	1·2	504	42	30
African bread (mixture of water, maize flour and bananas formed into a cake, wrapped in banana leaves and boiled until cooked)	0·3	102	37	70
African cake (mixture of water, maize flour and sugar baked in tin can)	1·2	297	26	45
Composite dishes – staples				
Hausa porridge (thin porridge of corn flour)	0·1	25	25	94
Porridge of corn grits	0·1	23	23	88
Banku (boiled mixture of corn dough and cassava dough)	0·7	107	16	73
Ga kenkey (corn dough made into dumplings and boiled in banana leaves)	0·8	172	19	71
Fanti kenkey (corn dough made into dumplings and boiled in plantain leaves)	0·7	118	21	72
Fufu (pounded boiled cassava and plantain)	0·4	96	24	69
Composite dishes – purchased meals				

Table 1 (*cont.*)

Food, and scientific name or recipe	Zn	Phy	Phy:Zn	% H$_2$O
Rice and stew (rice and standard ingredients‡)	0·6	118	21	68
Rice and beans (rice, cowpeas and standard ingredients)	0·5	107	18	70
Gari and beans (gari, cowpeas and standard ingredients)	0·9	178	22	59
Composite dishes – soups				
Palmnut soup (water, palmnut cream and standard ingredients)	0·4	n.a.	n.a.	86
Groundnut soup (water, groundnut paste and standard ingredients)	0·8	81	10	88
Composite dishes – stews				
Okra (okra and standard ingredients)	0·4	38	9	90
Bean (cowpeas and standard ingredients)	0·7	n.a.	n.a.	72

†, mg/100 g wet weight. n.a., not analysed.
‡, standard ingredients: tomato, red peppers, salt, onion, fish; palm oil in stews, rice and beans, and gari and beans.
Phy:Zn, [phytate]/[Zn] molar ratios. Phytate was analysed by the standard AOAC method (Harland & Oberleas, 1986).
All data from Ferguson *et al.* (1988, 1989*b*, 1993*a*).

AETIOLOGY OF ZINC DEFICIENCY IN DEVELOPING COUNTRIES

DIETARY FACTORS: LOW INTAKE AND POOR BIOAVAILABILITY OF DIETARY ZINC

The nutritional adequacy of dietary Zn depends on both its amount and bioavailability in the diet. Flesh foods are a rich source of Zn which is readily available because during their digestion certain L-amino acids and cysteine-containing peptides are released, which form soluble ligands with Zn (Sandström *et al.* 1980, 1989). In many developing countries, however, the content of flesh foods in rural diets is often low so that their contribution to total dietary Zn intake is small. Instead, diets are mainly plant based; cereals, starchy roots and/or tubers are often the major sources of Zn in rural diets. Of these staples, starchy roots and tubers generally have a lower Zn content than cereals, as shown by the Ghanaian and Malawian examples shown in Table 1. Hence, diets based on these staples tend to be correspondingly lower in Zn than cereal based diets (Gibson *et al.* 1991*a*; Ferguson *et al.* 1993*a*). Nevertheless, in certain geographical areas where Zn deficient soils exist, cereal staples will have a lower Zn content than when grown on Zn sufficient soils.

Plant based diets often contain high levels of phytic acid (myoinositol hexaphosphate) and dietary fibre, components known to inhibit the absorption of dietary Zn (Sandström, 1989). Of these antinutrients, phytic acid (Phy), the major storage form of phosphorus in cereals, legumes, and oleaginous seeds, is the most potent inhibitor of Zn absorption (Sandström & Lönnerdal, 1989). It forms insoluble chelates at a physiological pH. The lower inositol phosphates (i.e. tetra-, tri-, di-, and mono-inositol phosphates), formed by enzymic or non-enzymic hydrolysis of phytic acid, do not form insoluble complexes with Zn (Lönnerdal *et al.* 1989). The bioavailability of dietary Zn can be predicted from the ratio of phytic acid [Phy] to zinc [Zn] in diets. The critical [Phy]:[Zn] molar ratios associated with risk of Zn deficiency are equivocal; ratios above 15 have been associated with biochemical (Harland & Peterson, 1978; Oberleas & Harland, 1981; Turnlund *et al.* 1984; Bindra *et al.* 1986), and in some cases clinical signs of Zn deficiency in humans (Ferguson *et al.* 1989*a*).

Plant based staples such as unrefined maize flour, brown rice, sorghum and certain legumes (e.g. groundnuts, pigeon peas, kidney beans, and cowpeas) have elevated [Phy]:[Zn] molar ratios (Table 1; Ferguson, 1992). Hence, diets based on cereals and legumes have higher [Phy]:[Zn] molar ratios than those based on starchy roots and/or tubers (Ferguson et al. 1993a; Fitzgerald et al. 1993).

High levels of calcium potentiate the inhibitory effect of phytate on Zn absorption by forming a Ca:Zn:phytate complex that is even less soluble than phytate complexes formed by either ion alone (Wise, 1983). Hence, some authors have proposed that dietary [Phy][Ca]:[Zn] ratios may be a better predictor of Zn bioavailability than [Phy]:[Zn] ratios alone (Davies et al. 1985; Fordyce et al. 1987). To date, the critical [Ca][Phy]:[Zn] molar ratio that compromises Zn bioavailability in human diets has not been clearly defined. Retrospective calculations of experimental data from Cossack & Prasad (1983) suggest that molar ratios above 0.2 (200 mmol) may be associated with decreased Zn bioavailability in human diets. Most plant based diets are low in Ca, however, with the exception of those based on tortillas (Fitzgerald et al. 1993). The latter contain a relatively high concentration of Ca, derived from lime used to soak the maize in the preparation of nixtamal (soaking of corn kernels to liberate the husks) before being milled into masa (raw corn dough). Diets of lacto-ovo vegetarians may also have elevated [Ca][Phy]:[Zn] molar ratios (Bindra et al. 1986).

Dietary fibre, notably the insoluble fibres cellulose and lignin, may also inhibit Zn absorption to some degree, although their effects are equivocal, in part because fibre generally occurs concomitantly with phytic acid, making any independent inhibitory effect difficult to establish (Torre et al. 1991).

The bioavailability of Zn can also be affected by competitive interactions among certain micronutrients in the intestine, notably between Zn and non-haem Fe, and Zn and copper (Mills, 1985). The Fe and Cu contents of most human diets, however, are generally not high enough to compromise Zn bioavailability, unless high doses of supplemental non-haem iron are used (Solomons, 1986). In some cases, a negative Fe–Zn interaction has not been observed when the Fe is mixed with or is present as an intrinsic part of a food or meal (Valberg et al. 1984). Some (Milne et al. 1984; Mukherjee et al. 1984) but not all (Butterworth et al. 1988; Krebs et al. 1988) researchers have also observed a negative effect of high doses of folate supplements on Zn status, which could be of significance for pregnant women prescribed both supplemental folate and non-haem iron.

EXCESSIVE LOSSES

Additional factors that may exacerbate suboptimal Zn status in population groups living in developing countries include increased endogenous losses of Zn through perspiration; exfoliation of the skin as a result of the hot, humid climate; chronic haemolysis due to genetic factors (e.g. α-thalassaemia, sickle cell disease) and/or parasite infections (e.g. malaria, hookworm, schistosomiasis), and diarrhoea (Solomons, 1981; Ruz & Solomons, 1990). Ferguson (1992) estimated urinary Zn losses from haemolysis induced by schistosomiasis to range from 0.02 to 0.85 mg/d; faecal losses of Zn in infants with chronic diarrhoea can be as high as 300 μg/kg daily (Rothbaum et al. 1982). In areas where geophagia is practised, extensive faecal losses arising from poor absorption of dietary Zn may exacerbate Zn deficiency (Prasad et al. 1963).

HIGH PHYSIOLOGICAL REQUIREMENTS

The FAO/WHO/ILEA committee are currently revising the Zn requirements to include estimates to meet both basal and normative requirements (FAO/WHO/ILEA, unpublished observations, 1992). Basal requirements are the amount needed to prevent clinically detectable signs of functional impairment whereas the normative requirement represents the amount needed to maintain tissue stores or reserve capacity.

Physiological requirements of Zn are increased during periods of rapid growth because it has such a critical role in nucleic acid synthesis and protein metabolism. Hence, infants and children are especially vulnerable to Zn deficiency. In infants in developing countries, Zn stores at birth may be small as a consequence of their low birth weight and poor nutritional status of the mothers. Therefore, their dietary requirements for catch-up growth will be higher than those of infants from industrialized countries.

Male infants and children appear to have higher requirements for Zn than females, because of their higher growth rates and greater proportion of muscle/kg body weight; muscle contains a higher content of Zn than fat (Giugliano & Millward, 1984). In several double-blind supplementation studies, males have exhibited greater improvements in rate of linear growth and/or weight gain than their Zn supplemented female counterparts (Walravens & Hambidge, 1976; Walravens et al. 1983, 1989; Castillo-Duran et al. 1987; Schlesinger et al. 1992; M. Ruz, 1993, pers. comm.).

Requirements for Zn are also greater during pregnancy and lactation for the growth and development of the fetus and maternal tissues, and secretion of breast milk. The FAO/WHO/ILEA committee (unpublished, 1992) calculated the average individual physiological requirements for absorbed Zn during each trimester of pregnancy to be 0·8, 1·0, 1·4 mg/d for the basal requirements and 1·1, 1·4, and 2·0 mg/d for the normative requirements. These estimates do not take into account differences in the absorbability of dietary Zn or the varied intakes within the population. During the course of lactation, Zn concentrations in human milk decline (Casey et al. 1989). Hence, estimates of the average individual basal requirement range from 1·6 at 0–3 months and 1·5 at 3–6 months to 1·2 mg/d between 6 and 12 months; corresponding estimates for normative requirements are 1·9, 1·8, and 1·5 mg/d respectively (FAO/WHO/ILEA, unpublished observations, 1992).

ZINC INTAKES IN RELATION TO ESTIMATED REQUIREMENTS

In many developing countries, information on intakes and major food sources of Zn in local diets, as well as on the antinutrients dietary fibre and phytate, are limited, in part because of the paucity of data on the content of Zn and antinutrients in local foodstuffs. This is unfortunate because such data are essential for assessing the risk for inadequate intakes of dietary Zn, and for planning dietary strategies to improve its content and bioavailability in traditional diets.

Population groups consuming diets based predominantly on unrefined maize and rice generally have markedly higher intakes of phytate and elevated [Phy]:[Zn] molar ratios compared to those consuming diets based on starchy roots and/or tubers (Table 2) (Mbofung & Atinmo, 1987; Gibson et al. 1991a; Ferguson et al. 1993b; Fitzgerald et al. 1993). The latter, however, often have lower Zn intakes. Molar ratios of [Ca][Phy]:[Zn] in most of these plant based diets are low with the exception of those based on tortillas (Fitzgerald et al. 1993).

The adequacy of dietary Zn intakes can be evaluated by comparison with the newly revised requirements, provided an estimate of the bioavailability of Zn in the diet can be made. Diets can be categorized as high, moderate, or low in terms of Zn bioavailability, based on their content of animal or fish protein, calcium, and [Phy]:[Zn] molar ratios (FAO/WHO/ILEA, unpublished observations, 1992). Alternatively, more direct measure-

Table 2. *Dietary intakes (mean ±SD) of zinc, phytate, phytate:zinc molar ratios, and dietary fibre of children in some developing countries*

Country (n) Age in years Reference	Zinc (mg/day)	Phytate (mg/day)	Phy:Zn	Dietary fibre (g/day)
Papua New Guinea (67) 6–10 Gibson et al. 1991a	4·4 ± 1·3	646 ± 663	12	37·1 ± 11·4
Malawi (67) 4–6 Ferguson et al. 1993b	6·6 ± 1·7	1899 ± 590	25	24·9 ± 6·4†
Ghana (148) 3–6 Ferguson et al. 1993b	4·7 ± 1·1	591 ± 153	13	15·5 ± 3·8†
Egypt (96) 1·5–2·5 Murphy et al. 1992	5·2 ± 1·6	796 ± 249	16	17·4 ± 5·9
Kenya (100) 1·5–2·5 Murphy et al. 1992	3·7 ± 0·9	1066 ± 324	28	21·6 ± 5·5
Mexico (59) 1·5–2·5 Murphy et al. 1992	5·3 ± 1·3	1666 ± 650	30	15·3 ± 4·8
Guatemala (136) 6–8 Cavan et al. 1993a	9·0‡	962†	11	14·0‡
Canada (106) 4–6 Gibson et al. 1991b	6·9 ± 2·3	(300)‡	5	11·4 ± 5·5

† Non-starch polysaccharide.
‡ Median.

ments of the bioavailability of Zn in local diets can be made by using radioactive or stable isotope techniques (Sandström & Lönnerdal, 1989).

Some studies report average Zn intakes for a specific population group (Mbofung & Atinmo, 1987), often based on one day's intake per individual. Such data do not take into account the distribution of intakes among individuals and cannot be used to estimate the proportion of individuals within the population at risk for nutrient inadequacy. For the latter, food intake data based on at least two days' intake per person are required. If single days are used, prevalence estimates for risk of inadequacy are always too high (Beaton, 1985). To improve the reliability of the prevalence estimates for dietary inadequacy, they should be determined using the probability approach recommended by the Subcommittee for Criteria for Dietary Evaluation (National Research Council, 1986). When this approach has been used in studies of dietary Zn intakes of children in developing countries, a very high proportion of the children studied from Kenya and Malawi (> 90%), and more than two-thirds from Mexico and Ghana, were apparently at risk, assuming that the estimates used for both the bioavailability and basal requirements for Zn are valid (Murphy et al. 1992; Ferguson et al. 1993b).

Even the Zn intakes of exclusively breast fed infants may be inadequate during the first 4–6 months in some developing countries, especially if the infants are preterm and/or of low birth weight with high nutrient demands for catch-up growth. Moreover, their supply of Zn from breast milk may be compromised by the poor nutritional status of the lactating mothers, which may result in breast milk with an inherently low Zn content (Butte et al. 1992; Dorea, 1993) and/or low volume (Brown et al. 1986). To date, studies of the Zn concentrations of breast milk in poorly nourished lactating women with chronically inadequate Zn intakes have revealed inconsistent results. In some, breast milk Zn concentrations have been consistent with those reported for developed countries, and independent of maternal dietary Zn intakes (Kirsten et al. 1985; Karra et al. 1986; Moser et al. 1988; Simmer et al. 1990); others dispute this finding (Krebs et al. 1985; Shrimpton et al. 1985).

In many developing countries, breast milk output may also be compromised by the early introduction of weaning foods which replace rather than complement breast milk (Walker, 1990). Very often these weaning foods are prepared as thin porridges from staples with a low energy and nutrient density which fail to make up the nutrient deficit when breast milk no longer meets the infants' needs. If unrefined and unfermented cereals and/or legumes are used, the weaning foods will have a high phytic acid content; consequently Zn bioavailability will be low. During fermentation, hexa- and penta-inositol phosphates are hydrolysed enzymically to the lower inositol phosphates which do not inhibit Zn absorption (Lönnerdal et al. 1989). More work is required in developing countries to evaluate the adequacy of dietary Zn intakes for both exclusively breast fed infants and for weanlings. To date, no recommendations for the Zn content of weaning foods in developing countries exist (Royal Tropical Institute, Amsterdam, 1987). This is unfortunate because Zn deficiency impairs appetite, taste acuity, immune and intestinal function during infancy (Hambidge et al. 1972; Krebs et al. 1984; Castillo-Duran et al. 1987; Roy et al. 1992; Schlesinger et al. 1993; Tomkins et al. 1993) as well as growth (Hambidge, 1989). Such functional disturbances will have a further detrimental effect on the growth and development of the infants.

A high proportion of pregnant women from developing countries are probably also at risk through inadequate intakes of Zn. Although no data based on the probability approach are available in the literature, in a Guatemalan study 94 and 25% of the pregnant women had average Zn intakes below or less than two-thirds of the US Recommended Dietary Allowance for Zn (15 mg) respectively, assuming that 20% of Zn was absorbed from their diets. Mean Zn intakes for pregnant rural and urban women in Nigeria were 6·0 and 6·7 mg/d, respectively, whereas during lactation they ranged from 7·3 to 8·2 mg/d for rural women (Mbofung & Atinmo, 1987). Corresponding mean intakes for Nepalese (Moser et al. 1988) and Amazonian (Jackson et al. 1988) lactating women were 10·5 and 8·8 mg Zn/d respectively.

Comparison of Zn intakes with the current estimated requirements does not take into account the possibility that humans can adapt to chronically low Zn intakes and achieve Zn balance by increasing Zn absorption (King, 1986). Certainly, Amazonian lactating women maintained normal Zn balance in the presence of low intakes of Zn (and phytate) (Jackson et al. 1988), although there was evidence of functional impairment because breast milk Zn and retinol contents were abnormally low. Whether such adaptation also occurs in the presence of very high habitual intakes of phytate seems unlikely. Brune et al. (1989) reported that vegetarians did not adapt to their high phytate diet by increased absorption of ^{59}Fe.

Probability estimates for risk of Zn inadequacy do not identify actual individuals in the population who are deficient, or define the severity of the nutrient inadequacy. Such information can only be obtained when the dietary intake data are combined with laboratory and/or clinical indices of Zn status. This is especially important in developing countries where the coexistence of many other multifaceted health problems often confounds the diagnosis of Zn deficiency.

LABORATORY ASSESSMENT OF ZINC STATUS

BIOCHEMICAL INDICES OF ZINC STATUS

To date, no single, sensitive and specific index of Zn status exists (Golden, 1989). Serum/plasma Zn is the most frequently used index in human studies because it can be easily and accurately measured. Nevertheless, this index has several limitations. It can only be used when the serum samples are not haemolysed or contaminated, and conditions such as infection are absent. Erythrocytes have a high Zn content and in cases of Zn deficiency red cell fragility is increased (Bettger et al. 1978). Parasitaemia is prevalent in many developing countries, and its presence confounds the interpretation of serum Zn concentrations; during infection values are spuriously low because Zn is redistributed from the plasma to other tissues (Aggett, 1991; Filteau & Tomkins, 1994). Other important confounding factors which must be controlled when collecting blood samples for plasma Zn analysis include diurnal variation in circulating Zn level, fasting, meal consumption, the time interval between blood collection and separation of the plasma, and contamination of the blood sample from evacuated tubes with rubber stoppers and non-acid washed glassware (Gibson, 1989; Aggett, 1991; Wallock et al. 1993). In general, low plasma/serum Zn levels indicate deficiency or a redistribution of Zn, but normal levels do not necessarily preclude deficiency. For instance, in cases of chronic but mild Zn deficiency states, plasma concentrations are often normal (Gibson et al. 1989b; Ruz et al. 1991), making diagnosis difficult.

Alternative static biochemical indices of Zn status which have been investigated include the concentrations in hair, urine, leucocytes, neutrophils, platelets and saliva. Available evidence suggests that low concentrations in hair samples collected during infancy and childhood probably reflect chronic suboptimal Zn status when the confounding effects of severe protein–energy malnutrition and season are absent (Hambidge et al. 1972; Gibson et al. 1989b; Cavan et al. 1993a; Ferguson et al. 1993b). Clinical features of mild Zn deficiency in childhood, such as impairments in linear growth, appetite and taste acuity, have been associated with hair concentrations of less than 1·07 μmol/g (70 μg/g) (Hambidge et al. 1972; Krebs et al. 1984; Smit Vanderkooy & Gibson, 1987) in the summer, and less than 1·68 μmol/g (110 μg/g) in the winter (Gibson et al. 1989b; Cavan et al. 1993a). In some cases, the low hair concentrations have been related to poorly available dietary Zn (Smit Vanderkooy & Gibson, 1987; Ferguson et al. 1988; Gibson et al. 1991b; Cavan et al. 1993a).

Hair Zn cannot be used in cases of very severe malnutrition when the rate of growth of the hair shaft is often diminished. In such cases, hair Zn concentrations may be normal or even high (Erten et al. 1978; Bradfield & Hambidge, 1980). Standardized procedures must be used for sampling, washing, and analysing the hair samples (Hambidge, 1982). Supplementation trials must be undertaken over one year and all the subjects sampled at the same season of the year to minimize the confounding effects of seasonal variation (Gibson et al. 1989a).

Many investigators have failed to find any positive correlations between the Zn content of hair and serum/plasma Zn concentrations (Hambidge et al. 1972; Walravens et al. 1983, 1989; Gibson et al. 1989b). These findings are not unexpected. The Zn content of the hair shaft reflects the quantity of Zn available to the hair follicle over an earlier time interval. Positive correlations between hair and plasma Zn concentrations are only observed in chronic, severe deficiency states, in the absence of confounding factors.

Depletion of body Zn stores causes a reduction in urinary excretion, often before any detectable changes in serum/plasma Zn concentrations (Baer & King, 1984). Twenty-four hour urine collections are recommended because diurnal variation in urinary Zn excretion occurs, although casual urine samples can be used if Zn:creatinine ratios are determined (Zlotkin & Casselman, 1988). Several factors can affect urinary Zn concentrations, however, making interpretation of the results difficult. For example, despite the presence of suboptimal Zn status in sickle cell anaemia, hyperzincuria occurs. The absence of established interpretive criteria for urinary Zn levels further limits their use (Gibson, 1989).

The Zn contents of leucocytes or specific cellular types of leucocytes (e.g. neutrophils) have been used as an index of tissue Zn status; they are said to reflect soft tissue Zn (Jones *et al.* 1981) and correlate with retinal dark adaptation. They also have a shorter half-life than erythrocytes and hence should detect changes in Zn status over a shorter time period. Results, however, have been equivocal (Jones *et al.* 1981; Meadows *et al.* 1981; Prasad & Cossack, 1982; Thompson, 1991; Ruz *et al.* 1992). Relatively large volumes of blood are required and isolation of the leucocytes and specific cellular types, as well as their subsequent analysis, is lengthy and technically difficult, limiting their use in some countries. For example, Milne *et al.* (1985) have emphasised that the Zn content of leucocytes is a function of the type of separation used; contamination with Zn from the anticoagulant, reagents, density gradient system, and/or from erythrocytes and platelets may occur. Changes in the relative proportions of leucocyte subsets with physiological state (e.g. pregnancy) and haematological disorders must also be taken into account in the interpretation of the results. Finally, comparison of results among different studies is difficult because no consensus exists as to how to express Zn concentrations in the cell types.

Biochemical functional tests measure changes in the activities of certain enzymes or blood components dependent on Zn. Zinc is a constituent of over 200 metallo-enzymes which vary in their responses to Zn deficiency depending on the tissue examined, their Zn affinity, and rate of turnover of the enzyme. Of the Zn metallo-enzymes, activity of serum alkaline phosphatase has been most widely used to assess Zn status; its response has been inconsistent. In general, its activity is reduced in severe (Rothbaum *et al.* 1982) but not in mild (Ruz *et al.* 1991) Zn deficiency states. No significant changes in activity have been reported in mild Zn depletion–repletion studies of adults (Ruz *et al.* 1991), or in most (Hambidge *et al.* 1972; Walravens & Hambidge, 1976; Walravens *et al.* 1983, 1989; Gibson *et al.* 1989*b*; Cavan *et al.* 1993*b*), but not all (Udomkesmalee *et al.* 1992) of the Zn supplementation studies in infants and children.

The specificity of alkaline phosphatase as an index of Zn status is also poor; its activity is influenced by many factors other than Zn status such as low food intake, type of protein consumed, magnesium or manganese deficiency, season, and in states of increased bone turnover (Chesters & Will, 1978; Koo *et al.* 1989). Measurements of alkaline phosphatase activity in neutrophils (Ruz *et al.* 1991), leucocytes (Schiliro *et al.* 1987), and red blood cell membranes (Ruz *et al.* 1992; Cavan *et al.* 1993*b*) have also been investigated as indices of Zn status; more studies are needed before any definite conclusions can be reached. To date, there is no universally accepted Zn dependent enzyme which can be used to assess mild Zn deficiency.

Levels of the Zn binding protein metallothionein have been investigated in serum, urine, or erythrocytes as indices of Zn status (Golden, 1989). Levels fall in Zn deficiency as a result of impaired synthesis. Specificity is poor; levels are also affected by Fe deficiency, diurnal rhythm, and acute infection. Metallothionein is said to be much less responsive to stress and infection in erythrocytes than in plasma (Grider *et al.* 1990), and hence may provide a useful index of Zn status.

Serum thymulin has also been assessed as a potential index of Zn status. Thymulin is a Zn metallopeptide which controls cell mediated immune function (Prasad *et al.* 1988); its activity falls in mild Zn deficiency. Plasma somatomedin-C, a peptide of low molecular weight which is regulated by growth hormone, nutrition, and insulin, is increased in response to increases in Zn concentration in plasma and tibia of rats. Nevertheless, more work is required to establish the sensitivity, specificity, and validity of erythrocyte metallothionein, serum thymulin and somatomedin-C as indices of Zn status.

PHYSIOLOGICAL FUNCTIONAL INDICES OF ZINC STATUS

Physiological functions dependent on Zn, such as linear and ponderal growth, taste acuity, and immune competence, can also be used to assess Zn status. Such tests have greater biological significance than the biochemical tests because they measure the biological impact of Zn deficiency. Their specificity is low, and hence they must always be used in conjunction with biochemical indices.

Diminished taste acuity is a feature of mild Zn deficiency. Several methods for testing taste acuity have been used. In studies of Canadian (Gibson *et al.* 1989*b*) and Guatemalan (Cavan *et al.* 1993*a*) children, significant inverse relationships between recognition threshold for salt and hair Zn concentrations have been noted. These results suggest that impaired taste acuity can be used as a physiological functional test of suboptimal Zn nutriture in some children, provided a biochemical index of Zn status is also used. The test is not suitable, however, for infants and children less than five years of age.

Some changes in body composition have also been observed after Zn supplementation in some cases of deficiency in children. Specifically, increases in arm circumference were reported in Gambian children (Bates *et al.* 1993), whereas in Zn supplemented Jamaican children recovering from severe malnutrition, accretion of lean tissue was greater. The latter was attributed to an increased efficiency of nutrients for tissue synthesis after Zn supplementation. By contrast, triceps skin folds increased in Guatemalan Zn supplemented children (Cavan *et al.* 1993*b*), probably due to an increase in energy intake concomitant with improved appetite.

From the discussion above, it is evident that diagnosis of Zn deficiency is hampered by the lack of a single, specific, and sensitive index of status. A large number of indices have been proposed, but many are fraught with problems that affect their use and interpretation, especially in mild Zn deficiency states. Hence, it is not surprising that the true magnitude of Zn deficiency in developing countries is not yet known.

ZINC DEFICIENCY THROUGHOUT THE LIFE CYCLE

INFANCY AND CHILDHOOD

Cases of severe Zn deficiency in infancy and childhood are now rare but mild deficiency in infancy and childhood is not uncommon. Growth failure is the most prominent clinical feature of mild Zn deficiency, although impairments in body composition, taste acuity, appetite, immune function, dark adaptation, and delays in secondary sexual maturation have also been described (Hambidge, 1989). Growth failure is also a characteristic feature of childhood growth patterns in many developing countries, which has until recently been attributed to deficits in energy and/or protein. Inadequate Zn intakes are likely to be an important contributing factor because diets low in protein tend to be low in Zn (Golden & Golden, 1981*b*), and Zn has such a critical role in protein synthesis, cell replication, and appetite control.

The first cases of human Zn deficiency were reported in the Middle East among adolescent male dwarfs in the 1960s (Prasad *et al.* 1963). The syndrome was characterized by impaired growth and delayed sexual maturation, which were shown to respond to Zn supplementation in later studies (Ronaghy *et al.* 1969, 1974).

Since these first reports, nutritional Zn deficiency has been reported in infants and/or children living in some industrialized (Hambidge *et al.* 1972; Walravens & Hambidge, 1976; Arcasoy *et al.* 1978; Buzina *et al.* 1980; Walravens *et al.* 1983, 1989, 1992; Smit Vanderkooy & Gibson, 1987; Gibson *et al.* 1989*b*), and developing (Golden &

Table 3. *Double-blind zinc supplementation studies in infants*

Country, no. of subjects, age of subjects, experimental treatment, reference	Zinc suppl. Start	Zinc suppl. End	Control Start	Control End	Growth effects and other responses
USA. 68 normal healthy full term male infants at birth studied for 6 months. Double-blind study. Formula with 1·8 mg Zn/l v. 5·8 mg Zn/l. Walravens & Hambidge, 1976		11·9		11·0	Improved weight and length in males only.
France. 57 normal healthy infants 5·4 months old studied for 3 months. Double-blind placebo (32), 5 mg Zn/d (25). Walravens et al. 1992					Improved weight gain. Improved length gain in males only.
USA. 50 failure to thrive, 8–27 months old studied for 6 months. Randomized double-blind trial pair-matched. 5·7 mg Zn/d as syrup (25) and placebo (25). Walravens et al. 1989	10·7	9·8	10·7	10·4	Improved weight especially in boys. Tendency to increased activity of serum alkaline phosphatase in Zn group.
Chile. 32 marasmic infants, 7–8 months old, studied for 90 d. Randomized double-blind trial. 2 mg Zn/kg daily as solution (16). Placebo (16). Castillo-Duran et al. 1987	14·7	15·6	16·1	15·6	Weight for length effect. Decrease in % anergic infants, increase in serum IgA in Zn group.
Chile. 39 severely malnourished infants studied for 105 d. Double-blind trial 1·9 mg Zn/kg daily in formula (19) v. 0·35 mg Zn/kg daily in formula (20). Schlesinger et al. 1992	19·4	18·6	23·4	18·0	Linear length effect. Improved immune function.
Bangladesh. 60 severely malnourished infants 5–60 months old for 3 weeks. Rice based diet *ad lib.* and vitamins and minerals. 10 mg Zn/kg daily if < 6 kg or 50 mg Zn/d for those > 6 kg as $ZnSO_4$. Non-supplemented group (30). Khanum et al. 1988	8·2	18·5	7·9	10·6	Improved weight gain and weight for length.
Bangladesh. 65 children with AD 3–24 months old. 152 with PD 3–24 months old supplemented for 2 weeks. Followed for 2 and 3 months in a double-blind randomized study (placebo v. 15 mg Zn/kg daily). Roy et al. 1993					Improved length gain in AD group, and in PD with < 90 % weight/age and < 90 % height/age. Reduced no. of episodes of diarrhoea in AD and PD groups and attack rate of respiratory tract infections in AD group only.
Chile. 80 SGA neonates 38–41 weeks gestational age studied from birth for 6 months. Double-blind randomized study with placebo (41), 3 mg Zn/d (39). Rodriguez et al. 1991	12·6	10·5	12·1	8·9	Improved linear growth and weight gain. No difference in head circumference.

AD, acute diarrhoea; PD, persistent diarrhoea; SGA, small for gestational age.

Table 4. *Double-blind zinc supplementation studies in children*

Country, date, number of subjects, age of subjects, experimental treatment, reference	Dietary zinc intake (mg)	Zinc suppl. Start	Zinc suppl. End	Control Start	Control End	Growth effect and other responses
Egypt. 1965–6. 90 growth retarded school boys, 11–18 years old studied for 5·5 months. Randomized trial, placebo (30) and 14 mg Zn (30). Capsules given at school. Carter *et al.* 1969	14	10·7	19·2	11·7	13·3	No weight or height effects. No difference in sexual maturation. No serum alkaline phosphatase effect.
Iran. 1967–8. 60 growth retarded school boys 12–18 years old studied for 17 months (5 months trial, 7 months rest, 5 months trial). Controlled trial. 1st 5 months placebo (20), 28 mg Zn (20), 67 mg Fe (20). 2nd 5 months placebo (20), micronutrients (20), micronutrients+40 mg Zn (20). Capsules given at school. Ronaghy *et al.* 1969	12	17·2	14·7	11·6	14·1	No weight or height effects. Difference in sexual maturation.
Iran. 1969–71. 50 growth retarded school boys 13 years old studied for 17 months (5 months trial, 7 months rest, 5 months trial). Non-randomized trial placebo (10), micronutrients (20), micronutrients+40 mg Zn (20). Capsules given at school. Ronaghy *et al.* 1974	12	8·2	10·2	10·5	10·7	Weight and height effects. Difference in bone age. Tendency for faster sexual development. No serum alkaline phosphatase effect.
USA, Colorado. 40 growth retarded, low Zn status children 2–6 years old studied for 1 year. Randomized pair-matched trial with placebo (20) and 10 mg Zn/d (20). Syrup given by parents at home. Walravens *et al.* 1983	4·6	10·7	10·8	11·3	11·3	Height effect (especially boys). Increase in appetite.
Canada. 1986. 60 growth retarded boys 5–7 years old studied for 12 months. Randomized pair-matched trial with placebo (30) and 10 mg Zn/d (30). Fruit juice drink given by parents at home. Gibson *et al.* 1989*b*	6·4	15·6	16·2	16·5	16·4	Height effect only in subjects with low hair Zn (<1·68 μmol/g). Increase in appetite perceived by parents.

Thailand. 1989–90. 133 children 6–13 years old with suboptimal Zn and vitamin A nutriture for 6 months. Randomized pair-matched trial with placebo (35), 25 mg Zn/d (33), vit. A + Zn (32). Capsules taken on school days. Udomkesmalee *et al.* 1992	4·3	13·2	13·2	14·3	No weight or height effect. Increase in serum alkaline phosphatase activity. Improved dark adaption. Improved conjunctival integrity.
The Gambia. 1989–90. 109 apparently healthy children 1/2 to 3 years old for 15 months. Randomized group matched trial with placebo (54), and 70 mg Zn (55) as a drink twice a week at clinic. Bates *et al.* 1993			19·0		No weight or height effect. Increase in arm circumference. Less malaria. Improved intestinal permeability.
Guatemala. 1989. 162 school children 6–8 years old for 25 weeks. Randomized pair-matched trial. Micronutrients (82), micronutrients + 10 mg Zn/d (80). Chewable tablet given at school on weekdays. Cavan *et al.* 1993*b*	10	14·2	16·2	14·4	14·9	No weight or height effect. Increase in triceps skinfold. Smaller decrease in mid arm circumference. No increase in serum alkaline phosphatase.
Chile. 1991. 46 short stature school children, 6–12 years old, consuming diets providing 50–60% of normal daily Zn intake. 12 month randomized study involving placebo *v.* 10 mg Zn/d. Castillo-Duran *et al.* 1987					No weight effect. Height effect in males only. No difference in plasma Zn.
Chile. 1993. 98 healthy preschool children studied for 14 months. Placebo *v.* 10 mg Zn/d. Ruz, 1992					Height effect in males. Trend in improved immune function and reduced giardiasis.

Golden, 1981a; Xue-Cun et al. 1985; Castillo-Duran et al. 1987; Khanum et al. 1988; Simmer et al. 1988; Udomkesmalee et al. 1990, 1992; Schlesinger et al. 1992; Bates et al. 1993; Cavan et al. 1993a, b; Roy et al. 1993; M. Ruz, 1993, pers. comm.; Smith et al. 1993) countries (Tables 3 & 4). In most of these studies, clinical signs of severe Zn deficiency were not apparent. Instead, mild Zn deficiency existed, characterized by reductions in linear and/or ponderal growth, and/or impairments in taste acuity, appetite, immune and intestinal function, and dark adaptation, some of which have responded positively to Zn supplementation in double-blind studies. Biochemical evidence of Zn deficiency has not been a consistent finding. This is not unexpected; physiological functional consequences (e.g. growth retardation) of mild Zn deficiency are often apparent before the Zn concentrations in plasma and/or tissues are significantly reduced (Gibson et al. 1989b; Ruz et al. 1991), emphasizing the importance of confirming mild Zn deficiency by a positive response to a supplement in double-blind studies.

PREGNANCY

Animal studies have clearly demonstrated the teratogenic effects of Zn deficiency (Hurley & Swenerton, 1966), but results of human studies have been inconsistent. In severe Zn deficiency in humans arising from acrodermatitis enteropathica, abortions and gross congenital malformations (e.g. anencephaly) have been reported (Hambidge et al. 1975). The existence of mild Zn deficiency during pregnancy and its effect on pregnancy outcome is less clear, in part because of difficulties in establishing the existence of marginal Zn status during pregnancy and/or inadequacies in the experimental designs. No double-blind Zn supplementation studies during pregnancy have been carried out in developing countries.

Serum Zn has been the most frequently used index of Zn status during pregnancy; it declines during pregnancy even in the presence of optimal maternal Zn nutriture (Swanson & King, 1987), attributed in part to expansion in plasma volume. Nevertheless, in women with inadequate Zn intakes, the decline in serum Zn during pregnancy may be abnormally large (Hambidge et al. 1983; Cherry et al. 1989).

Relationships between maternal plasma Zn and pregnancy outcome have been inconsistent, and have varied with both the stage of gestation and the outcome variable measured (Swanson & King, 1987). For example, plasma Zn correlated weakly with birth weight when sampled at mid pregnancy (McMichael et al. 1982), more strongly in early rather than in later pregnancy, i.e. third trimester (Neggers et al. 1990), or not at all (Arcasoy et al. 1978; Buzina et al. 1980; Hambidge et al. 1983; Campbell-Brown et al. 1985; Hunt et al. 1985; Tuttle et al. 1985; Mahomed et al. 1989). Plasma Zn has also been reported to correlate with pregnancy complications such as prolonged labour, hypertension, postpartum haemorrhage, spontaneous abortions, and/or congenital malformations by some (Jameson, 1976; Çavdar et al. 1980, 1991; Cherry et al. 1981; McMichael et al. 1982; Soltan & Jenkins, 1982; Hunt et al. 1985) but not all (Breskin et al. 1983; Mukherjee et al. 1984) investigators. In some double-blind Zn supplementation studies, significant reductions in pregnancy complications have been observed in the Zn treated compared to the placebo group (Hunt et al. 1984; Cherry et al. 1989; Jameson et al. 1990; Simmer et al. 1991) (Table 5), associated in some cases with alterations in prostaglandin metabolism (O'Dell et al. 1977).

Some relationships have also been reported between low maternal Zn concentrations in leucocytes and/or lymphocytes and intrauterine growth retardation (Meadows et al. 1981; Simmer & Thompson, 1985), low birth weight (Wells et al. 1987; Malhotra et al. 1990), and neural tube defects (Hinks et al. 1989). In two double-blind Zn supplementation studies

Table 5. *Zinc supplementation studies in pregnant women*

Country, date, no. of subjects, type of subjects, experimental treatment, reference	Dietary zinc intake (mg)	Responses
UK. 1985–6. 494 middle class women studied for last 4 months of pregnancy. Randomized double-blind trial with placebo (248) and 20 mg Zn/d (246). Capsules taken at home. Mahomed et al. 1989	9	No effect on birth weight. No difference in leucocyte Zn in supplemented and placebo group.
USA, New Orleans. 556 low income adolescent women studied for last 3 months of pregnancy. Randomized double-blind trial with placebo (288) and 30 mg Zn/d (268). Tablets taken at home. Cherry et al. 1989	?	No effect on birth weight. Reduced rates of prematurity and neonatal morbidity.
USA, Los Angeles. 1981–2. 138 Hispanic teenagers studied for last 4 months of pregnancy. Randomized double-blind trial with micronutrients (68) and micronutrients + 20 mg Zn/d (70). Capsules taken at home. Hunt et al. 1985	9·8	No effect on birth weight.
USA, Los Angeles. 213 Hispanic low income women enrolled < 27 week gestation age. Randomized double-blind trial with micronutrients (106) and micronutrients + 20 mg Zn/d (107). Hunt et al. 1984	9·3	No effect on birth weight. Reduced incidence of pregnancy induced hypertension.
UK. 56 pregnant females at risk of SGA infants. Studied last 15–25 weeks. Randomized double-blind trial with placebo (26) and 22·5 mg Zn/d (30). Simmer et al. 1991	?	Lower incidence of IUGR. Labour induced less often. C-section less often.
USA. 46 pregnant middle income females studied for 7–9 months. Not randomized double-blind study. Placebo (36) v. 15 mg Zn/d (10). Tablet taken 2 h after dinner. Hambidge et al. 1983	11	No effect on birth weight. No other effects observed.
Sweden. 1983–6. 1231 pregnant women. Not randomized double-blind study. 15–90 mg Zn/d (depending on serum Zn) given to 598 subjects; 633 given no Zn supplement. Jameson et al. 1990	9·4	Fewer preterm deliveries before 33rd week of gestation. Reduction in perinatal deaths; fewer spontaneous abortions.

C-section, Caesarian section; IUGR, intra-uterine growth retardation; SGA, small for gestational age.

(Mahomed et al. 1989; Thauvin et al. 1992), however, no differences in leucocyte Zn concentrations between the two groups were observed.

Several adaptive mechanisms exist during pregnancy to help meet the increased demands for Zn, including an increase in absorption, a reduction in endogenous losses, redistribution of tissue Zn, and an efficient maternal–fetal transfer (Swanson & King, 1987). Although such adaptive mechanisms may be adequate to prevent Zn deficiency in women in developed countries, they may not be sufficient for pregnant women from developing countries, whose Zn status may be especially low because of frequent reproductive cycling, excessive losses of endogenous Zn, combined with diets low in readily available Zn. Unfortunately, however, investigations of the Zn status of pregnant women in developing countries are limited (Çavdar et al. 1980; Prema, 1980; Okonofua et al. 1989, 1990); none has involved double-blind Zn supplementation trials.

In view of the inconsistencies noted above, the precise nature of the association between Zn status and pregnancy outcome remains unclear. Existing evidence suggests that the prevalence of deficiency in women during pregnancy in developing countries is likely to be

Table 6. *Double-blind zinc supplementation studies in lactating women*

Country, no. of subjects, type of subjects, experimental treatment, reference	Dietary zinc intake (mg)	Response
Brazil, Amazon region. 65 poor urban women studied for first 5 months of lactation. Randomized trial with placebo (28) and group consuming 15 mg Zn/d (37). Capsules taken at home. Shrimpton *et al.* 1983 and Shrimpton *et al.* 1985	23	No effect on milk Zn levels. Milk vitamin A levels increased. Less diarrhoea in infants.
USA, Colorado. 53 middle income lactating women, for varying durations up to 9 months. Controlled (8) trial with placebo (39) and group consuming 15 mg Zn/d (14). Tablets taken at home. Krebs *et al.* 1985	12·2	Decreased fall in milk Zn levels.
USA, Indiana. 49 middle income mothers studied during first 6 months of lactation. Controlled trial with groups consuming micronutrients (25) and micronutrients+25 mg Zn (24). Different commercial supplements taken at home. Karra *et al.* 1986	11·2	Higher milk Zn levels.
USA, Maryland. 40 middle income women studied during the first 6 months of lactation. Randomized double-blind trial with groups consuming micronutrients (20) and micronutrients+25 mg Zn/d (20). Tablets taken at home. Moser-Veillon & Reynolds, 1990	12	No effect on milk Zn levels.

higher than that in developed countries, but large, well designed, double-blind Zn supplementation trials are required to confirm the existence of nutritional deficiency and its precise impact on pregnancy outcome.

LACTATION

Studies of maternal Zn status during lactation are limited (Table 6). Some have documented low plasma concentrations in the presence of normal concentrations in hair, urine (Jackson *et al.* 1988), and/or breast milk, even in poorly nourished lactating women with chronically inadequate intakes of dietary Zn (Kirsten *et al.* 1985; Karra *et al.* 1986; Simmer *et al.* 1990). Two Zn supplementation studies during lactation (Krebs *et al.* 1985; Shrimpton *et al.* 1985) documented a reduction in the abnormally steep decline in breast milk Zn content during late lactation, although the numbers of subjects in these studies were small. Furthermore, the incidence of diarrhoea in the infants decreased, and milk retinol content was maintained at a higher level throughout lactation in the Zn supplemented Amazonian women (Shrimpton *et al.* 1983, 1985).

By contrast, in a US study in Indiana (Karra *et al.* 1986) in which 25 mg Zn/d were given, Zn levels of breast milk apparently increased. Such increases were not observed by Moser-Veillon & Reynolds (1990), despite a comparable daily Zn supplement to US Maryland lactating women. The study of Karra *et al.* (1986), unlike the Maryland study (Moser-Veillon & Reynolds, 1990), was not a double-blind randomized trial.

Even in malnourished women from developing countries whose breast milk Zn concentrations are not compromised, their volume of breast milk may be reduced (Brown *et al.* 1986), thus contributing to growth failure in early infancy. Traditional weaning foods used in many developing countries are often based on unrefined cereals and/or legumes, low in bioavailable Zn. If these weaning foods are not processed to reduce their phytic acid content, their use may further compromise infant growth, especially if they replace rather than complement breast milk (Walker, 1990). Strategies which can be used in developing countries to reduce the phytic acid content of traditional staple foods, including weaning foods, are outlined below.

NUTRITION INTERVENTION STRATEGIES TO PREVENT ZINC DEFICIENCY IN DEVELOPING COUNTRIES

Both short term and long term nutrition intervention strategies can be used to prevent Zn deficiency in developing countries: (1) supplementation; (2) fortification; and (3) dietary modification/diversification using traditional household techniques. For pregnant women, supplementation or fortification is appropriate because a relatively short term response is required to improve their Zn status before the end of pregnancy. Moreover, requirements for Zn during pregnancy, like Fe, cannot be met from dietary sources alone. Such approaches can also be used to provide several micronutrients simultaneously. They do, however, rely on a stable infrastructure and require financial support on a long standing economic basis if they are to be successful. All too often such programmes have been suspended for economic, political, and logistical reasons.

The third approach, dietary modification/diversification, involves changes in food selection patterns and/or traditional household methods for preparing and processing indigenous foods. It is a more economically feasible, culturally acceptable, and sustainable intervention for alleviating Zn deficiency in developing countries. Possible dietary changes to improve both the content and bioavailability of Zn include increasing the consumption of flesh foods, rich sources of readily available Zn, when economically feasible, and making modifications to food preparation and processing practices to reduce the level of the higher inositol phosphates in plant based staples. Higher inositol phosphates can be hydrolysed to lower inositol phosphates enzymically *via* fermentation and/or germination (Svanberg & Sandberg, 1988). Alternatively, in some cases non-enzymic hydrolysis of the higher inositol phosphates can be achieved by thermal processing, or soaking, provided the phytic acid is present as the soluble potassium salt (Reddy *et al.* 1989). The extent of the hydrolysis of the higher inositol phosphates can be monitored using the HPLC method for phytic acid analysis (Lehrfeld, 1989). The latter, unlike the AOAC method (Harland & Oberleas, 1986), differentiates the hexaphosphate and pentaphosphate from the lower inositol phosphates. Only the former inhibit the bioavailability of Zn (Tao *et al.* 1986; Lönnerdal *et al.* 1989).

To be successful, these dietary modifications/diversifications must be introduced using well designed educational and social marketing projects aimed to change attitudes and dietary behaviours. To enhance their effectiveness and sustainability, they should be integrated into ongoing national health and nutrition programmes in developing countries which emphasize the broader health consequences of micronutrient deficiencies. This approach has been highly successful in the Philippines for controlling vitamin A deficiency (Solon, 1986). Implementation of these dietary strategies could have far reaching consequences for both maternal and infant health in many developing countries, decreasing morbidity and complications in pregnancy, reducing mortality during childbirth, risk of prematurity and low birth weight, and enhancing growth and development in infancy and childhood.

REFERENCES

Aggett, P. J. (1991). The assessment of zinc status: a personal view. *Proceedings of the Nutrition Society* **50**, 9–17.
Arcasoy, A., Çavdar, A. O. & Babacan, E. (1978). Decreased iron and zinc absorption in Turkish children with iron deficiency and geophagia. *Acta Hematologica* **60**, 76–84.
Baer, M. T. & King, J. C. (1984). Tissue zinc levels and zinc excretion during experimental zinc depletion in young men. *American Journal of Clinical Nutrition* **39**, 556–570.
Bates, C. J., Evans, P. H., Dardenne, M., Prentice, A., Lunn, P. G., Northrop-Clewes, C. A., Hoare, S., Cole, T. J., Horan, S. J., Longman, S. C., Stirling, D. & Aggett, P. J. (1993). A trial of zinc supplementation in young rural Gambian children. *British Journal of Nutrition* **69**, 243–255.
Beaton, G. H. (1985). Uses and limits of the use of the Recommended Dietary Allowances for evaluating dietary intake data. *American Journal of Clinical Nutrition* **41**, 155–164.
Bettger, W. J., Fish, T. J. & O'Dell, B. L. (1978). Effects of copper and zinc status of rats on erythrocyte stability and superoxide dismutase activity. *Proceedings of the Society for Experimental Biology and Medicine* **158**, 279–282.
Bindra, G. S., Gibson, R. S. & Thompson, L. U. (1986). [Phytate][calcium]/[zinc] ratios in Asian immigrant lacto-ovo vegetarian diets and their relationship to zinc nutriture. *Nutrition Research* **6**, 475–483.
Bradfield, R. B. & Hambidge, K. M. (1980). Problems with hair zinc as an indicator of body zinc status. *Lancet* **i**, 363.
Breskin, M. W., Worthington-Roberts, B. S., Knopp, R. H., Brown, Z., Plovie, B., Mottet, N. K. & Mills, J. L. (1983). First trimester serum zinc concentrations in human pregnancy. *American Journal of Clinical Nutrition* **38**, 943–953.
Brown, K., Akhtar, N. A., Robertson, A. D. & Ahmed, M. G. (1986). Lactational capacity of marginally nourished mothers: relationships between maternal nutritional status and quantity and proximate composition of milk. *Pediatrics* **78**, 909–919.
Brune, M., Rossander, L. & Hallberg, L. (1989). Iron absorption: no intestinal adaptation to a high-phytate diet. *American Journal of Clinical Nutrition* **49**, 542–545.
Butte, N. F., Villalpando, S., Wong, W. W., Flores-Huerta, S., Hernandez-Beltran, M. de J., O'Brian Smith, E. & Garza, C. (1992). Human milk intake and growth faltering of rural Mesoamerindian infants. *American Journal of Clinical Nutrition* **55**, 1109–1116.
Butterworth, C. E., Hatch, K., Cole, P., Sauberlich, H. E., Tamura, T., Cornwell, P. E. & Soong, S.-J. (1988). Zinc concentration in plasma and erythrocytes of subjects receiving folic acid supplementation. *American Journal of Clinical Nutrition* **47**, 484–486.
Buzina, R., Jušić, M., Sapunar, J. & Milanović, N. (1980). Zinc nutrition and taste acuity in school children with impaired growth. *American Journal of Clinical Nutrition* **33**, 2262–2267.
Campbell-Brown, M., Ward, R. J., Haines, A. P., North, W. R. S., Abraham, R. & McFadyen, I. R. (1985). Zinc and copper in Asian pregnancies – is there evidence for a nutritional deficiency? *British Journal of Obstetrics and Gynaecology* **92**, 875–885.
Carter, J. P., Grivetti, L. E., Davis, J. T., Nasiff, S., Mansour, A., Mousa, W. A., Atta, A., Patwardhan, V. N., Moneim, M. A., Abdou, I. A. & Darby, W. J. (1969). Growth and sexual development of adolescent Egyptian village boys. Effects of zinc, iron, and placebo supplementation. *American Journal of Clinical Nutrition* **22**, 59–78.
Casey, C. E., Neville, M. C. & Hambidge, K. M. (1989). Studies in human lactation: secretion of zinc, copper, and manganese in human milk. *American Journal of Clinical Nutrition* **49**, 773–785.
Castillo-Duran, C., Heresi, G., Fisberg, M. & Uauy, R. (1987). Controlled trial of zinc supplementation during recovery from malnutrition: effects on growth and immune function. *American Journal of Clinical Nutrition* **45**, 602–608.
Cavan, K. R., Gibson, R. S., Grazioso, C. F., Isalgue, A. M., Ruz, M. & Solomons, N. W. (1993a). Growth and body composition of periurban Guatemalan children in relation to zinc status: a cross-sectional study. *American Journal of Clinical Nutrition* **57**, 334–343.
Cavan, K. R., Gibson, R. S., Grazioso, C. F., Isalgue, A. M., Ruz, M. & Solomons, N. W. (1993b). Growth and body composition of periurban Guatemalan children in relation to zinc status: a zinc intervention trial. *American Journal of Clinical Nutrition* **57**, 344–352.
Çavdar, A. O., Babacan, E. & Arcasoy, A. (1980). Effect of nutrition on serum zinc concentration during pregnancy in Turkish women. *American Journal of Clinical Nutrition* **33**, 542–544.
Çavdar, A. O., Bahceci, M., Akar, N., Erten, J. & Yavuz, H. (1991). Effect of zinc supplementation in a Turkish woman with two previous anencephalic infants. *Gynecologic and Obstetrical Investigation* **32**, 123–125.
Cherry, F. F., Bennett, E. A., Bazzano, G. S., Johnson, L. K., Fosmire, G. J. & Batson, H. K. (1981). Plasma zinc in hypertension/toxemia and other reproductive variables in adolescent pregnancy. *American Journal of Clinical Nutrition* **34**, 2367–2375.
Cherry, F. F., Sandstead, H. H., Rojas, P., Johnson, L. K., Batson, H. K. & Wang, X. B. (1989). Adolescent pregnancy: associations among body weight, zinc nutriture, and pregnancy outcome. *American Journal of Clinical Nutrition* **50**, 945–954.
Chesters, J. K. & Will, M. (1978). The assessment of zinc status of an animal from the uptake of ^{65}Zn by the cells of whole blood in vitro. *British Journal of Nutrition* **38**, 297–306.

Cossack, Z. T. & Prasad, A. S. (1983). Effect of protein source on the bioavailability of zinc in human subjects. *Nutrition Research* **3**, 23–31.

Davies, N. T., Carswell, A. J. P. & Mills, C. F. (1985). The effect of variations in dietary calcium intake on the phytate-zinc interaction in rats. In *Trace Elements in Man and Animals* (*5th International Symposium*), pp. 456–457 [C. F. Mills, I. Bremner and J. K. Chesters, editors]. Farnham Royal: Commonwealth Agricultural Bureaux.

Dorea, J. G. (1993). Is zinc a first limiting nutrient in human milk? *Nutrition Research* **13**, 659–666.

Erten, J., Arcasoy, A., Çavdar, A. O. & Cin, S. (1978). Hair zinc levels in healthy and malnourished children. *American Journal of Clinical Nutrition* **31**, 1172–1174.

Ferguson, E. L. (1992). A comparison of food consumption patterns and zinc status of preschool children from Southern Malawi and Ghana. Ph.D. thesis, University of Guelph.

Ferguson, E. L., Gibson, R. S., Opare-Obisaw, C., Osei-Opare, F., Stephen, A. M., Lehrfeld, J. & Thompson, L. U. (1993*a*). The zinc, calcium, copper, manganese, nonstarch polysaccharide and phytate content of seventy-eight locally grown and prepared African foods. *Journal of Food Composition and Analysis* **6**, 87–99.

Ferguson, E. L., Gibson, R. S., Opare-Obisaw, C., Ounpuu, S., Thompson, L. U. & Lehrfeld, J. (1993*b*). The zinc nutriture of preschool children living in two African countries. *Journal of Nutrition* **123**, 1487–1496.

Ferguson, E. L., Gibson, R. S., Thompson, L. U. & Ounpuu, S. (1989*a*). Dietary calcium, phytate, and zinc intakes and the calcium, phytate, and zinc molar ratios of the diets of a selected group of East African children. *American Journal of Clinical Nutrition* **50**, 1450–1456.

Ferguson, E. L., Gibson, R. S., Thompson, L. U., Ounpuu, S. & Berry, M. (1988). Phytate, zinc, and calcium contents of 30 East African foods and their calculated phytate:Zn, Ca:phytate, and [Ca][phytate]/[Zn] molar ratios. *Journal of Food Composition and Analysis* **1**, 316–325.

Ferguson, E. L., Gibson, R. S., Weaver, S. D., Heywood, P., Heywood, A. & Yaman, C. (1989*b*). The mineral content of commonly consumed Malawian and Papua New Guinean foods. *Journal of Food Composition and Analysis* **2**, 260–272.

Filteau, S. M. & Tomkins, A. M. (1994). Micronutrients and tropical infections. *Transactions of the Royal Society of Tropical Medicine and Hygiene* **88**, 1–3.

Fitzgerald, S. L., Gibson, R. S., Quan de Serrano, J., Portocarrero, L., Vasquez, A., de Zepeda, E., Lopez-Palacios, C. Y., Thompson, L. U., Stephen, A. M. & Solomons, N. W. (1993). Trace element intakes and dietary phytate/Zn and Ca × phytate/Zn millimolar ratios of periurban Guatemalan women during the third trimester of pregnancy. *American Journal of Clinical Nutrition* **57**, 195–201.

Fordyce, E. J., Forbes, R. M., Robins, K. R. & Erdman, J. W. (1987). Phytate × calcium/zinc molar ratios: are they predictive of zinc bioavailability? *Journal of Food Science* **52**, 440–444.

Gibson, R. S. (1989). Assessment of trace element status in humans. *Progress in Food and Nutrition Science* **13**, 67–111.

Gibson, R. S., Ferguson, E. F., Smit Vanderkooy, P. D. & MacDonald, A. C. (1989*a*). Seasonal variations in hair zinc concentrations in Canadian and African children. *Science of the Total Environment* **84**, 291–298.

Gibson, R. S., Heywood, A., Yaman, C., Sohlström, A., Thompson, L. U. & Heywood, P. (1991*a*). Growth in children from the Wosera subdistrict, Papua New Guinea, in relation to energy and protein intakes and zinc status. *American Journal of Clinical Nutrition* **53**, 782–789.

Gibson, R. S., Smit Vanderkooy, P. D., MacDonald, A. C., Goldman, A., Ryan, B. A. & Berry, M. (1989*b*). A growth-limiting, mild zinc-deficiency syndrome in some Southern Ontario boys with low height percentiles. *American Journal of Clinical Nutrition* **49**, 1266–1273.

Gibson, R. S., Smit Vanderkooy, P. D. & Thompson, L. U. (1991*b*). Dietary phytate × calcium/zinc millimolar ratios and zinc nutriture in some Ontario preschool children. *Biological Trace Element Research* **30**, 87–94.

Giugliano, R. & Millward, D. J. (1984). Growth and zinc homeostasis in the severely Zn-deficient rat. *British Journal of Nutrition* **52**, 545–560.

Golden, M. H. N. (1989). The diagnosis of zinc deficiency. In *Zinc in Human Biology* (*International Life Sciences Institute of Human Nutrition Reviews*), pp. 324–333 [C. F. Mills, editor]. Berlin: Springer-Verlag.

Golden, M. H. N. & Golden, B. E. (1981*a*). Effect of zinc supplementation on the dietary intake, rate of weight gain, and energy cost of tissue deposition in children recovering from severe malnutrition. *American Journal of Clinical Nutrition* **34**, 900–908.

Golden, M. H. N. & Golden, B. E. (1981*b*). Trace elements: potential importance in human nutrition with particular reference to zinc and vanadium. *British Medical Bulletin* **37**, 31–36.

Grider, A., Bailey, L. B. & Cousins, R. J. (1990). Erythrocyte metallothionein as an index of zinc status in humans. *Proceedings of the National Academy of Sciences, USA* **87**, 1259–1262.

Hambidge, K. M. (1982). Hair analyses: worthless for vitamins, limited for minerals. *American Journal of Clinical Nutrition* **36**, 943–949.

Hambidge, K. M. (1989). Mild zinc deficiency in children. In *Zinc in Human Biology* (*International Life Sciences Institute of Human Nutrition Reviews*), pp. 285–295 [C. F. Mills, editor]. Berlin: Springer-Verlag.

Hambidge, K. M., Hambidge, C., Jacobs, M. & Baum, J. D. (1972). Low levels of zinc in hair, anorexia, poor growth, and hypogeusia in children. *Pediatric Research* **6**, 868–874.

Hambidge, K. M., Krebs, N. F., Jacobs, M. A., Favier, A., Guyette, L. & Ikle, D. N. (1983). Zinc nutritional status during pregnancy: a longitudinal study. *American Journal of Clinical Nutrition* **37**, 429–442.

Hambidge, K. M., Neldner, K. H. & Walravens, P. A. (1975). Zinc, acrodermatitis enteropathica, and congenital malformations. *Lancet* **i**, 577–578.

Harland, B. F. & Oberleas, D. (1986). Anion-exchange method for determination of phytate in foods: collaborative study. *Journal of the Association of Official Analytical Chemists* **69**, 667–670.

Harland, B. F. & Peterson, M. (1978). Nutritional status of lacto-ovo vegetarian Trappist monks. *Journal of the American Dietetic Association* **72**, 259–264.

Hinks, L. J., Ogilvy-Stuart, A., Hambidge, K. M. & Walker, V. (1989). Maternal zinc and selenium status in pregnancies with a neural tube defect or elevated plasma α-fetoprotein. *British Journal of Obstetrics and Gynaecology* **96**, 61–66.

Hunt, I. F., Murphy, N. J., Cleaver, A. E., Faraji, B., Swendseid, M. E., Browdy, B. L., Coulson, A. H., Clark, V. A., Settlage, R. H. & Smith, J. C. (1985). Zinc supplementation during pregnancy in low-income teenagers of Mexican descent: effects on selected blood constituents and on progress and outcome of pregnancy. *American Journal of Clinical Nutrition* **42**, 815–828.

Hunt, I. F., Murphy, N. J., Cleaver, A. E., Faraji, B., Swendseid, M. E., Coulson, A. H., Clark, V. A., Browdy, B. L., Cabalum, M. J. & Smith, J. C. (1984). Zinc supplementation during pregnancy: effects on selected blood constituents and on progress and outcome of pregnancy in low-income women of Mexican descent. *American Journal of Clinical Nutrition* **40**, 508–521.

Hurley, L. S. & Swenerton, H. (1966). Congenital malformations resulting from zinc deficiency in rats. *Proceedings of the Society for Experimental Biology and Medicine* **123**, 692–696.

Jackson, M. J., Giugliano, R., Giugliano, L. G., Oliveira, E. F., Shrimpton, R. & Swainbank, I. G. (1988). Stable isotope metabolic studies of zinc nutrition in slum-dwelling lactating women in the Amazon valley. *British Journal of Nutrition* **59**, 193–203.

Jameson, S. (1976). Effects of zinc deficiency in human reproduction. *Acta Medica Scandinavica* Suppl. 593, 5–89.

Jameson, S., Burtsröm, M. & Hellsing, K. (1990). Zinc status in pregnancy. The effect of zinc therapy on perinatal mortality. In *International Symposium on Trace Elements in Man and Animals VII*, pp. 4.8–4.9 [B. Momcilovic, editor].

Jones, R. B., Keeling, P. W. N., Hilton, P. J. & Thompson, R. P. H. (1981). The relationship between leucocyte and muscle zinc in health and disease. *Clinical Science* **60**, 237–239.

Karra, M. V., Udipi, S. A., Kirksey, A. & Roepke, J. L. B. (1986). Changes in specific nutrients in breast milk during extended lactation. *American Journal of Clinical Nutrition* **43**, 495–503.

Khanum, S., Alam, A. N., Anwar, I., Akbar Ali, M. & Mujibur Rahaman, M. (1988). Effect of zinc supplementation on the dietary intake and weight gain of Bangladeshi children recovering from protein-energy malnutrition. *European Journal of Clinical Nutrition* **42**, 709–714.

King, J. C. (1986). Assessment of techniques for determining human zinc requirements. *Journal of the American Dietetic Association* **86**, 1523–1528.

Kirsten, G. F., Heese, H. de V., Watermeyer, S., Dempster, W. S., Pocock, F. & Varkvisser, H. (1985). Zinc and copper levels in the breast-milk of Cape Town mothers. *South African Medical Journal* **68**, 402–405.

Koo, W. W. K., Succop, P. & Hambidge, K. M. (1989). Serum alkaline phosphatase and serum zinc concentrations in preterm infants with rickets and fractures. *American Journal of Diseases of Children* **143**, 1342–1345.

Krebs, N. F., Hambidge, K. M., Hagerman, R. J., Peirce, P. L., Johnson, K. M., English, J. L., Miller, L. L. & Fennessey, P. V. (1988). Effects of pharmacologic doses of folate on zinc absorption and zinc status. In *Nutrient Availability: Chemical and Biological Aspects* (*Royal Society of Chemistry Special Publication No. 72*), pp. 226–228 [D. A. T. Southgate, I. T. Johnson and G. R. Fenwick, editors]. Cambridge: Royal Society of Chemistry.

Krebs, N. F., Hambidge, K. M., Jacobs, M. A. & Rasbach, J. O. (1985). The effects of a dietary zinc supplement during lactation on longitudinal changes in maternal zinc status and milk zinc concentrations. *American Journal of Clinical Nutrition* **41**, 560–570.

Krebs, N. F., Hambidge, K. M. & Walravens, P. A. (1984). Increased food intake of young children receiving a zinc supplement. *American Journal of Diseases of Children* **138**, 270–273.

Lehrfeld, J. (1989). High-performance liquid chromatography analysis of phytic acid on a pH-stable, macroporous polymer column. *Cereal Chemistry* **66**, 510–515.

Lönnerdal, B., Sandberg, A.-S., Sandström, B. & Kunz, C. (1989). Inhibitory effects of phytic acid and other inositol phosphates on zinc and calcium absorption in suckling rats. *Journal of Nutrition* **119**, 211–214.

McMichael, A. J., Dreosti, I. E., Gibson, G. T., Hartshorne, J. M., Buckley, R. A. & Colley, D. P. (1982). A prospective study of serial maternal serum zinc levels and pregnancy outcome. *Early Human Development* **7**, 59–69.

Mahomed, K., James, D. K., Golding, J. & McCabe, R. (1989). Zinc supplementation during pregnancy: a double blind randomised controlled trial. *British Medical Journal* **299**, 826–830.

Malhotra, A., Fairweather-Tait, S. J., Wharton, P. A. & Gee, H. (1990). Placental zinc in normal and intra-uterine growth-retarded pregnancies. *British Journal of Nutrition* **63**, 613–621.

Mbofung, C. M. F. & Atinmo, T. (1987). Trace element nutriture of Nigerians. *World Review of Nutrition and Dietetics* **51**, 105–139.

Meadows, N. J., Ruse, W., Smith, M. F., Day, J., Keeling, P. W. N., Scopes, J. W., Thompson, R. P. H. & Bloxam, D. L. (1981). Zinc and small babies. *Lancet* **ii**, 1135–1137.

Mills, C. F. (1985). Dietary interactions involving the trace elements. *Annual Review of Nutrition* **5**, 173–193.
Milne, D. B., Canfield, W. K., Mahalko, J. R. & Sandstead, H. H. (1984). Effect of oral folic acid supplements on zinc, copper, and iron absorption and excretion. *American Journal of Clinical Nutrition* **39**, 535–539.
Milne, D. B., Ralston, N. V. C. & Wallwork, J. C. (1985). Zinc content of cellular components of blood: methods for cell separation and analysis evaluated. *Clinical Chemistry* **31**, 65–69.
Monsen, E. R. (1988). Iron nutrition and absorption: dietary factors which impact iron bioavailability. *Journal of the American Dietetic Association* **88**, 786–791.
Moser, P. B., Reynolds, R. D., Acharya, S., Howard, M. P., Andon, M. B. & Lewis, S. A. (1988). Copper, iron, zinc, and selenium dietary intake and status of Nepalese lactating women and their breast-fed infants. *American Journal of Clinical Nutrition* **47**, 729–734.
Moser-Veillon, P. B. & Reynolds, R. D. (1990). A longitudinal study of pyridoxine and zinc supplementation of lactating women. *American Journal of Clinical Nutrition* **52**, 135–141.
Mukherjee, M. D., Sandstead, H. H., Ratnaparkhi, M. V., Johnson, L. K., Milne, D. B. & Stelling, H. P. (1984). Maternal zinc, iron, folic acid, and protein nurture and outcome of human pregnancy. *American Journal of Clinical Nutrition* **40**, 496–507.
Murphy, S. P., Beaton, G. H. & Calloway, D. H. (1992). Estimated mineral intakes of toddlers: predicted prevalence of inadequacy in village populations in Egypt, Kenya, and Mexico. *American Journal of Clinical Nutrition* **56**, 565–572.
National Academy of Sciences. (1991). *Nutrition During Pregnancy*. Washington, DC: National Academy Press.
National Research Council. (1986). *Nutrient Adequacy: Assessment Using Food Consumption Surveys*. Washington, DC: National Academy Press.
Neggers, Y. H., Cutter, G. R., Acton, R. T., Alvarez, J. O., Bonner, J. L., Goldenberg, R. L., Go, R. C. P. & Roseman, J. M. (1990). A positive association between maternal serum zinc concentration and birth weight. *American Journal of Clinical Nutrition* **51**, 678–684.
O'Dell, B. L., Reynolds, G. & Reeves, P. G. (1977). Analogous effects of zinc deficiency and aspirin toxicity in the pregnant rat. *Journal of Nutrition* **107**, 1222–1228.
Oberleas, D. & Harland, B. F. (1981). Phytate content of foods: effect on dietary zinc bioavailability. *Journal of the American Dietetic Association* **79**, 433–436.
Okonofua, F. E., Amole, F. A., Emofurieta, W. O. & Ugwu, N. C. (1989). Zinc and copper concentration in plasma of pregnant women in Nigeria. *International Journal of Gynecology and Obstetrics* **29**, 19–23.
Okonofua, F. E., Isinkaye, A., Onwudiegwu, U., Amole, F. A., Emofurieta, W. A. & Ugwu, N. C. (1990). Plasma zinc and copper in pregnant Nigerian women at term and their newborn babies. *International Journal of Gynecology and Obstetrics* **32**, 243–245.
Prasad, A. S. & Cossack, Z. T. (1982). Neutrophil zinc: an indicator of zinc status in man. *Transactions of the Association of American Physicians* **95**, 165–176.
Prasad, A. S., Meftah, S., Abdallah, J., Kaplan, J., Brewer, G. J., Bach, J. F. & Dardenne, M. (1988). Serum thymulin in human zinc deficiency. *Journal of Clinical Investigation* **82**, 1202–1210.
Prasad, A. S., Miale, A., Farid, Z., Sandstead, H. H. & Schulert, A. R. (1963). Zinc metabolism in patients with syndrome of iron deficiency anemia, hepatosplenomegaly, dwarfism, and hypogonadism. *Journal of Laboratory and Clinical Medicine* **61**, 537–549.
Prema, K. (1980). Predictive value of serum copper and zinc in normal and abnormal pregnancy. *Indian Journal of Medical Research* **71**, 554–560.
Reddy, N. R., Pierson, M. D., Sathe, S. K. & Salunkhe, D. K. (1989). *Phytates in Cereals and Legumes*. Boca Raton, FL: CRC Press.
Rodriguez, A., Venegas, G. V. & Torres, S. (1991). Zinc supplementation of infants fetal malnourished. In *Reunión de la Sociedad Latinoamericana de Oncologia Pediatrica VIII*.
Ronaghy, H. S., Reinhold, J. G., Mahloudji, M., Ghavami, P., Spivey Fox, M. R. & Halsted, J. A. (1974). Zinc supplementation of malnourished schoolboys in Iran: increased growth and other effects. *American Journal of Clinical Nutrition* **27**, 112–121.
Ronaghy, H. S., Spivey Fox, M. R., Garn, S. M., Israel, H., Harp, A., Moe, P. G. & Halsted, J. A. (1969). Controlled zinc supplementation for malnourished school boys: a pilot experiment. *American Journal of Clinical Nutrition* **22**, 1279–1289.
Rothbaum, R. J., Maur, P. R. & Farrell, M. K. (1982). Serum alkaline phosphatase and zinc undernutrition in infants with chronic diarrhea. *American Journal of Clinical Nutrition* **35**, 595–598.
Roy, S. K., Behrens, R. H., Haider, R., Akramuzzaman, S. M., Mahalanabis, D., Wahed, M. A. & Tomkins, A. M. (1992). Impact of zinc supplementation on intestinal permeability in Bangladeshi children with acute diarrhoea and persistent diarrhoea syndrome. *Journal of Pediatric Gastroenterology and Nutrition* **15**, 289–296.
Roy, S. K., Tomkins, A. M., Haider, R., Behrens, R. H. & Akramuzzaman, S. M. (1993). The importance of zinc deficiency in stunting and morbidity. *International Congress of Nutrition XV*, p. 713 (Abstr.)
Royal Tropical Institute, Amsterdam. (1987). *Weaning Food – a New Approach to Small-scale Weaning Food Production from Indigenous Raw Materials in Tropical Countries*, 2nd edn. Amsterdam: Royal Tropical Institute.
Ruz, M., Cavan, K. R., Bettger, W. J. & Gibson, R. S. (1992). Erythrocytes, erythrocyte membranes, neutrophils and platelets as biopsy materials for the assessment of zinc status in humans. *British Journal of Nutrition* **68**, 515–527.

Ruz, M., Cavan, K. R., Bettger, W. J., Thompson, L. U., Berry, M. & Gibson, R. S. (1991). Development of a dietary model for the study of mild zinc deficiency in humans and evaluation of some biochemical and functional indices of zinc status. *American Journal of Clinical Nutrition* **53**, 1295–1303.

Ruz, M. & Solomons, N. W. (1990). Mineral excretion during acute dehydrating diarrhea treated with oral rehydration therapy. *Pediatric Research* **27**, 170–175.

Sandström, B. (1989). Dietary pattern and zinc supply. In *Zinc in Human Biology (International Life Sciences Institute Human Nutrition Reviews)*, pp. 351–363 [C. F. Mills, editor]. Berlin: Springer-Verlag.

Sandström, B., Almgren, A., Kivistö, B. & Cederblad, Å. (1989). Effect of protein level and protein source on zinc absorption in humans. *Journal of Nutrition* **119**, 48–53.

Sandström, B., Arvidsson, B., Cederblad, Å. & Björn-Rasmussen, E. (1980). Zinc absorption from composite meals. 1. The significance of wheat extraction rate, zinc, calcium, and protein content in meals based on bread. *American Journal of Clinical Nutrition* **33**, 739–745.

Sandström, B. & Lönnerdal, B. (1989). Promoters and antagonists of zinc absorption. In *Zinc in Human Biology (International Life Sciences Institute Human Nutrition Reviews)*, pp. 57–78 [C. F. Mills, editor]. Berlin: Springer-Verlag.

Schiliro, G., Russo, A., Azzia, N., Mancuso, G. R., Di Gregorio, F. D., Romeo, M. A., Fallico, R. & Sciacca, S. (1987). Leucocyte alkaline phosphatase (LAP): a useful marker of zinc status in β-thalassemic patients. *American Journal of Pediatric Hematology/Oncology* **9**, 149–152.

Schlesinger, L., Arevalo, M., Arredondo, S., Diaz, M., Lönnerdal, B. & Stekel, A. (1992). Effect of a zinc-fortified formula on immunocompetence and growth of malnourished infants. *American Journal of Clinical Nutrition* **56**, 491–498.

Shrimpton, R., Alencar, F. H., Vasconcelos, J. C. & Rocha, Y. R. (1985). Effect of maternal zinc supplementation on the growth and diarrhoeal status of breast fed infants. *Nutrition Research* Suppl. 1, 338S–342S.

Shrimpton, R., Marinho, H. A., Rocha, Y. S. & Alencar, F. H. (1983). Zinc supplementation in urban Amazonian mothers: concentrations of Zn and retinol in maternal serum and milk. *Proceedings of the Nutrition Society* **42**, 122A.

Simmer, K., Ahmed, S., Carlsson, L. & Thompson, R. P. H. (1990). Breast milk zinc and copper concentrations in Bangladesh. *British Journal of Nutrition* **63**, 91–96.

Simmer, K., Khanum, S., Carlsson, L. & Thompson, R. P. H. (1988). Nutritional rehabilitation in Bangladesh – the importance of zinc. *American Journal of Clinical Nutrition* **47**, 1036–1040.

Simmer, K., Lort-Phillips, L., James, C. & Thompson, R. P. H. (1991). A double-blind trial of zinc supplementation in pregnancy. *European Journal of Clinical Nutrition* **45**, 139–144.

Simmer, K. & Thompson, R. P. H. (1985). Maternal zinc and intrauterine growth retardation. *Clinical Science* **68**, 395–399.

Smit-Vanderkooy, P. D. & Gibson, R. S. (1987). Food consumption patterns of Canadian preschool children in relation to zinc and growth status. *American Journal of Clinical Nutrition* **45**, 609–616.

Smith, J. C., Udomkesmalee, E. & Dhanamitta, S. (1993). Effect of vitamin A and zinc supplementation of children in Thailand and Belize, Central America. *International Congress of Nutrition XV*, p. 618 (Abstr.).

Solomons, N. W. (1981). Zinc and copper in human nutrition. In *Nutrition in the 1980s: Constraints on Our Knowledge (Progress in Clinical and Biological Research* Vol. 67), pp. 97–127 [N. Selvey and P. L. White, editors]. New York: Alan R. Liss.

Solomons, N. W. (1986). Competitive interaction of iron and zinc in the diet: consequences for human nutrition. *Journal of Nutrition* **116**, 927–935.

Solon, M. A. (1986). Control of vitamin A deficiency by education and the public health approach. In *Vitamin A Deficiency and its Control*, pp. 285–318 [J. C. Bauernfeind, editor]. New York: Academic Press.

Soltan, M. H. & Jenkins, D. M. (1982). Maternal and fetal plasma zinc concentration and fetal abnormality. *British Journal of Obstetrics and Gynaecology* **89**, 56–58.

Svanberg, U. & Sandberg, A. S. (1988). Improved iron availability in weaning foods. In *Improved Young Child Feeding in Eastern and Southern Africa: Household Level Food Technology*, pp. 366–373 [D. Alnwick, S. Moses and O. G. Schmidt, editors]. Ottawa: International Development Research Center.

Swanson, C. A. & King, J. C. (1987). Zinc and pregnancy outcome. *American Journal of Clinical Nutrition* **46**, 763–771.

Tao, S.-H., Spivey Fox, M. R., Phillippy, B. Q., Fry, B. E., Johnson, M. L. & Johnston, M. R. (1986). Effects of inositol phosphates on mineral utilization. *Federation Proceedings* **45**, 819.

Thauvin, E., Fusselier, M., Arnaud, J., Faure, H., Favier, M., Coudray, C., Richard, M.-J. & Favier, A. (1992). Effects of a multivitamin mineral supplement on zinc and copper status during pregnancy. *Biological Trace Element Research* **32**, 405–414.

Thompson, R. P. H. (1991). Assessment of zinc status. *Proceedings of the Nutrition Society* **50**, 19–28.

Tomkins, A., Behrens, R. & Roy, S. (1993). The role of zinc and vitamin A deficiency in diarrhoeal syndromes in developing countries. *Proceedings of the Nutrition Society* **52**, 131–142.

Torre, M., Rodriguez, A. R. & Saura-Calixto, F. (1991). Effects of dietary fiber and phytic acid on mineral availability. *CRC Critical Reviews in Food Science and Nutrition* **30**, 1–22.

Turnlund, J. R., King, J. C., Keyes, W. R., Gong, B. & Michel, M. C. (1984). A stable isotope study of zinc absorption in young men: effects of phytate and α-cellulose. *American Journal of Clinical Nutrition* **40**, 1071–1077.

Tuttle, S., Aggett, P. J., Campbell, D. & MacGillivray, I. (1985). Zinc and copper in human pregnancy: a longitudinal study in normal primigravidae and in primigravidae at risk of delivering a growth retarded baby. *American Journal of Clinical Nutrition* **41**, 1032–1041.

Udomkesmalee, E., Dhanamitta, S., Sirisinha, S., Charoenkiatkul, S., Tuntipopipat, S., Banjong, O., Rojroongwasinkul, N., Kramer, T. R. & Smith, J. C. (1992). Effect of vitamin A and zinc supplementation on the nutriture of children in Northeast Thailand. *American Journal of Clinical Nutrition* **56**, 50–57.

Udomkesmalee, E., Dhanamitta, S., Yhoung-Aree, J., Rojroongwasinkul, N. & Smith, J. C. (1990). Biochemical evidence suggestive of suboptimal zinc and vitamin A status in schoolchildren in Northeast Thailand. *American Journal of Clinical Nutrition* **52**, 564–567.

United Nations. (1991). Some options for improving nutrition in the 1990s. *SCN News* No. 7, Suppl. 16–18.

Valberg, L. S., Flanagan, P. R. & Chamberlain, M. J. (1984). Effects of iron, tin, and copper on zinc absorption in humans. *American Journal of Clinical Nutrition* **40**, 536–541.

Walker, A. F. (1990). The contribution of weaning foods to protein–energy malnutrition. *Nutrition Research Reviews* **3**, 25–47.

Wallock, L. M., King, J. C., Hambidge, K. M., English-Westcott, J. E. & Pritts, J. (1993). Meal-induced changes in plasma, erythrocyte, and urinary zinc concentrations in adult women. *American Journal of Clinical Nutrition* **58**, 695–701.

Walravens, P. A., Chakar, A., Mokni, R., Denise, J. & Lemonnier, D. (1992). Zinc supplements in breastfed infants. *Lancet* **340**, 683–685.

Walravens, P. A. & Hambidge, K. M. (1976). Growth of infants fed a zinc supplemented formula. *American Journal of Clinical Nutrition* **29**, 1114–1121.

Walravens, P. A., Hambidge, K. M. & Koepfer, D. M. (1989). Zinc supplementation in infants with a nutritional pattern of failure to thrive: a double-blind controlled study. *Pediatrics* **83**, 532–538.

Walravens, P. A., Krebs, N. F. & Hambidge, K. M. (1983). Linear growth of low income preschool children receiving a zinc supplement. *American Journal of Clinical Nutrition* **38**, 195–201.

Wells, J. L., James, D. K., Luxton, R. & Pennock, C. A. (1987). Maternal leucocyte zinc deficiency at start of third trimester as a predictor of fetal growth retardation. *British Medical Journal* **294**, 1054–1056.

Wise, A. (1983). Dietary factors determining the biological activities of phytate. *Nutrition Abstracts and Reviews* **53**, 791–806.

Xue-Cun, C., Tai-An, Y., Jin-Sheng, H., Qui-Yan, M., Zhi-Min, H. & Li-Xiang, L. (1985). Low levels of zinc in hair and blood, pica, anorexia, and poor growth in Chinese preschool children. *American Journal of Clinical Nutrition* **42**, 694–700.

Zlotkin, S. H. & Casselman, C. (1988). Diurnal variation in urinary zinc excretion and the use of zinc/Cr ratio from random urine samples to monitor zinc status. *Canadian Federation of Biological Sciences*, Quebec, p. 624 (Abstr.).

Printed in Great Britain

ANTICARCINOGENIC FACTORS IN PLANT FOODS: A NEW CLASS OF NUTRIENTS?

I. T. JOHNSON[1]*, G. WILLIAMSON[2] AND S. R. R. MUSK[1]

Departments of [1]Nutrition, Diet and Health and [2]Food Molecular Biochemistry, Institute of Food Research, Norwich Laboratory, Norwich Research Park, Colney, Norwich, NR4 7UA

CONTENTS

INTRODUCTION	175
MECHANISMS OF CARCINOGENESIS	176
BLOCKING AGENTS	178
PHASE I METABOLISM – ACTIVATION BY MONOOXYGENASES	178
Free radical mediated damage	180
Alternative route to phase I metabolism – quinone reductase (QR)	181
CONJUGATION BY PHASE II ENZYMES	183
INDIRECT EFFECTS *VIA* ENTERIC BACTERIAL METABOLISM	184
INDUCERS OF DNA REPAIR	185
SUPPRESSING AGENTS	186
INHIBITORS OF CELL PROLIFERATION	186
Modification of intracellular signalling	187
Inhibition of oncogene expression	188
Polyamine metabolism	188
Oestrogen metabolism	189
DIRECT ACTING MODULATORS OF CELL DAMAGE	190
Suppression of free radical production	190
Selective cytotoxins	191
INDUCERS OF CELLULAR DIFFERENTIATION	191
ANTIMETASTATIC AGENTS	192
IMMUNOMODULATION	192
A NEW CLASS OF NUTRIENTS?	192
REFERENCES	194

INTRODUCTION

In their now famous and widely quoted study of the role of environmental factors in the aetiology of cancer, Doll & Peto (1981) estimated that the proportion of cancer deaths attributable to an adverse effect of diet was approximately 30% in the USA, the same as that due to tobacco. However, unlike tobacco, for which the risk was well defined and, at

---

* To whom correspondence should be addressed.

least in theory, easily avoidable, the effects of diet were considered so complex that their mechanisms of action and their means of manipulation could only be guessed at. One of the strongest signals to emerge from epidemiological studies of diet and cancer incidence over the last decade has been the protective effect of diets rich in vegetables and fruits. Block *et al.* (1992) reviewed nearly 200 studies of the influence of fruit and vegetable intake on the risk of cancer in man. In 156 of these studies the results were expressed in terms of relative risk, and of these 128 showed a statistically significant protective effect. The results were particularly consistent for cancers of the alimentary tract, lung, breast and female reproductive organs, and were relatively weak only for cancer of the prostate. It is probable therefore that in industrialized societies many of the important effects of diet are protective, and that dietary strategies for the avoidance of cancer require the maximization of these protective effects. Initiatives such as the '5 A Day for Better Health' campaign, by which the National Cancer Institute has recently sought to promote the consumption of vegetables and fruit in the USA, are a practical reflection of this principle, but the mechanisms through which the various components of these foods may act remain uncertain.

Interest in the anticarcinogenic effects of foods has a relatively long history. Crabtree (1947) defined an anticarcinogen as any factor "which delays or prevents the emergence of malignant characters in any tissue of any species of organism". Such a broad definition encompasses nutrients as well as other biologically active compounds which fall outside any currently accepted definition of nutrients. Certainly vegetables and fruits are a rich source of micronutrients, including several which are intimately involved in cell proliferation and the maintenance of tissue integrity. These include the folates, and the carotenoids which, together with vitamin E and ascorbate, are thought also to protect against oxidative damage to DNA and other cellular components (Diplock, 1991). There is good epidemiological evidence to suggest that levels of carotenoids (Stähelin *et al.* 1991; Ziegler, 1991) and vitamin E in the serum (Murphy *et al.* 1990; Comstock *et al.* 1991) are inversely related to risk of cancer. Nevertheless, the protective effects of these nutrients remain to be established, and it may be that high serum levels are merely markers for a high intake of plant foods.

In this paper we are concerned with the very large number of biologically active 'non-nutrient' compounds in plant foods for which potentially anticarcinogenic effects have been demonstrated experimentally. The growing realization that such compounds exert biochemical and physiological effects in humans raises important theoretical and practical issues. Many of these substances have previously been regarded as potential toxicants. If their biological activity contributes to the protective effects of fruit and vegetables against cancer, what is the balance of risk and benefit? Furthermore, if their protective value is proven, should they be considered for classification as micronutrients in their own right? In this review we will consider these questions in the light of the known mechanisms of action of the biologically active non-nutrients.

MECHANISMS OF CARCINOGENESIS

One difficulty in any discussion of anticarcinogenic components of plant foods is the development of a satisfactory classification scheme. The number of secondary metabolites present in fruits and vegetables is very large and many of them appear to inhibit carcinogenesis by more than one mechanism. They are therefore best classified in terms of their biological effects rather than their chemistry, but such a scheme requires a knowledge of their mechanisms of action which in many cases does not yet exist. Much insight has been gained from studies in which vegetables, fruits, or compounds isolated from them

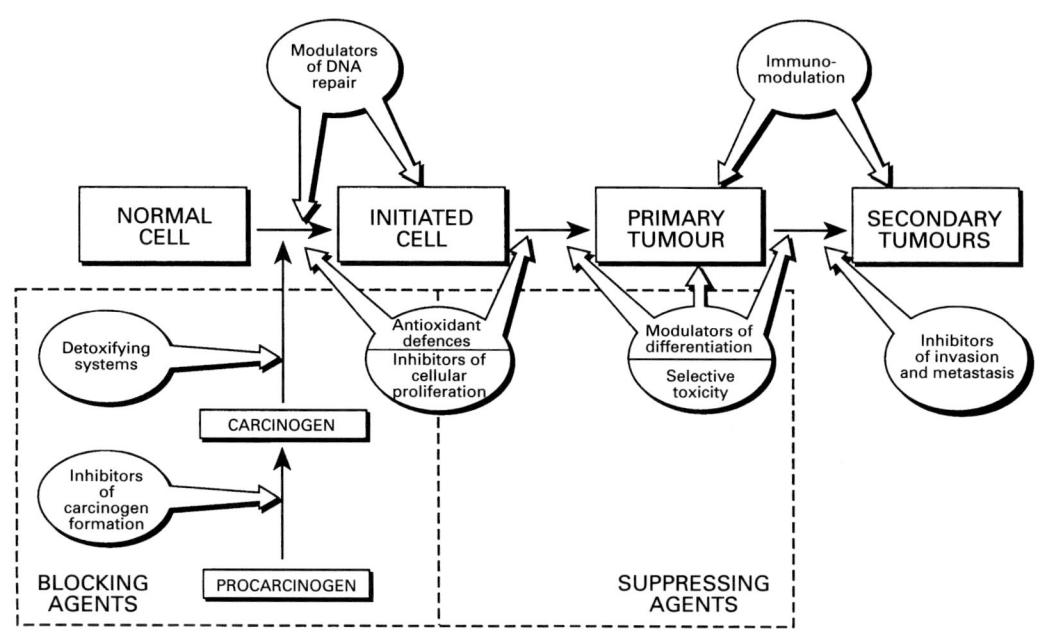

Fig. 1. Mechanisms and sites of interaction whereby protective factors may inhibit the carcinogenic process.

have been administered to experimental animals treated with chemical carcinogens. Before discussing this and other approaches to classification, however, the nature of tumours and the process of carcinogenesis will be briefly reviewed.

Tumour cells are typified by a loss of responsiveness to some or all of the factors regulating cell growth, differentiation and programmed death (apoptosis) in the tissue from which they have arisen. There is usually reduced structural and functional specialization, and malignant tumour cells have the capacity to invade adjacent mesenchyme, and ultimately to migrate to distant sites, giving rise to secondary tumours (metastases). In the simplest model of carcinogenesis the process is assumed to occur in two stages – initiation and promotion. Initiation is the primary event in which cellular DNA undergoes damage which remains unrepaired or becomes misrepaired. The resulting acquired somatic mutation is reproduced at mitosis, giving rise to a clonal population of initiated cells. The initiating event in carcinogenesis is often the result of DNA adduct formation with a genotoxic chemical, but endogenous production of free radicals may be equally important.

Initiated cells do not necessarily give rise to a malignant tumour until they have undergone 'promotion', a process which facilitates their full transformation to an invasive state. Chemical substances which function as promoters are not always genotoxic, but they are often mitogenic, and they may interfere with the expression of genes controlling differentiation and growth. However, some compounds, including polycyclic aromatic hydrocarbons and nitrosamines, are classified as 'complete carcinogens' because they are capable of inducing tumours in experimental animals without the need for exposure to chemical promoters.

Much of the pioneering work on the anticarcinogenic properties of naturally occurring compounds has been carried out by Wattenberg and his coworkers, who proposed a system of classification based on the stage of carcinogenesis at which they act (Wattenberg, 1985). According to this scheme, anticarcinogens are subdivided into two major classes defined operationally as 'blocking agents' and 'suppressing agents'. Blocking agents are typically compounds which have been found to be effective when given immediately before or during treatment with chemical carcinogens. They are thought to prevent initiation, either by inhibiting the formation of carcinogens from precursor compounds, or by preventing the active carcinogenic species from acting upon its cellular target. In contrast, suppressing agents are thought to act by preventing the progression of initiated cells to fully transformed tumour cells. Such compounds inhibit the emergence of tumours even when given after treatment with a complete carcinogen or a combination of incomplete carcinogen and promoting agent. We will retain here the general concepts of blocking and suppressing agents, while exploring the extent to which the mechanisms of blocking and suppression have been defined. Fig. 1 summarizes the various possible mechanisms for the inhibition of initiation, promotion and metastasis.

BLOCKING AGENTS

The first line of defence against chemical carcinogenesis is the ability of tissues such as those of the liver and intestinal mucosa to intercept and detoxify potentially damaging environmental substances. Many carcinogenic compounds enter cells by passive diffusion (Landers & Bunce, 1991), and are detoxified by several mechanisms. Fig. 2 charts the detoxification pathways which lead to an excretable metabolite of a carcinogen which is represented in this example by a quinone. This figure will form the basis for much of the subsequent discussion. For convenience the metabolic pathway is usually divided into Phase I (which often involves carcinogen activation), Phase II (conjugation) and Phase III (transport out of the cell: Prochaska *et al.* 1985; Ishikawa, 1992). Paradoxically, the metabolic pathways involved in detoxification may also serve to generate potentially damaging chemical species. Products of phase I reactions in particular may lead to free radical mediated damage of lipids, protein, carbohydrate and DNA, and many of the activated compounds can form DNA adducts. One of the earliest blocking agents to be identified was disulfiram, which inhibits induction of large bowel tumours by 1,2-dimethylhydrazine in rodents by preventing its metabolic conversion to an active form (Fiala *et al.* 1977). Subsequently, a variety of naturally occurring non-nutrient compounds, including aromatic isothiocyanates and indoles from cruciferous vegetables and organo-sulphur compounds found particularly in allium species, have been found to block chemical carcinogenesis in animals (Wattenberg, 1993). In this section we will review both non-nutrient and some nutrient blocking agents, and show how they play a vital role in regulating the efficiency of carcinogen metabolism. We will also review mechanisms whereby the adverse effects of endogenous free radicals may be blocked by naturally occurring plant constituents.

PHASE I METABOLISM – ACTIVATION BY MONOOXYGENASES

Hydrophobic carcinogens partition into the cell membrane, where they are activated by membrane bound monooxygenase enzymes, primarily cytochrome P450 (Black & Coon, 1987; Guengerich, 1992*a*), and flavin dependent enzymes (Ziegler, 1991). The product of this reaction is an oxygenated compound such as an epoxide which forms a substrate for further metabolism by phase II enzymes (Guengerich, 1992*a*). The final product of the P450 catalysed reaction may itself be highly carcinogenic (McManus & McKinnon, 1991). For example, aflatoxin B_1 is metabolized to aflatoxin-B_1-8,9-epoxide, which is capable of forming an adduct with the N-7 atom of guanine in DNA (Ishikawa, 1992). Benzo(*a*)pyrene,

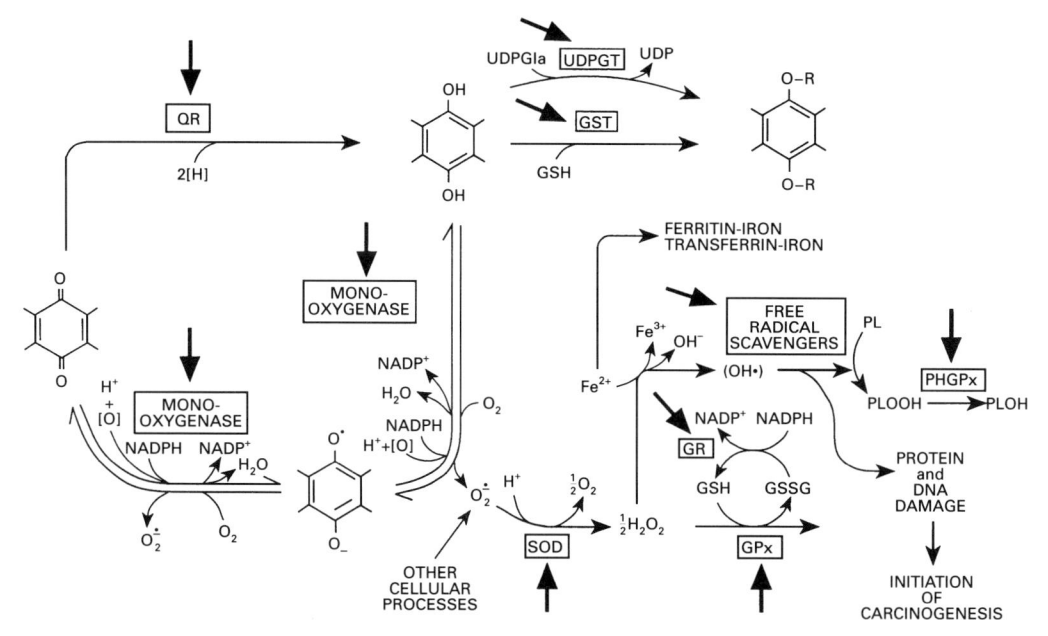

Fig. 2. The inter-relationship between detoxification pathways and free radical mediated oxidative damage, using a quinone xenobiotic as an example. Possible sites for dietary intervention in the detoxification of 2, 3, 5, 6-tetramethyl-*p*-benzoquinone are shown by filled arrows. UDPGla, UDP-glucuronic acid; PL, phospholipid, its hydroperoxide (PLOOH) and alcohol (PLOH).

a polycyclic aromatic hydrocarbon, is converted to many genotoxic products including phenols and epoxides (Guengerich, 1992a).

Phase I reactions may also give rise to problems for the cell by the production of free radicals. The role of cytochrome P450 in free radical damage has been investigated by several groups (Castillo *et al.* 1992; Dai *et al.* 1993; Ohmori *et al.* 1993). With reconstituted enzyme systems, in the absence of substrate, reduction of P450 by NADPH or NADH causes the production of reactive oxygen species (Aust *et al.* 1985). Even in the presence of substrate, coupling is so loose that 50% of the reducing equivalents appear as superoxide and peroxide, rather than causing oxidation of substrate (Hochstein *et al.* 1988). These results indicate that the P450 monooxygenase system may be responsible for the production of reactive oxygen species as a consequence of normal cellular metabolism, although such extreme conditions may not occur *in vivo*. Certain cytochrome P450 isoenzymes are more prone to the production of reactive oxygen species than others. Thus in animal studies, P450 2E1, an enzyme involved in glycerol oxidation, is poorly coupled, and liver microsomes prepared from animals treated with P450 2E1 inducers exhibit higher rates of H_2O_2 production (Clejan & Coderbaum, 1991). P450 are therefore often considered as pro-oxidants (Dai *et al.* 1993).

The number of P450 isoenzymes so far found in human tissues exceeds 30 (Guengerich, 1992a) and a full discussion is beyond the scope of this review. The nomenclature of the P450 isoenzymes is updated in Nelson *et al.* (1993). The P450 which metabolize xenobiotics belong to classes I, II and III (Guengerich, 1992b). Nutrient status influences phase I metabolism by affecting the transcription, translation or protein stability of cytochrome P450 (Yang *et al.* 1992). In rats, vitamin E deficiency gives a slight decrease in total P450 (Williams *et al.* 1992), whereas in the case of vitamin C, both high doses and a deficiency lead to lowered monooxygenase activities (Yang *et al.* 1992).

Many dietary non-nutrients bring about much greater changes in P450 levels. Indole-3-carbinol, a compound found in cruciferous vegetables, gives rise to a 7·5-fold increase in P450 1A1 in the liver, and a 14-fold increase in the small intestine, when fed to rats at a very high level (500 mg/kg body weight) for 14 d (Wortelboer et al. 1992). Quercetin, a flavonoid which occurs widely in the diet, has been shown to induce P450 levels in vitro. For example it can give rise to increases in P450 1A1 as high as 12-fold in mouse hepatoma cells in culture (De Long et al. 1986). The relevance of such effects in the context of real human diets remains to be established.

There are many examples of drugs and xenobiotics (Forrester et al. 1992), including caffeine (Gandhi & Khanduja, 1992) and ethanol (Lucas et al. 1992), which increase some P450 involved in phase I metabolism. Other non-nutrient components of human foods which have been shown to have this effect include the flavonoids tangeretin, flavone and nobiletin (Yang et al. 1992). Induction of P450 may, in theory, increase the risk of carcinogenesis. On the other hand diallyl sulphone, a metabolite of diallyl sulphide which is a major flavour component of garlic oil, irreversibly inhibits P450 2E1 (Brady et al. 1991), and this has been proposed as a mechanism for the blocking activity of diallyl sulphone (Brady et al. 1991; Horie et al. 1992). Phenethyl isothiocyanate, a breakdown product from many cruciferous vegetables (Yang et al. 1992), with demonstrated blocking activity (Morse et al. 1989), and naringenin, from grapefruit juice (Fuhr et al. 1993), also inhibit P450 2E1 and 1A2 respectively, and thus both have the potential to decrease phase I metabolism. It has been suggested that inhibition of P450 activity by phenethyl isothiocyanate occurs both through chemical inactivation and by a competitive mechanism (Smith et al. 1993). Again, it must be emphasized that the significance of results obtained in vitro or at very high doses in vivo remains to be properly assessed.

The mechanism of control of P450 1A1 transcription has been extensively studied (Saatcioglu et al. 1990; Fujii-Kuriyama et al. 1992; Gonzalez et al. 1993; Wu & Whitlock, 1993). The key protein is the Ah receptor, which binds to xenobiotics including dietary compounds, translocates into the nucleus, and then binds to a specific sequence of DNA (the xenobiotic responsive element) upstream of the P450 1A1 gene, which enhances transcription (Gonzalez et al. 1993). The increase in transcription is directly related to the specificity of the Ah receptor protein as shown for indole-3-carbinol, 3,3′-diindolylmethane (the main gastric conversion product of indole-3-carbinol) and dioxin (Jellinck et al. 1993).

Free radical mediated damage

Free radical mediated damage occurs as a consequence of normal cellular metabolism, and is exacerbated by poorly coupled redox reactions, including those mediated by cytochrome P450, by free iron released from storage proteins (Minotti et al. 1991; Winterbourn et al. 1991; Reif, 1992), by tissue injury (Powell & Tortollani, 1992), by reperfusion (Kilgore & Lucchesi, 1993), and by u.v. light (Godar et al. 1993). Free radical mediated damage is one of the mechanisms involved in initiation of carcinogenesis and such damage is reduced by antioxidants. Nutrient antioxidants such as vitamin E and carotenes play an important role in protecting the membrane and LDL (Bowry et al. 1992; Packer, 1991, 1992; Rousseau et al. 1992; Krinsky, 1993). The amount of vitamin E in membranes is apparently low: about 1 molecule per 1000–2000 phospholipids (Packer, 1992). Vitamin C is also usually considered to be an antioxidant, and may be involved in regeneration of vitamin E from tocopheryl radicals (Buettner, 1993). Vitamin E and C are chain breaking antioxidants since they react very poorly with oxygen, can be regenerated by enzymic systems, and their respective radicals are relatively harmless (Buettner, 1993). Oxidized vitamin C (dehydroascorbic acid) is converted back to vitamin C by protein disulphide-isomerase (*EC* 5.3.4.1) and thioltransferase, also called glutaredoxin (Wells et al. 1990), by a reaction which results in oxidized glutathione. This provides an important metabolic link between vitamin C and glutathione, which plays a vital role in both detoxification (phase II) reactions (see later) and in protection against free radical mediated damage (Meister, 1991).

The properties of non-nutrient food borne antioxidants have been studied *in vitro* by several groups, but few have been proven to work *in vivo*. In assessing a potential biological antioxidant, several methods are necessary (Halliwell, 1990), and all compounds classed as antioxidants, including vitamins E and C, can be pro-oxidant under certain conditions (Halliwell, 1990; Maiorino *et al.* 1993; Mukai *et al.* 1993). There are many examples in the literature of non-nutrient antioxidants (for a review see Pratt, 1992). Some examples from a long list of candidate antioxidants are carnosol and carnosic acid from the herb rosemary (Aruoma *et al.* 1992), caffeine (Shi *et al.* 1991), cinnamic acids, especially ferulic, *p*-coumaric and caffeic acids which are found in many plant foods (Howie *et al.* 1990; Scott *et al.* 1993), the flavouring agent, vanillin (Liu & Mori, 1993), flavonoids, especially catechin, found in high concentration in tea, and quercetin, found in many plant foods (Morel *et al.* 1993; Scott *et al.* 1993), and diallyl polysulphides from 'aged' garlic extracts (Horie *et al.* 1992). Pro-oxidant activities of some non-nutrients have also been described. Examples include the flavour cinnamaldehyde (Raveendran *et al.* 1993) and carnosol/carnosic acid (Aruoma *et al.* 1992).

In addition to compounds which act as antioxidants by virtue of their redox chemistry, there are also endogenous antioxidant systems (Fig. 2). These include the enzymes glutathione peroxidase (EC 1.11.1.9; Ladenstein, 1984), the α form of glutathione transferase (GST, EC 2.5.1.18; Mannervik & Danielson, 1988), glutathione reductase (EC 1.6.4.2; Chow, 1988), superoxide dismutase (EC 1.15.1.1; Hirose *et al.* 1993), catalase (EC 1.11.1.6; Deisseroth & Dounce, 1970), phospholipid hydroperoxide glutathione peroxidase (Schuckelt *et al.* 1991), and (see Fig. 2) metal binding proteins such as ferritin and transferrin (Bomford & Munro, 1992; Testa *et al.* 1993). The relative importance of some of these enzymes has been estimated (Remacle *et al.* 1992). In addition, there are enzymes which remove the products of free radical mediated damage such as macro-oxyproteinase and other proteinase systems, which remove damaged proteins (Pacifici & Davies, 1990), DNA repair enzymes (see below), and the glutathione transferases and peroxidases which inactivate lipid hydroperoxides (Ladenstein, 1984; Ursini *et al.* 1985; Ursini & Bindoli, 1987; Mannervik & Danielson, 1988) and remove lipid breakdown products (Mannervik *et al.* 1985).

Diet exerts influence over many of these repair mechanisms (Fig. 2). For example the nutrient selenium is essential for glutathione peroxidase and phospholipid hydroperoxide glutathione peroxidase (Schrauzer, 1992) and copper, zinc and manganese are essential for synthesis of CuZn superoxide dismutase and Mn superoxide dismutase respectively (Donnelly & Robinson, 1991). The prosthetic group of glutathione reductase is a flavin, and so dietary riboflavin is essential for the active enzyme. Indole-3-carbinol decreases the levels of superoxide dismutase and glutathione peroxidase in rat liver (Shertzer & Sainsbury, 1991) but, with the exception of GST discussed below, generally little is known about the effect of non-nutrients on the expression of these enzymes.

Alternative route to phase I metabolism – quinone reductase (QR)

QR (EC 1.6.99.2) is an enzyme which catalyses the 2-electron reduction of many quinones, without a 1-electron reduced free radical intermediate (Lind *et al.* 1982), and consequently there is less likelihood of free radical mediated damage (Fig. 2). Theoretically QR is able to compete with phase I enzymes for substrates, but being mainly a cytosolic enzyme its activity toward hydrophobic substrates *in vivo* is probably limited. The main isoenzyme of QR requires a flavin prosthetic group (Edwards *et al.* 1980), and so dietary riboflavin is essential. However, its expression is markedly influenced by many non-nutrient compounds.

A rapid and convenient assay for QR has been developed (Prochaska *et al.* 1985; Prochaska & Santamaria, 1988) and a range of inducers has been identified. These include quercetin, coumarin, α-angelicalactone (De Long *et al.* 1986), benzyl isothiocyanate (Talalay & Prochaska, 1987) and sulphoraphane (Zhang *et al.* 1992). Extracts from many vegetables, especially brassicas, also induce QR. This is dependent on processing (Kore *et al.* 1993; Tawfiq *et al.* 1994) because endogenous plant enzymes in the vegetables break down glucosinolates into more biologically active species such as isothiocyanates (Fenwick *et al.* 1989). Experiments on animals have also indicated a wide range of inducers of QR, such as phenethyl isothiocyanate (Guo *et al.* 1993), eugenol (Verhagen *et al.* 1993) and erucin (Zhang *et al.* 1992). Diet can intervene at more than one level in the induction process, including the stabilization of protein or mRNA, but it is most effective at the transcription stage. Certain inducers of QR, including aromatic isothiocyanates and flavonoids, are known to possess blocking activity (Wattenberg, 1993).

QR is often considered as a phase II enzyme, and a unifying theory for induction of phase I and phase II enzymes has been presented (Prochaska *et al.* 1985), postulating that inducers fall into two classes – bifunctional (induce phase I and II enzymes) and monofunctional (induce phase II only). The former include compounds such as benzo(*a*)pyrene and aflatoxin (Talalay & Prochaska, 1987). The latter include diphenols such as hydroquinone and isothiocyanates (Talalay & Prochaska, 1987). Bifunctional inducers require activation by phase I enzymes to redox labile molecules before they are able to induce phase II enzymes. This hypothesis suggests that the redox signal is the most important for induction of QR, but does not satisfactorily explain induction by compounds such as isothiocyanates (Daniel, 1993).

The control of QR by redox signals is *via* a sequence of DNA in the 5′ flanking region of the gene, which is called the ARE (antioxidant responsive element; Rushmore *et al.* 1991). Depending on the species, these responsive elements may or may not contain within them an element responsive to phorbol-12-*O*-tetradecanoate-13-acetate (Li & Jaiswal, 1992; Xanthoudakis *et al.* 1992; Daniel, 1993; Rushmore & Pickett, 1993) which binds to the transcription factors *fos* and *jun*. These are regulated *via* a protein called *ref*-1, which controls the redox state of sulphydryl residues, essential for DNA binding, on *fos* and *jun* (Abate *et al.* 1990; Xanthoudakis *et al.* 1992). The sulphydryl–disulphide redox state of the cell is communicated *via* thioredoxin. This implies that a wide range of antioxidants will influence expression of QR *via* thioredoxin, and indeed butylated hydroxyanisole, butylated hydroxytoluene, and ethoxyquin do affect expression (De Long *et al.* 1986; Derbel *et al.* 1993). However, there is a growing body of evidence that transcription factors other than *fos* and *jun* are involved in controlling expression (Rushmore & Pickett, 1993). Nguyen & Pickett (1992) have identified two transcription factors of 28000 and 45000 Daltons, that are constitutive and bind to the ARE sequence. In identifying the ARE, Pickett's group showed that this element responded to redox cycling phenolics, such as hydroquinone and catechol, but not to resorcinol which cannot redox cycle (Rushmore *et al.* 1991). By implication, redox cycling phenolics in the diet may affect binding to the ARE. It is clear that control of expression *via* the ARE is a subject for future research. Indeed, a variety of chemicals with a large range of structures is able to induce QR, and this property may correlate with their reactivity as Michael acceptors (Riley & Workman, 1992).

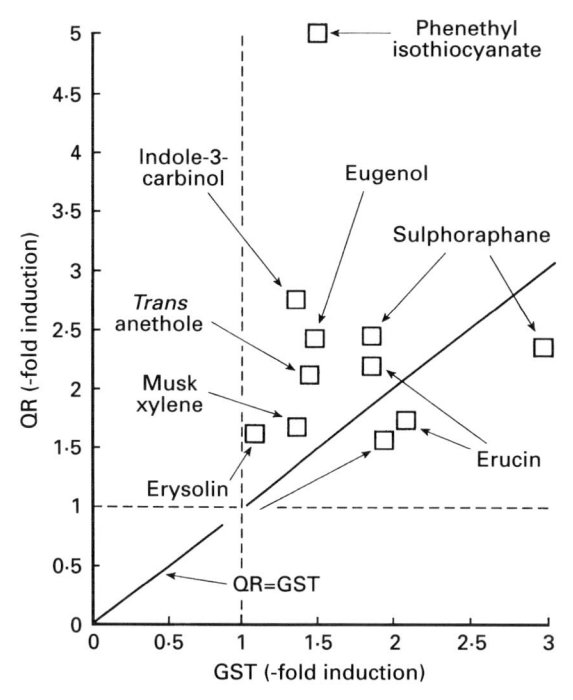

Fig. 3. Co-ordinate induction of GST and QR in animals? Induction (-fold) of GST and QR (equal induction represented by ——) is compared for several dietary compounds using animal models: sulphoraphane, erucin and erysolin in mouse liver and stomach (Zhang et al. 1992); phenethyl isothiocyanate in rat liver (Guo et al. 1993); indole-3-carbinol in rat liver (Wortelboer et al. 1992); eugenol and transanethole in rat liver (Verhagen et al. 1993); musk xylene in rat liver (Iwata et al. 1993).

CONJUGATION BY PHASE II ENZYMES

Conjugation results in a xenobiotic which is more hydrophilic, and can therefore be transported more readily out of the cell. Metabolites from the action of monooxygenase or QR reduction are conjugated with glutathione by GST, with UDP-glucuronic acid by UDPglucuronosyltransferase (UDPGT, 2.4.1.17; Mannervik, 1985; Brierley & Burchell, 1993), with 3′-phosphoadenosine-5′-phosphosulphate by sulphotransferases (EC 2.8.2.x) or by amino acid transferases (commonly with glycine). The route of conjugation depends on the chemical nature of the compound and varies from one species to another. GST and UDPGT are the most studied systems. GST requires no nutrients as prosthetic groups, but requires glutathione as a cofactor: synthesis of glutathione requires dietary cysteine (Meister, 1991). There are a large number of isoforms of GST in humans, and all GST can be classified as α, μ, π, θ or microsomal (Black & Coon, 1987; Mannervik & Danielson, 1988; Black & Wolf, 1991). The isoforms have individual but overlapping specificity profiles for a wide range of electrophilic carcinogens. Only GST α is able to reduce organic hydroperoxides (but not H_2O_2) and so contributes to reducing free radical mediated damage (Mannervik & Danielson, 1988). Most of the work on control of expression of GST has been performed on animal models, and many compounds have been tested for their ability to induce GST in rats (Nijhoff et al. 1993 and Fig. 3). In one study, quercetin, flavone, ferulic acid, ellagic acid, coumarin, α-angelicalactone and curcumin all induced GST α in rat liver, although in contrast to other studies (Bradfield et al. 1985) large doses of Brussels sprouts (20% of diet) did not. All of the above, except for quercetin and curcumin, also increased liver glutathione levels, as did Brussels sprouts (Nijhoff et al.

1993). Although the genes have been sequenced (Klone et al. 1992; Rozen et al. 1992), control of GST α in humans is poorly understood. The rat and mouse α classes are controlled via the ARE/EpRE/TRE mechanism (EpRE, electrophile responsive element; TRE, phorbol-12-O-tetradecanoate 13-acetate responsive element) as described above for QR (Friling et al. 1992; Nguyen & Pickett, 1992; Daniel, 1993), and there is some evidence for coordinate induction of GST and QR in these species (Talalay et al. 1988). Although many compounds which are inducers of QR are also inducers of GST, Fig. 3 shows that the induction is not quantitatively coordinated for a range of non-nutrient compounds. It will be very interesting to see if regulation of QR and GST are coordinately controlled in humans. Compounds with the ability to induce both QR and GST in humans would presumably exhibit potent activities as blocking agents, especially if this were combined with antioxidant properties.

UDPGT also exists as a family of enzymes (Brierley & Burchell, 1993), again with different but overlapping specificities. The induction of UDPGT by drugs and other xenobiotics has been relatively well studied (Sutherland et al. 1993), and compounds such as dipyridyls induce UDPGT as well as GST, but not sulphotransferase in rats (Franklin, 1991). Dietary lipid affects the UDPGT activity. In a study in which rats were fed a diet supplemented with fish oil there was a 3-fold increase in UDPGT after 4 weeks compared to a control group given a fat-free diet (Dannenberg & Yang, 1992). Cytochrome P450 1A2 and 2E1 are also induced by fish oils (Yoo et al. 1990), and regulation of expression of UDPGT, like P450 1A1/2, is influenced by the Ah receptor (Brierley & Burchell, 1993). Expression is also influenced by the peroxisome proliferator receptor (Brierley & Burchell, 1993). The effect of non-nutrients on UDPGT expression has been poorly studied in humans. Subjects fed a diet high in cabbage and Brussels sprouts showed a modest increase in UDPGT (Miners & Mackenzie, 1991). Given the wide range of drugs that induce UDPGT, it will be of interest to determine which, if any, non-nutrients affect the expression of UDPGT in humans.

In the final phase of detoxification (Phase III) glutathione conjugates are actively removed from the cell by the glutathione-S-conjugate export pump, which is dependent on ATP (Ishikawa, 1992). Some carcinogens are removed by P-glycoprotein, without the need for conjugation and this reaction also requires ATP (Ishikawa, 1992). We are unaware of any non-nutrient food constituents which influence this process.

INDIRECT EFFECTS *VIA* ENTERIC BACTERIAL METABOLISM

Mallett & Rowland (1990) have pointed out that bacteria present in the gut lumen express enzymes which may influence the metabolism of xenobiotics in host tissues. For instance, when potentially carcinogenic compounds are detoxified by UDPGT (Dutton, 1980), the conjugates are secreted via the bile into the gut lumen. The activity of bacterial β-D-glucuronidase (*EC* 3.2.1.31; Cole et al. 1985; Gadelle et al. 1985) may result in the hydrolysis of the conjugates, and their consequent reabsorption into the circulation in their active, deconjugated form (Rowland et al. 1985). β-Glucuronidase activity has been implicated in the activation of benzo(a)pyrene and 1,2-dimethylhydrazine conjugates in the colon (Nanno et al. 1986; Mallett & Rowland, 1990). It has been demonstrated that the yield of colorectal tumours is lower in germ free rats exposed to 1,2-dimethylhydrazine than in similarly treated conventional rats (Reddy et al. 1974) and that a β-glucuronidase inhibitor can protect rats with a conventional flora from the induction of tumours by azoxymethane, a metabolite of 1,2-dimethylhydrazine (Takada et al. 1982).

Gut bacteria also express β-glycosidase activity, one of the effects of which is the liberation of biologically active flavonoid aglycones from their glycoside conjugates

(Brown, 1988). It has been shown that the presence of the dietary flavonoid rutin (itself a substrate for β-glycosidase) in the diet of rats leads to a 20-fold increase in the level of β-glycosidase activity in the caecum and that this is associated with changes in the ability of hepatic microsomes obtained from those rats to activate bacterial promutagens *in vitro* (Mallett *et al.* 1989). Modulation of β-glucuronidase and β-glycosidase activities in enteric bacteria remains a hypothetical blocking mechanism and in theory would remove the requirement that a compound taken orally would need to traverse the gut wall and enter the circulation in an active form in order to modulate the expression of xenobiotic metabolizing enzymes.

INDUCERS OF DNA REPAIR

Wattenberg does not include agents that modify DNA repair in his classification scheme but Morse & Stoner (1993) classified such compounds as blocking agents. Clearly they cannot be described as suppressors for they exert their influence prior to promotion. However, they do not share the common feature of all other blocking agents, which act to prevent the occurrence of DNA lesions rather than to reduce the deleterious effects of such lesions. Furthermore, while DNA repair may be involved in the conversion of various forms of DNA damage into mutagenic lesions, and thus could be said to be involved in initiation, certain lesions such as O^6-alkylguanine adducts are thought to be capable of inducing mutations without further modification. Thus their repair could be considered a postinitiatory event. Given these complicating factors, we propose that modifiers of DNA repair should be included as a category in their own right, acting after blocking agents, as strictly defined, but prior to suppressing agents.

Much of the evidence for the existence of naturally occurring inducers of DNA repair has been obtained using prokaryotic cells. For example it has been reported that the natural flavouring vanillin can inhibit the induction of mutations in *Escherichia coli* by the promotion of *RecA* dependent recombinational repair (Ohta *et al.* 1986, 1988). Similar antimutagenic properties have been reported for the flavouring agents cinnamaldehyde and coumarin (Ohta *et al.* 1983 *a*, *b*) while Shimoi *et al.* (1985) have reported that tannic acid, which is found in tea, protects *E. coli* from the mutagenic effects of u.v. light. The fact that protection was only conferred by tannin components (gallic acid, (-)-epicatechin gallate, (-)-epigallocatechin and (-)-epigallocatechin gallate) to uvr^+ but not $uvrA^-$ strains led them to propose that the antimutagenic activity was due to stimulation of $uvrA$ dependent excision repair (Shimoi *et al.* 1986).

The relevance of these findings to human nutrition may be questioned, given that prokaryotic and eukaryotic DNA repair processes are known to differ markedly in their sensitivity to modifying agents such as caffeine. However, vanillin, anisaldehyde, cinnamaldehyde and coumarin have been shown to protect CHO cells from the clastogenic and mutagenic effects of u.v. light and X-rays *in vitro* (Sasaki *et al.* 1990*a*). Vanillin has also been shown to protect CHO cells from the clastogenic effects of mitomycin C (Sasaki *et al.* 1987) and from the mutagenic effects of u.v. and X-rays (Imanishi *et al.* 1990), and has been shown to have protective activity against the induction in mice of micronuclei by mitomycin C or by X-rays (Inouye *et al.* 1988; Sasaki *et al.* 1990*b*) and of mutations by ethylnitrosourea (Imanishi *et al.* 1990). It has been proposed that vanillin promotes DNA repair *via* DNA nucleotidyltransferase β (EC 2.7.7.7; Sasaki *et al.* 1990*a*). The complexity of this field is illustrated by the work of Imanishi *et al.* (1991) who showed that components of tea tannin can inhibit mammalian DNA repair when unmodified, but that metabolites resulting from pretreatment with a metabolically active hepatic microsome extract (S9) can actually promote excision repair activity in the same test systems.

SUPPRESSING AGENTS

The concept of a suppressing agent was introduced by Wattenberg to account for the ability of compounds such as sodium cyanate, tert-butylisocyanate and benzyl isothiocyanate to inhibit the appearance of tumours, even when fed considerably after the chemical treatment which initiated the neoplasia (Wattenberg, 1981). Benzyl isothiocyanate is found in cruciferous vegetables, and Wattenberg's group went on to feed brassicas to rats previously treated with 7,12-dimethylbenz(*a*)anthracene in order to test their ability to inhibit the emergence of mammary tumours. Cabbage or broccoli was fed to rats for an 18 week period, starting 1 week after cessation of treatment with the carcinogen. For both vegetables the numbers of animals with mammary tumours in the treatment group were 40%–50% lower than in the control groups and the average number of tumours per rat was about 60% lower (Wattenberg *et al*. 1989*a*, *b*). Similar observations were made for other foods including green coffee beans, green Brazilian cocoa beans and orange oil (Wattenberg, 1983). Recently many more naturally occurring suppressing agents have been identified (Wattenberg, 1993).

Any logical classification of suppressing agents must be based upon the identification of their mechanisms of action, but in most cases this is unknown. Even for the aromatic isothiocyanates, first identified as suppressing agents by Wattenberg, the mechanism of action remains uncertain and is probably multifactorial (Wattenberg, 1993). For example, Sugie *et al*. (1993) have reported that oral administration of benzyl isothiocyanate and benzyl thiocyanate to male F344 rats resulted in the suppression of proliferative activity in hepatocyte primary cultures derived from them. However Musk & Johnson (1993*b*) have shown that aromatic isothiocyanates are selectively toxic against transformed colorectal tumour cells. Thus the same compounds may exert quite different effects at different stages of carcinogenesis. The difficulty of achieving a satisfactory classification of suppressing agents reflects our current incomplete understanding of tumour biology. By definition, suppressing agents act by inhibiting tumour promotion. In Fig. 1 we have used a simple classification which reflects the possible mechanisms whereby compounds may inhibit the progression of cells initiated by genotoxic damage.

INHIBITORS OF CELL PROLIFERATION

Cellular hyperproliferation is now recognized as a potentially important mechanism of carcinogenesis (Preston-Martin *et al*. 1990). Cell populations undergoing a high rate of division are more susceptible to DNA damage, both by endogenous oxidative mutagenesis and by exogenous chemical mutagens, than quiescent cells (Tong *et al*. 1980). Single stranded DNA is intrinsically more vulnerable to damage than double stranded, and the time available for DNA repair is reduced in rapidly proliferating cells (Ames & Gold, 1991). Furthermore, rapid proliferation is thought to favour somatic recombination, allowing cells which are heterozygous for a mutation favouring transformation to become homozygous and give rise to neoplastic progeny (Groden *et al*. 1990). Somatic recombination may be an important aspect of carcinogenesis during both initiation and promotion. Environmental factors which can inhibit proliferation could therefore be classified as either blocking or suppressing agents, depending upon the tissue concerned, and the particular mechanism of carcinogenesis which is being inhibited. This illustrates once more the difficulty of fitting the growing list of food borne protective factors into any rigid system of classification. Those cellular control mechanisms that have been identified as potentially susceptible to modification by food borne protective factors will now be described briefly and examples will be given. In view of the incomplete understanding of promotion, the relationships between the mechanisms remain undefined and, in the case of multifunctional agents, their relative importance cannot yet be assessed.

Modification of intracellular signalling

The proliferation of normal cells is regulated by a complex system of membrane receptors and intracellular signal pathways, some of which may be susceptible to modulation by plant cell constituents derived from food. Receptor mediated hydrolysis of membrane bound phosphatidylinositol-4,5-biphosphate yields two intracellular second messengers, inositol 1,4,5-triphosphate and 1,2-diacylglycerol. This sequence of events forms part of a signal transduction pathway which is involved in the regulation of cellular proliferation in a range of mammalian tissues (Berridge & Irvine, 1984). Under normal circumstances, release of 1,2-diacylglycerol leads to activation of protein kinase C (PKC, EC 2.7.1.37), a step which is known to stimulate cell proliferation by mechanisms which include activation of ornithine decarboxylase (ODC), while inositol 1,4,5-triphosphate mobilizes intracellular calcium.

Inositol hexaphosphate, usually referred to as phytate, is widely distributed in plant foods and is present at particularly high levels in cereals and legume seeds. It has been proposed that the supposed anticarcinogenic properties of diets rich in cereals may be due to the presence of phytate rather than dietary fibre, and evidence for a protective effect of isolated inositol hexaphosphate has been obtained in animal experiments (Shamsuddin & Sakamoto, 1992). When the compound was administered orally to rats 2 weeks or 5 weeks after treatment with the chemical carcinogen azoxymethane, a significant reduction was observed in the number of animals with tumours and the number and size of tumours in the affected animals (Shamsuddin & Ullah, 1989). This is strong evidence to suggest that exogenous inositol hexaphosphate can function as a suppressing agent, perhaps by entering and modifying the inositol phosphate metabolism pathways of the cell. Further work will be needed to clarify the mechanisms involved and assess the significance in relation to the human diet.

The direct inhibition of PKC might also provide a mechanism for suppression as stimulation of PKC leads to the activation of proteins that affect cellular proliferation (Blumberg, 1988). Glycyrrhetic acid is a breakdown product of glycyrrhizin, a sweet constituent of licorice. This compound has been demonstrated to inhibit PKC (O'Brian *et al.* 1990) and to possess antitumorigenic activity, preventing the inflammatory and toxic effects of 12-*O*-tetradecanoylphorbol-13-acetate (Wang *et al.* 1991; Nishino, 1992). PKC may also be involved in the control of intercellular communication in that tumour promoters which have been shown to activate PKC have also been demonstrated to inhibit the transfer of fluorescent dye between adjacent cells in culture (Enomoto & Yamasaki, 1985). It has been proposed that compounds that inhibit such communication may function as tumour promoters by interfering with cell–cell growth regulatory mechanisms and thereby encouraging cellular replication (Trosko & Chang, 1984). Epidemiological evidence suggests that green tea may be protective against cancer (Oguni *et al.* 1989). An extract of green tea consisting largely of various catechins (Maeda & Nakagawa, 1977) previously shown to possess antimutagenic and anticarcinogenic properties (Cheng *et al.* 1986; Ruch *et al.* 1989) apparently prevents the inhibition of intercellular communication induced in cultured mammalian cells by the promoters phenobarbital and 12-*O*-tetradecanoylphorbol-13-acetate (Ruch *et al.* 1989).

Proteinase inhibitors have been reported to suppress tumour promotion in rodents at a variety of sites including skin, colon, breast and liver (Yavelow *et al.* 1983; Troll *et al.* 1984, 1992; Billings *et al.* 1990). They inhibit the transformation of mammalian cells *in vitro* by radiation and by oncogenes (Kennedy & Little, 1981; Yavelow et al. 1983, 1985; Garte et al. 1987) and inhibit the production of oxygen radicals by tumour promoters (Frenkel et al. 1987; Troll et al. 1987). Troll et al. (1992) have proposed that proteinase inhibitors might exert their antitumorigenic effect by inhibiting the proteolytic modification of PKC which is necessary for its activation as a tumour promoter (Murray et al. 1987). Many experiments using proteinase inhibitors have involved the topical application of inhibitors to skin. However, it has been demonstrated that inclusion of a soyabean extract containing the Bowman–Birk proteinase inhibitor in the diet can protect mice against the induction of lung tumours by 3-methylcholanthrene (Witschi & Kennedy, 1989) and against induction of colonic adenomas by 1,2-dimethylhydrazine (Weed et al. 1985). Furthermore, epidemiological evidence suggests that diets containing high levels of foods rich in proteinase inhibitors are protective against cancers of the breast, colon and prostate (Correa, 1981). It has been proposed by Schelp & Pongpaew (1988) that chemoprevention may be possible by a nutritionally induced increase in the production of endogenous inhibitors.

Inhibition of oncogene expression

The progressive loss of proliferative control which characterizes cellular transformation is accompanied by an accumulation of acquired genetic defects involving proto-oncogenes, many of which code for proteins involved in the regulation of cell proliferation. One mechanism by which dietary agents might suppress promotion is by inhibiting the post-translational modification of oncoproteins. For example the *ras* family of proto-oncogenes code for a group of small guanosine 5′-triphosphatases, including p21^{ras}, which regulate various aspects of cellular growth and differentiation. *Ras* mutations are associated with many types of tumour. To exert its transforming properties the mutated p21^{ras} must be translocated to the plasma membrane, and for this to occur it must undergo farnesylation (Casey et al. 1989; Kato et al. 1992). Limonene is a monoterpene component of orange peel oil which inhibits the development of mammary carcinomas at the promotional stage of induction by the carcinogen 7,12-dimethylbenz(*a*)anthracene in the rat (Elegbede et al. 1986) and has also demonstrated antitumorigenic properties in a murine model (Wattenberg et al. 1989b). It has been proposed that the antitumorigenic properties of *d*-limonene reflect its ability to reduce the farnesylation of p21^{ras} (Crowell et al. 1991). Although exposure to other monoterpenes has been shown to bring about a decrease in expression of an enzyme involved in the synthesis of farnesyl moieties (3-hydroxy-3-methylglutaryl coenzyme A reductase; *EC* 1.1.1.88) in the liver of rats (Clegg et al. 1982), this reduction does not seem to play a role in influencing the farnesylation of p21^{ras} (Crowell et al. 1991). This mechanism may be relevant to other aspects of cellular control mediated by isoprenylated proteins. It is interesting to note that limonene is one example of a substance which, though it is capable of suppressing tumorigenesis in certain models, can also be shown to be carcinogenic in other circumstances. In this instance, the induction of kidney tumours in the rat by limonene appears to be species and sex specific (Dietrich & Swenberg, 1991).

Polyamine metabolism

The intracellular level of polyamines plays a central role in the control of proliferation (Williams-Ashman & Canellakis, 1979; Pegg, 1988). Tumour promoters such as 12-*O*-tetradecanoylphorbol-13-acetate have been found to increase the activity of ODC, which catalyses the formation of the polyamine precursor putrescine from ornithine (McCann et al. 1992), and so increases the concentration of polyamines in affected tissues (Slaga, 1983). It has been demonstrated that an increase in ODC activity and polyamine levels is an absolute requirement for the proliferation of a variety of human tumour cells both *in vitro* and *in vivo* (Luk, 1992), and inhibitors of ODC have been shown to deplete polyamine levels and block proliferation in cultured cells (Mamont *et al.* 1976) and in tumours *in vivo* (Pegg, 1988). It has been proposed that the ability of the flavonoid apigenin to suppress skin tumorigenesis in mice, and of curcumin, a constituent of turmeric, to suppress early markers of tumorigenesis in the rat colon, may be related to reduced ODC levels in the target tissues (Wei *et al.* 1990; Rao *et al.* 1993). Other flavonoids that have been reported to inhibit promoter induced increases in ODC include kaempferol, luteolin, morin and fisetin (Nakadate *et al.* 1984; Fujiki *et al.* 1986).

Inhibition of arachidonic acid metabolism is closely linked to the direct inhibition of ODC. Increased metabolism of arachidonate is commonly seen in experimental models of tumour promotion (Earnest *et al.* 1992). Both the cyclo-oxygenase and the lipoxygenase (*EC* 1.13.11.12) pathways of arachidonic acid metabolism (Moncada *et al.* 1980) are implicated in this promoting effect, which may operate *via* the up-regulation of ODC activity (Yamamoto & Kato, 1992). Blocking the metabolism of arachidonate has been shown to inhibit the growth of human cells both *in vitro* and *in vivo* (Levine *et al.* 1972; Hial *et al.* 1977; Bayer *et al.* 1979; Sato *et al.* 1983; Goodlad *et al.* 1989). Curcumin, an inhibitor of promotion induced by croton oil and 12-*O*-tetradecanoylphorbol-13-acetate in the murine skin model (Huang *et al.* 1988; Soudamini & Kuttan, 1989), has been shown to inhibit the activities of cyclo-oxygenases and lipoxygenases (Huang *et al.* 1991; Rao *et al.* 1993). The ability of the flavonoids quercetin, fisetin, kaempferol and morin to suppress promotion (Kato *et al.* 1983; Nakadate *et al.* 1984) may correlate with their ability to inhibit lipoxygenase (Nakadate *et al.* 1984).

Oestrogen metabolism

Certain tumour types are responsive to oestrogens and this presents another mechanism by which promotion could theoretically be inhibited. 16β-Hydroxyoestrone functions as a promoter of mammary cell transformation (Telang *et al.* 1992) and enhances the expression of oncogenes in human cancer cells (Hsu *et al.* 1991), whilst 2-hydroxyoestrone is anti-oestrogenic in cell culture (Schneider *et al.* 1984). Thus the induction of enzymes that increase the 2-hydroxylation of oestrogens, relative to their 16β-hydroxylation, may be anticarcinogenic (Jellinck *et al.* 1993). Such induction has been reported in mouse and rat liver following feeding with indole-3-carbinol, which, as we have seen, is derived from cruciferous vegetables (Baldwin & LeBlanc, 1992; Jellinck *et al.* 1993). In human volunteers the rate of 2-hydroxylation of oestradiol in an *in vivo* radiometric assay was increased by 50% after daily exposure to 6–7 mg indole-3-carbinol/kg for 7 d and the urinary excretion of 2-hydroxyoestrone was increased relative to that of oestriol (Michnovicz & Bradlow, 1990, 1991). Indole-3-carbinol forms condensation products under acid conditions; the ability of these products to induce similar effects to indole-3-carbinol and the inactivity of the parent compound when administered intraperitoneally has led to the suggestion that the condensation products formed in the acid milieu of the stomach might be the active moieties in this induction (Bradfield & Bjeldanes, 1991).

Caffeine has been shown to reduce the number of mammary tumours induced in mice by a combination of 17β-oestradiol and progesterone (VanderPloeg & Welsch, 1991). In other studies however, caffeine has been shown to enhance the development of mammary tumours in mice, spontaneous or induced by 7,12-dimethylbenz(*a*)anthracene (Nagasawa & Konishi, 1988; Welsch *et al.* 1988), and to enhance the development of pancreatic tumours in hamsters exposed to *N*-nitroso-bis-(2-oxopropyl)amine at a postinitiation stage (Nishikawa *et al.* 1992). The mechanism(s) by which caffeine exerts its effects remain a matter for conjecture, though Alldrick & Rowland (1988) have demonstrated that it can inhibit the ability of mouse hepatic microsomes to metabolize 2-amino-3,4-dimethylimidazo[4,5-*f*]quinoline, Trp-P-2 and 2-amino-3,8-dimethylimidazo[4,5-*f*]quinoxaline to bacterial mutagens and Welsch *et al.* (1988) have observed an enhancing effect of caffeine on the responsiveness of murine mammary organ cultures to mammotrophic growth hormone.

Human blood, faeces and urine commonly contain varying quantities of oestrogenic compounds derived from precursors found in plant foods. Lignans are compounds with a dibenzylbutane structure that were first identified as secondary metabolites of plants. The two most important mammalian lignans, enterodiol and enterolactone, are derived from lignans of plant origin by the activity of bacteria in the alimentary tract (Axelson *et al.* 1982). Isoflavonoids are a group of plant diphenols, of which many have oestrogenic activity. They occur extensively in plants used as human foods, and many of them have been detected in human and animal urine. These compounds include equol, methylequol, daidzen, dihydrodaidzen, 3′,7-dihydroxyisoflavan and others (Adlercreutz, 1990).

The mammalian lignans and the isoflavonoids exhibit weak oestrogen-like activity, together with a variety of other physiological activities including antiproliferative effects and cytotoxicity (Setchell & Adlercreutz, 1988). Many of these effects are apparently mediated *via* the interaction of phyto-oestrogens with mammalian oestrogen receptors. For example lignans and isoflavonoids compete with endogenous oestrogens for type II oestrogen binding sites and may thereby regulate the growth of oestrogen dependent tissues (Markaverich *et al.* 1988). A second important effect of these compounds is the stimulation of endogenous production of sex hormone binding globulin (SHBG) which modulates the biological activity of oestrogen in humans. Adlercreutz and co-workers have observed that enterolactone causes a dose dependent stimulation of SHBG synthesis in HepG2 cells in culture. A positive relationship between plasma SHBG levels and urinary lignan and isoflavonoid excretion in humans has also been observed, and ascribed to stimulation of hepatic SHBG synthesis by dietary phyto-oestrogens (Adlercreutz *et al.* 1993).

Epidemiological evidence can be interpreted in favour of a protective role for the phyto-oestrogens against sex hormone dependent tumours, principally carcinoma of the breast and prostate cancer. Urinary excretion of these compounds is reported to be higher in vegetarians than in omnivores, and lower in breast cancer patients. Japanese women consuming a traditional diet have both a low incidence of breast cancer and a relatively low mortality from this disease (Nomura *et al.* 1978), and Japanese men experience relatively low mortality from prostate cancer, although the incidence of this common but often slow growing tumour is apparently similar to that of western populations. Adlercreutz *et al.* (1991) proposed that the slow growth of both breast and prostate cancers in Japanese patients consuming traditional diets may be due to inhibitory effects of isoflavanoids derived primarily from soya products. This proposal is supported by the observation that a high intake of soya products is associated with reduced risk of breast cancer amongst women in Singapore (Lee *et al.* 1991), but there is little direct experimental evidence for the putative mechanism.

DIRECT ACTING MODULATORS OF CELL DAMAGE

Suppression of free radical production

Oxidative damage by free radicals appears to be important at the promotion stage as well as the initiation stages of carcinogenesis discussed earlier. Many food borne compounds that can inhibit the generation of oxidative damage may exert an antipromotional effect. For example epigallocatechin gallate, which is found in teas, has been shown to reduce both the induction of lung tumours, and the formation of 8-hydroxydeoxyguanosine (a marker for oxidative DNA damage) in mouse lung tissue exposed to 4-(methylnitrosamino)-1-(3-pyridyl)-1-butanone, without having any effect on the initial level of alkylation in the DNA (Xu et al. 1992). It has also shown antipromoter activity in skin, duodenum and liver (Fujiki et al. 1992). The catechin-rich extract of green tea referred to above as an inhibitor of intercellular communication has also been shown to protect cultured cells against cytotoxicity induced by oxygen radicals (Ruch et al. 1989).

Selective cytotoxins

Chemotherapy and radiotherapy involve exposure to high levels of cytotoxic agents with the aim of inhibiting the growth of tumour cells and ultimately of killing them. Ideally therapeutic agents would be, if not absolutely specific for tumour cells, then at least highly selective in their mode of action. It is not conceivable that the human diet normally contains biologically significant concentrations of cytotoxic agents of a similar non-specificity to commonly used therapeutic agents but many secondary plant metabolites are cytotoxic, and some of these compounds may selectively target tumours or precancerous cells.

Evidence for selective toxicity against transformed cells has been presented for polyunsaturated fatty acids (Bégin et al. 1986), and also for the compounds 1-cyano-2-hydroxy-3-butene, which is found in cruciferous vegetables (Wallig et al. 1993) and quercetin (Larocca et al. 1991), the latter having been shown to be active *in vivo* (Castillo et al. 1989). We have recently shown that the dietary compound allyl isothiocyanate, again derived from cruciferous vegetables and a major constituent of mustard, is more toxic towards transformed HT29 human colorectal adenocarcinoma cells than towards cells which have been experimentally detransformed *in vitro* (Musk & Johnson, 1993a). A similar selective effect of quercetin has also been observed (S. R. R. Musk & I. T. Johnson, unpublished observations). It will clearly be of interest to determine whether other dietary compounds with demonstrated cytotoxic activity (Mori et al. 1988; Babich et al. 1993) might also be selective and if so, whether such compounds might have a practical role to play as food borne protective factors in chemoprevention, or as chemotherapeutic agents.

INDUCERS OF CELLULAR DIFFERENTIATION

Tumour promoters are known to influence cell differentiation, either by inhibiting normal differentiation or by inducing inappropriate differentiation programmes (Yamasaki, 1984). The second class of suppressing agent comprises compounds which act by modulating the differentiation of tumorous or pretumorous cells, thereby slowing or even halting their growth. Retinoids, calcium and vitamin D have been shown to modify cell differentiation both *in vitro* and *in vivo* (Lotan 1992a; Moon et al. 1992), and to exert antipromotive effects in animal models of tumorigenesis (Lotan, 1980; Dawson & Okamura, 1990) although it has also been proposed that these compounds may act by modifying the expression of ODC (Dawson et al. 1987) or by enhancing intercellular communication (Hossain et al. 1989). Certain dietary carotenoids such as astaxanthin, canthaxanthin and fucoxanthin, which do not demonstrate provitamin A activity in mammals, are known to exhibit anticarcinogenic properties in animal models (Mathews-Roth, 1982; Grubbs et al. 1991; Okuzumi et al. 1993; Tanaka et al. 1994). An effect on differentiation remains a theoretical mode of action for these compounds. Fucoxanthin suppressed the growth of, and expression of N-*myc* in, tumour cells *in vitro* (Okuzumi et al. 1990) and canthaxanthin inhibited the growth of tumour cells and the transformation of 10T1/2 cells (Pung et al. 1988; Huang et al. 1992).

ANTIMETASTATIC AGENTS

The final stages of epithelial tumorigenesis involve invasion of the basal lamina by fully transformed cells and migration, *via* the blood or lymph systems, to new tissue sites where secondary tumours can develop. Bracke *et al.* (1989) have reported that the invasive behaviour of MO$_4$ cells (fetal mouse cells transformed with Kirsten murine sarcoma virus) towards embryonic chick heart fragments *in vitro* can be greatly reduced by exposure of the cultures to the flavonoid tangeretin. Inhibition was noticeable at a concentration of 0·01 mM and was completely reversible after removal of the tangeretin. Similar observations have been reported for another flavonoid, (+)-catechin (Bracke *et al.* 1984, 1987) and for retinoids (Lotan, 1992b). Such an effect might contribute to the inhibitory action of dietary cabbage on the yield of pulmonary metastases in mice injected with BALB/c tumour cells (Scholar *et al.* 1989). Thus dietary components may modify the carcinogenic process even after the completion of promotion.

IMMUNOMODULATION

Certain compounds that possess antitumorigenic properties have also been shown to enhance the immune response. Examples include β-carotene and the carotenoids canthaxanthin and astaxanthin (Bendich & Shapiro, 1986; Jyonouchi *et al.* 1991, 1993). It has been proposed that the antitumorigenic action of these compounds may be causally linked to their immunoenhancing properties (Shklar & Schwartz, 1988; Bendich, 1989). Given that high levels of prostaglandins E_1, E_2, A_1 and A_2 have been shown to suppress the killing of tumour cells by natural killer cells and that inhibitors of prostaglandin synthesis can restore natural killer functions to mice with experimentally depressed activity (Brunda *et al.* 1980; Taffet & Russell, 1981), it is also possible that inhibitors of arachidonic acid metabolism might exert an antitumour effect *via* this mechanism (Goodwin, 1984).

A NEW CLASS OF NUTRIENTS?

Clearly, a variety of biologically active constituents other than those conventionally recognized as nutrients may contribute to the protective effects of diets rich in fruits and vegetables against cancer. There is still much to be done to clarify the mechanisms by which plant constituents act, and there is an urgent need to assess their relative importance so that the practical implications for human diets can be properly assessed. Although an unprecedented variety of plant foods is now available to modern western consumers, the bulk of fruit and vegetable consumption involves relatively few species and varieties. Moreover the varieties most readily available tend to be determined by commercial factors related to agricultural production and distribution. It is inevitable that the increasing application of molecular biology will lead to the manipulation of food crops to improve pest resistance, yield, keeping qualities and flavour. The implications for health of such trends can only be assessed if the ensuing compositional changes and their biological significance are properly understood.

One obvious practical issue which must be addressed is the balance of risk and benefit conferred by the consumption of compounds which in many cases express more than one type of potent biological activity. Many of the substances that we have described as potential anticarcinogens have also been shown to possess potentially hazardous properties. For example, indole-3-carbinol has been shown to promote carcinogenesis in certain models (Pence *et al.* 1986; Bailey *et al.* 1987) and its toxicology will need to be thoroughly

Table 1. *Examples of dietary anticarcinogens with genotoxic or tumorigenic effects*

Compound	Assay	Reference
Allyl isothiocyanate	Positive in Ames test	Yamaguchi, 1980
Allyl isothiocyanate	Induction of aberrations in mammalian cells	Kasamaki *et al.* 1982
Allyl isothiocyanate	Induction of bladder tumours by feeding	Dunnick *et al.* 1982
Allyl isothiocyanate	Transformation of mammalian cells	Kasamaki *et al.* 1987
Anisaldehyde	Induction of aberrations in mammalian cells	Kasamaki *et al.* 1982
Benzyl isothiocyanate	Positive in Ames test	Yamaguchi, 1980
Benzyl isothiocyanate	Induction of aberrations in mammalian cells	Musk & Johnson, 1993*b*
Benzyl isothiocyanate	Induction of SCE in mammalian cells	Musk & Johnson, 1993*b*
trans r-Cinnamaldehyde	Induction of aberrations in mammalian cells	Kasamaki *et al.* 1982
trans r-Cinnamaldehyde	Transformation of mammalian cells	Kasamaki *et al.* 1987
Coumarin	Induction of bile duct tumours by feeding	Griepentrog, 1973
Curcumin	Induction of aberrations in mammalian cells	Ishidate *et al.* 1988
Diallyl sulphide	Enhancement of DEN induced hepatocarcinogenesis	Takahashi *et al.* 1992
Indole-3-carbinol	Promotion of AFB_1 induced carcinogenesis	Bailey *et al.* 1987
Indole-3-carbinol	Enhancement of DMH induced carcinogenesis	Pence *et al.* 1986
Kaempferol	Induction of mutations in mammalian cells	Maruta *et al.* 1979
Phenethyl isothiocyanate	Induction of aberrations in mammalian cells	Musk & Johnson, 1993*b*
Phenethyl isothiocyanate	Induction of SCE in mammalian cells	Musk & Johnson, 1993*b*
Quercetin	Positive in Ames test and SOS chromotest	Rueff *et al.* 1992
Quercetin	Induction of aberrations in mammalian cells	Ishidate *et al.* 1988
Quercetin	Induction of recombination in mammalian cells	Suzuki *et al.* 1991
Tannic acid	Induction of liver tumours by s.c. administration	Korpássy & Mosonyi, 1950
Vanillin	Induction of aberrations in mammalian cells	Jansson & Zech, 1987
Vanillin	Induction of SCE in mammalian cells	Jansson & Zech, 1987

studied before an increase in its consumption can be recommended (Bradfield & Bjeldanes, 1991). Other examples of mutagenic, clastogenic, carcinogenic and cocarcinogenic effects of dietary constituents previously referred to are given in Table 1. Clearly the message of epidemiology is that vegetables and fruits are overwhelmingly beneficial in their effects, but the prospect of manipulating the composition of commercial varieties inevitably raises issues of safety. It remains unclear whether any of the compounds which have been identified experimentally as potential anticarcinogens would represent a real hazard to humans if their intake were increased, but the caveat of Bradfield & Bjeldanes (1991) regarding the need for a thorough investigation of the toxicology of indole-3-carbinol should be extended to the other putatively anticarcinogenic agents.

Can it be argued that any of the biologically active substances discussed in this review should be classified as micronutrients? The Oxford English Dictionary defines a *nutrient* simply as a substance serving as nourishment. *To nourish* is "to supply (a thing) with whatever is necessary to promote its growth or formation or to maintain it in proper condition". The last part of this definition would certainly encompass the prevention of cancer, but nutritional science tends to regard nutrients more narrowly as substances which are essential in the sense that a specific deficiency disease results if they are absent from the diet. It is improbable that any of the biologically active substances discussed here meet this criterion, but it is conceivable that human beings have become adapted to a cocktail of food borne plant metabolites which help to maintain resistance to neoplasia. The fact that many of these compounds are potentially toxic if consumed above a threshold dose need not necessarily preclude their classification as nutrients. At first sight this is something of a paradox but several recognized micronutrients have been shown to be mutagenic under appropriate conditions *in vitro* (Chow, 1990).

Although human beings are highly adaptable omnivores, it is reasonable to assume that a high intake of fruits and vegetables is a biological norm. Hunter-gathering has probably been the dominant means of food provision throughout most of human evolutionary history. Modern hunter gatherers obtain more than half of their food energy from plant sources (Lee, 1967) and it has been calculated that even in industrialized societies human beings are exposed to about 1·5 g of potentially toxic plant constituents/d (Ames & Gold, 1990). Many of these compounds are themselves mutagenic, and may be carcinogenic at high doses. Natural selection has probably ensured that the chemical defence mechanisms of the human body are both highly effective and inducible (Ames & Gold, 1991). It would then follow that the maintenance of such defence mechanisms is a normal, diet dependent physiological function, and it may be appropriate to regard the constant provision of the necessary food borne factors as an aspect of normal nutrition. If this hypothesis survives further rigorous research it will be necessary to recognize the existence of a class of dietary substances which, although distinct from micronutrients defined in the conventional sense, are necessary for the maintenance of optimum health. The term 'dietary phytoprotectants' may be suitable as a collective term for such substances.

REFERENCES

Abate, C., Patel, L., Rauscher, F. J. & Curran, T. (1990). Redox regulation of fos and jun DNA binding activity in vitro. *Science* **249**, 1157–1161.

Adlercreutz, H. (1990). Western diet and Western diseases. Some hormonal and biochemical mechanisms and associations. *Scandinavian Journal of Clinical and Laboratory Investigation* **50**, Suppl. 201, 3–23.

Adlercreutz, H., Carson, M., Mousavi, Y., Palotie, A., Booms, S., Loukovaara, M., Mäkelä, T., Wähälä, K., Brunow, G. & Hase, T. (1993). Lignans and isoflavanoids of dietary origin and hormone-dependent cancer. In *Food and Cancer Prevention: Chemical and Biological Aspects*, pp. 348–352 [K. W. Waldron, I. T. Johnson & G. R. Fenwick, editors]. Cambridge: Royal Society of Chemistry.

Adlercreutz, H., Honjo, H., Higashi, H., Fotsis, T., Hamalainen, E., Hasegawa, T. & Okada, H. (1991). Urinary excretion of lignans and isoflavanoid phytoestrogens in Japanese men and women consuming a traditional Japanese diet. *American Journal of Clinical Nutrition* **54**, 1093–1100.

Alldrick, A. J. & Rowland, I. R. (1988). Caffeine inhibits hepatic-microsomal activation of some dietary genotoxins. *Mutagenesis* **3**, 423–427.

Ames, B. N. & Gold, L. S. (1990). Chemical carcinogenesis: too many rodent carcinogens. *Proceedings of the National Academy of Sciences, USA* **87**, 7772–7776.

Ames, B. N. & Gold, L. S. (1991). Mitogenesis, mutagenesis and rodent cancer tests. In *Origins of Human Cancer: a Comprehensive Review*, pp. 125–135 [J. Brugge, T. Curran, E. Harlow and F. McCormick, editors]. New York: Cold Spring Harbor Laboratory Press.

Aruoma, O. I., Halliwell, B., Aeschbach, R. & Loligers, J. (1992). Antioxidant and pro-oxidant properties of active rosemary constituents – carnosol and carnosic acid. *Xenobiotica* **22**, 257–268.

Aust, S. D., Morehouse, L. A. & Thomas, C. E. (1985). Role of metals in oxygen radical reactions. *Journal of Free Radicals in Biology and Medicine* **1**, 3–25.

Axelson, M., Sjövall, J., Gustafsson, B. E. & Setchell, K. D. R. (1982). Origin of lignans in mammals and identification of a precursor from plants. *Nature* **298**, 659–660.

Babich, H., Borenfreund, E. & Stern, A. (1993). Comparative cytotoxicities of selected minor dietary non-nutrients with chemopreventive properties. *Cancer Letters* **73**, 127–133.

Bailey, G. S., Hendricks, J. D., Shelton, D. W., Nixon, J. E. & Pawlowski, N. E. (1987). Enhancement of carcinogenesis by the natural anticarcinogen indole-3-carbinol. *Journal of the National Cancer Institute* **78**, 931–934.

Baldwin, W. S. & LeBlanc, G. A. (1992). The anti-carcinogenic plant compound indole-3-carbinol differentially modulates P450-mediated steroid hydroxylase activities in mice. *Chemico-Biological Interactions* **83**, 155–169.

Bayer, B. M., Kruth, H. S., Vaughan, M. & Beaven, M. A. (1979). Arrest of cultured cells in the G_1 phase of the cell cycle by indomethacin. *Journal of Pharmacology and Experimental Therapeutics* **210**, 106–111.

Bégin, M. E., Ells, G., Das, U. N. & Horrobin, D. F. (1986). Differential killing of human carcinoma cells supplemented with n-3 and n-6 polyunsaturated fatty acids. *Journal of the National Cancer Institute* **77**, 1053–1062.

Bendich, A. (1989). Carotenoids and the immune response. *Journal of Nutrition* **119**, 112–115.

Bendich, A. & Shapiro, S. S. (1986). Effect of β-carotene and canthaxanthin on the immune responses of the rat. *Journal of Nutrition* **116**, 2254–2262.

Berridge, M. J. & Irvine, R. F. (1984). Inositol triphosphate, a novel second messenger in cellular signal transduction. *Nature* **312**, 315–321.
Billings, P. C., Newberne, P. M. & Kennedy, A. R. (1990). Protease inhibitor suppression of colon and anal gland carcinogenesis induced by dimethylhydrazine. *Carcinogenesis* **11**, 1083–1086.
Black, S. D. & Coon, M. J. (1987). P-450 cytochromes: structure and function. *Advances in Enzymology* **60**, 35–87.
Black, S. M. & Wolf, C. R. (1991). The role of glutathione-dependent enzymes in drug resistance. *Pharmacology and Therapeutics* **51**, 139–154.
Block, G., Patterson, B. & Subar, A. (1992). Fruit, vegetables, and cancer prevention. A review of the epidemiological evidence. *Nutrition and Cancer* **18**, 1–29.
Blumberg, P. M. (1988). Protein kinase C as the receptor for the phorbol ester tumor promoters: sixth Rhoads Memorial Award lecture. *Cancer Research* **48**, 1–8.
Bomford, A. B. & Munro, H. N. (1992). Ferritin gene expression in health and malignancy. *Pathobiology* **60**, 10–18.
Bowry, V. W., Ingold, K. U. & Stocker, R. (1992). Vitamin E in human low-density lipoprotein—when and how this antioxidant becomes a pro-oxidant. *Biochemical Journal* **288**, 341–344.
Bracke, M. E., Castronovo, V., Van Cauwenberge, R. M. L., Coopman, P., Vakaet, L., Strojny, P., Foidart, J.-M. & Mareel, M. M. (1987). The anti-invasive flavonoid (+)-catechin binds to laminin and abrogates the effect of laminin on cell morphology and adhesion. *Experimental Cell Research* **173**, 193–205.
Bracke, M. E., Van Cauwenberge, R. M.-L. & Mareel, M. M. (1984). (+)-Catechin inhibits the invasion of malignant fibrosarcoma cells into chick heart *in vitro*. *Clinical and Experimental Metastasis* **2**, 161–170.
Bracke, M. E., Vyncke, B. M., Van Larebeke, N. A., Bruyneel, E. A., De Bruyne, G. K., De Pestel, G. H., De Coster, W. J., Espeel, M. F. & Mareel, M. M. (1989). The flavonoid tangeretin inhibits invasion of MO$_4$ mouse cells into embryonic chick heart *in vitro*. *Clinical and Experimental Metastasis* **7**, 283–300.
Bradfield, C. A. & Bjeldanes, L. F. (1991). Modification of carcinogen metabolism by indolylic autolysis products of *Brassica oleraceae*. In *Nutritional and Toxicological Consequences of Food Processing*, pp. 153–163 [M. Friedman, editor]. New York: Plenum Press.
Bradfield, C. A., Chang, Y. & Bjeldanes, L. F. (1985). Effects of commonly consumed vegetables on hepatic xenobiotic metabolising enzymes in the mouse. *Food and Chemical Toxicology* **23**, 899–904.
Brady, J. F., Ishizaki, H., Fukuto, J. M., Lin, M. C., Fadel, A., Gapac, J. M. & Yang, C. S. (1991). Inhibition of cytochrome P-450 2E1 by diallyl sulfide and its metabolites. *Chemical Research in Toxicology* **4**, 642–647.
Brierley, C. H. & Burchell, B. (1993). Human UDP-glucuronosyl transferases – chemical defence, jaundice and gene therapy. *Bioessays* **15**, 749–754.
Brown, J. P. (1988). Hydrolysis of glycosides and esters. In *Role of the Gut Flora in Toxicity and Cancer*, pp. 109–144 [I. R. Rowland, editor]. London: Academic Press.
Brunda, M. J., Herberman, R. B. & Holden, H. T. (1980). Inhibition of murine natural killer cell activity by prostaglandins. *Journal of Immunology* **124**, 2682–2687.
Buettner, G. R. (1993). The pecking order of free radicals and antioxidants – lipid peroxidation, alpha-tocopherol, and ascorbate. *Archives of Biochemistry and Biophysics* **300**, 535–543.
Casey, P. J., Solski, P. A., Der, C. J. & Buss, J. E. (1989). p21 ras is modified by a farnesyl isoprenoid. *Proceedings of the National Academy of Sciences, USA* **86**, 8323–8327.
Castillo, M. H., Perkins, E., Campbell, J. H., Doerr, R., Hassett, J. M., Kandaswami, C. & Middleton, E. (1989). The effects of the bioflavonoid quercetin on squamous cell carcinoma of head and neck origin. *American Journal of Surgery* **158**, 351–355.
Castillo, T., Koop, D. R., Kamimura, S., Triadafilopoulos, G. & Tsukamoto, H. (1992). Role of cytochrome-P-450 2E1 in ethanol-dependent, carbon tetrachloride-dependent and iron-dependent microsomal lipid peroxidation. *Hepatology* **16**, 992–996.
Cheng, S.-J., Ho, C.-T., Lou, H.-Z., Bao, Y.-D., Jian, Y.-Z., Li, M.-H., Gao, Y.-N., Zhu, G.-F., Bai, J.-F., Guo, S.-P. & Li, X.-Q. (1986). A preliminary study on the antimutagenicity of green tea antioxidants. *Acta Biologiae Experimentalis Sinica* **19**, 427–431.
Chow, C. K. (1988). Interrelationships of cellular antioxidant defense systems. In *Cellular Antioxidant Defense Mechanisms*, vol. 2, pp. 217–237 [C. K. Chow, editor]. Boca Raton, FL: CRC Press.
Chow, C. K. (1990). Mutagenesis and micronutrients relationship. *Food Additives and Contaminants* **7**, Suppl. 1, S44-S47.
Clegg, R. J., Middleton, B., Bell, G. D. & White, D. A. (1982). The mechanism of cyclic monoterpene inhibition of hepatic 3-hydroxy-3-methylglutaryl coenzyme A reductase *in vivo* in the rat. *Journal of Biological Chemistry* **257**, 2294–2299.
Clejan, L. A. & Cederbaum, A. I. (1991). Role of iron, hydrogen peroxide and reactive oxygen species in microsomal oxidation of glycerol to formaldehyde. *Archives of Biochemistry and Biophysics* **285**, 83–89.
Cole, C. B., Fuller, R., Mallett, A. K. & Rowland, I. R. (1985). The influence of the host on expression of intestinal microbial enzyme activities involved in metabolism of foreign compounds. *Journal of Applied Bacteriology* **59**, 549–553.
Comstock, G. W., Helzlsouer, K. J. & Bush, T. L. (1991). Prediagnostic serum levels of carotenoids and vitamin E as related to subsequent cancer in Washington County, Maryland. *American Journal of Clinical Nutrition* **53**, 260S-264S.

Correa, P. (1981). Epidemiological correlations between diet and cancer frequency. *Cancer Research* **41**, 3685–3690.
Crabtree, H. G. (1947). Anti-carcinogenesis. *British Medical Bulletin* **4**, 345–348.
Crowell, P. L., Chang, R. R., Ren, Z., Elson, C. E. & Gould, M. N. (1991). Selective inhibition of isoprenylation of 21–26 kDa proteins by the anticarcinogen d-limonene and its metabolites. *Journal of Biological Chemistry* **266**, 17679–17685.
Dai, Y., Rashbastep, J. & Cederbaum, A. I. (1993). Stable expression of human cytochrome-P4502E1 in HepG2 cells. Characterization of catalytic activities and production of reactive oxygen intermediates. *Biochemistry* **32**, 6928–6937.
Daniel, V. (1993). Glutathione S-transferases. Gene structure and regulation of expression. *CRC Critical Reviews in Biochemistry and Molecular Biology* **28**, 173–207.
Dannenberg, A. J. & Yang, E. K. (1992). Effect of dietary lipids on levels of UDP-glucuronosyltransferase in liver. *Biochemical Pharmacology* **44**, 335–340.
Dawson, M. I., Chao, W.-R. & Helmes, C. T. (1987). Inhibition by retinoids of anthralin-induced mouse epidermal ornithine decarboxylase activity and anthralin-promoted skin tumor formation. *Cancer Research* **47**, 6210–6215.
Dawson, M. I. & Okamura, W. H. (1990). *Chemistry and Biology of Synthetic Retinoids*. Boca Raton, FL: CRC Press.
Deisseroth, A. & Dounce, A. L. (1970). Catalase: physical and chemical properties, mechanism of catalysis, and physiological role. *Physiological Reviews* **50**, 319–375.
De Long, M. J., Prochaska, H. J. & Talalay, P. (1986). Induction of NAD(P)H: quinone reductase in murine hepatoma cells by phenolic antioxidants, azo dyes, and other chemoprotectors: a model system for the study of anticarcinogens. *Proceedings of the National Academy of Sciences, USA* **83**, 787–791.
Derbel, M., Igarashi, T. & Satoh, T. (1993). Differential induction of glutathione S-transferase subunits by phenobarbital, 3-methylcholanthrene and ethoxyquin in rat liver and kidney. *Biochimica et Biophysica Acta* **1158**, 175–180.
Dietrich, D. R., & Swenberg, J. A. (1991). The presence of α2u-globulin is necessary for d-limonene promotion of male rat kidney tumors. *Cancer Research* **51**, 3512–3521.
Diplock, A. T. (1991). Antioxidant nutrients and disease prevention: an overview. *American Journal of Clinical Nutrition* **53**, 189S–193S.
Doll, R. & Peto, R. (1981). The causes of cancer: quantitative estimates of avoidable risks of cancer in the United States today. *Journal of the National Cancer Institute* **66**, 1191–1308.
Donnelly, J. K. & Robinson, D. S. (1991). Superoxide dismutase. In *Oxidative Enzymes in Foods*, pp. 49–91 [D. S. Robinson and N. A. M. Eskin, editors]. London: Elsevier Applied Science.
Dunnick, J. K., Prejean, J. D., Haseman, J., Thompson, R. B., Giles, H. D. & McConnell, E. E. (1982). Carcinogenesis bioassay of allyl isothiocyanate. *Fundamental and Applied Toxicology* **2**, 114–120.
Dutton, G. J. (1980). *Glucuronidation of Drugs and Other Compounds*. Boca Raton, FL: CRC Press.
Earnest, D. L., Hixson, L. J., Finley, P. R., Blackwell, G. G., Einspahr, J., Emerson, S. S. & Alberts, D. S. (1992). Arachidonic acid cascade inhibitors in chemoprevention of human colon cancer: preliminary studies. In *Cancer Chemoprevention*, pp. 165–180 [L. Wattenberg, M. Lipkin, C. W. Boone & G. J. Kelloff, editors]. Boca Raton, FL: CRC Press.
Edwards, Y. H., Potter, J. & Hopkinson, D. A. (1980). Human FAD-dependent NAD(P)H diaphorase. *Biochemical Journal* **187**, 429–436.
Elegbede, J. A., Elson, C. E., Tanner, M. A., Qureshi, A. & Gould, M. N. (1986). Regression of rat primary mammary tumors following dietary d-limonene. *Journal of the National Cancer Institute* **76**, 323–325.
Enomoto, T. & Yamasaki, H. (1985). Rapid inhibition of intercellular communication between BALB/c 3T3 cells by diacylglycerol, a possible endogenous functional analog of phorbol esters. *Cancer Research* **45**, 3706–3710.
Fenwick, G. R., Heaney, R. K. & Mawson, R. (1989). Glucosinolates. In *Toxicants of Plant Origin*, vol. 2, *Glycosides*, pp. 1–41 [P. R. Cheeke, editor]. Boca Raton, FL: CRC Press.
Fiala, E. S., Bobotas, G., Kulakis, C., Wattenberg, L. W. & Weisburger J. H. (1977). Effects of disulfiram and related compounds on the metabolism *in vivo* of the colon carcinogen, 1,2-dimethylhydrazine. *Biochemical Pharmacology* **26**, 1763–1768.
Forrester, L. M., Henderson, C. J., Glancey, M. J., Back, D. J., Park, B. K., Ball, S. E., Kitteringham, N. R., McLaren, A. W., Miles, J. S., Skett, P. & Wolf, C. R. (1992). Relative expression of cytochrome P450 isoenzymes in human liver and association with the metabolism of drugs and xenobiotics. *Biochemical Journal* **281**, 359–368.
Franklin, M. R. (1991). Drug metabolizing enzyme induction by simple diaryl pyridines; 2-substituted isomers selectively increase only conjugation enzyme activities, 4-substituted isomers also induce cytochrome P450. *Toxicology and Applied Pharmacology* **111**, 24–32.
Frenkel, K., Chrzan, K., Ryan, C. A., Wiesner, R. & Troll, W. (1987). Chymotrypsin-specific protease inhibitors decrease hydrogen peroxide formation by activated human polymorphonuclear leukocytes. *Carcinogenesis* **8**, 1207–1212.
Friling, R. S., Bergelson, S. & Daniel, V. (1992). Two adjacent AP-1 like binding sites form the electrophile responsive element of the murine glutathione S-transferase Ya subunit gene. *Proceedings of the National Academy of Sciences, USA* **89**, 668–672.

Fuhr, U., Klittich, K. & Staib, A. H. (1993). Inhibitory effect of grapefruit juice and its bitter principal, naringenin, on CYP1A2 dependent metabolism of caffeine in man. *British Journal of Clinical Pharmacology* **35**, 431–436.

Fujii-Kuriyama, Y., Imataka, H., Sogawa, K., Yasumoto, K.-I. & Kikuchi, Y. (1992). Regulation of CYP1A1 expression. *FASEB Journal* **6**, 706–710.

Fujiki, H., Horiuchi, T., Yamashita, K., Hakii, H., Suganuma, M., Nishino, H., Iwashima, A., Hirata, Y. & Sugimura, T. (1986). Inhibition of tumor promotion by flavonoids. *Progress in Clinical and Biological Research* **213**, 429–440.

Fujiki, H., Suganuma, M., Yoshizawa, S., Yatsunami, J., Nishikawa, S., Furuya, H., Okabe, S., Nishiwaki-Matsushima, R., Matsunaga, S., Muto, Y., Okuda, T. & Sugimura, T. (1992). Sarcophytol A and (-)epigallocatechin gallate (EGCG), nontoxic inhibitors of cancer development. In *Cancer Chemoprevention*, pp. 393–405 [L. Wattenberg, M. Lipkin, C. W. Boone & G. J. Kelloff, editors]. Boca Raton, FL: CRC Press.

Gadelle, D., Raibaud, P. & Sacquet, E. (1985). β-Glucuronidase activities of intestinal bacteria determined both in vitro and in vivo in gnotobiotic rats. *Applied and Environmental Microbiology* **49**, 682–685.

Gandhi, R. K. & Khanduja, K. L. (1992). Action of caffeine in altering the carcinogen-activating and carcinogen-detoxifying enzymes in mice. *Journal of Clinical Biochemistry and Nutrition* **12**, 19–26.

Garte, S. J., Currie, D. D. & Troll, W. (1987). Inhibition of H-*ras* oncogene transformation of NIH 3T3 cells by protease inhibitors. *Cancer Research* **47**, 3159–3162.

Godar, D. E., Thomas, D. P., Miller, S. A. & Lee, W. (1993). Long-wavelength UVA radiation induces oxidative stress, cytoskeletal damage and hemolysis. *Photochemistry and Photobiology* **57**, 1018–1026.

Gonzalez, F. J., Liu, S. Y. & Yano, M. (1993). Regulation of cytochrome-P450 genes – molecular mechanisms. *Pharmacogenetics* **3**, 51–57.

Goodlad, R. A., Madgwick, A. J., Moffatt, M. R., Levin, S., Allen, J. L. & Wright, N. A. (1989). Prostaglandins and the gastric epithelium: effects of misoprostol on gastric epithelial cell proliferation in the dog. *Gut* **30**, 316–321.

Goodwin, J. S. (1984). Immunologic effects of nonsteroidal anti-inflammatory drugs. *American Journal of Medicine* **77** (4B), 7–15.

Griepentrog, F. (1973). [Pathological-anatomical results on the effect of coumarin in animal experiments.] *Toxicology* **1**, 93–102.

Groden, J., Nakamura, Y. & German, J. (1990). Molecular evidence that homologous recombination occurs in proliferating human somatic cells. *Proceedings of the National Academy of Sciences, USA* **87**, 4315–4319.

Grubbs, C. J., Eto, I., Juliana, M. M. & Whitaker, L. M. (1991). Effect of canthaxanthin on chemically induced mammary carcinogenesis. *Oncology* **48**, 239–245.

Guengerich, F. P. (1992a). Metabolic activation of carcinogens. *Pharmacology & Therapeutics* **54**, 17–61.

Guengerich, F. P. (1992b). Human cytochrome-P-450 enzymes. *Life Sciences* **50**, 1471–1478.

Guo, Z., Smith, T. J., Wang, E., Eklind, K. I., Chung, F. L. & Yang, C. S. (1993). Structure-activity relationships of arylalkyl isothiocyanates for the inhibition of 4-(methylnitrosamino)-1-(3-pyridyl)-1-butanone metabolism and the modulation of xenobiotic-metabolizing enzymes in rats and mice. *Carcinogenesis* **14**, 1167–1173.

Halliwell, B. (1990). How to characterize a biological antioxidant. *Free Radical Research Communications* **9**, 1–32.

Hial, V., De Mello, M. C. F., Horakova, Z. & Beaven, M. A. (1977). Antiproliferative activity of anti-inflammatory drugs in two mammalian cell culture lines. *Journal of Pharmacology and Experimental Therapeutics* **202**, 446–454.

Hirose, K., Longo, D. L., Oppenheim, J. J. & Matsushima, K. (1993). Overexpression of mitochondrial manganese superoxide dismutase promotes the survival of tumor cells exposed to interleukin-1, tumor necrosis factor, selected anticancer drugs, and ionizing radiation. *FASEB Journal* **7**, 361–368.

Hochstein, P., Atallah, A. S. & Ernster, L. (1988). DT diaphorase and the toxicity of quinones: status and perspectives. In *Cellular Antioxidant Defense Mechanisms*, vol. 2, pp. 123–131 [C. K. Chow, editor]. Boca Raton, FL: CRC Press.

Horie, T., Awazu, S., Itakura, Y. & Fuwa, T. (1992). Identified diallyl polysulfides from an aged garlic extract which protects the membranes from lipid peroxidation. *Planta Medica* **58**, 468–469.

Hossain, M. Z., Wilkens, L. R., Mehta, P. P., Loewenstein, W. & Bertram, J. S. (1989). Enhancement of gap junctional communication by retinoids correlates with their ability to inhibit neoplastic transformation. *Carcinogenesis* **10**, 1743–1748.

Howie, A. F., Forrester, L. M., Glancey, M. J., Schlager, J. J., Powis, G., Beckett, G. J., Hayes, J. D. & Wolf, C. R. (1990). Glutathione S-transferase and glutathione peroxidase expression in normal and tumour human tissues. *Carcinogenesis* **11**, 451–8.

Hsu, C.-J., Kirkman, B. R. & Fishman, J. (1991). Differential expression of oncogenes c-fos, c-myc and neu/Her 2 induced by estradiol and 16α-hydroxyestrone in human cancer cell line. *Seventh Annual Meeting of the Endocrine Society* Abstract 586.

Huang, D. S., Odeleye, O. E. & Watson, R. R. (1992). Inhibitory effects of canthaxanthin on *in vitro* growth of murine tumor cells. *Cancer Letters* **65**, 209–213.

Huang, M.-T., Lysz, T., Ferraro, T., Abidi, T. F., Laskin, J. D. & Conney, A. H. (1991). Inhibitory effects of curcumin on *in vitro* lipoxygenase and cyclooxygenase activities in mouse epidermis. *Cancer Research* **51**, 813–819.

Huang, M.-T., Smart, R. C., Wong, C.-Q. & Conney, A. H. (1988). Inhibitory effect of curcumin, chlorogenic acid, caffeic acid, and ferulic acid on tumor promotion in mouse skin by 12-O-tetradecanoylphorbol-13-acetate. *Cancer Research* **48**, 5941–5946.

Imanishi, H., Sasaki, Y. F., Matsumoto, K., Watanabe, M., Ohta, T., Shirasu, Y. & Tutikawa, K. (1990). Suppression of 6-TG-resistant mutations in V79 cells and recessive spot formations in mice by vanillin. *Mutation Research* **243**, 151–158.

Imanishi, H., Sasaki, Y. F., Ohta, T., Watanabe, M., Kato, T. & Shirasu, Y. (1991). Tea tannin components modify the induction of sister-chromatid exchanges and chromosome aberrations in mutagen-treated cultured mammalian cells and mice. *Mutation Research* **259**, 79–87.

Inouye, T., Sasaki, Y. F., Imanishi, H., Watanabe, M., Ohta, T. & Shirasu, Y. (1988). Suppression of mitomycin C-induced micronuclei in mouse bone marrow cells by post-treatment with vanillin. *Mutation Research* **202**, 93–95.

Ishidate, M., Harnois, M. C. & Sofuni, T. (1988). A comparative analysis of data on the clastogenicity of 951 chemical substances tested in mammalian cell cultures. *Mutation Research* **195**, 151–213.

Ishikawa, T. (1992). The ATP-dependent glutathione S-conjugate export pump. *Trends in Biochemical Sciences* **17**, 463–468.

Iwata, N., Minegishi, K., Suzuki, K., Ohno, Y., Igarashi, T., Satoh, T. & Takahashi, A. (1993). An unusual profile of musk xylene-induced drug-metabolizing enzymes in rat liver. *Biochemical Pharmacology* **45**, 1659–1665.

Jansson, T. & Zech, L. (1987). Effects of vanillin on sister-chromatid exchanges and chromosome aberrations in human lymphocytes. *Mutation Research* **190**, 221–224.

Jellinck, P. H., Forkert, P. G., Riddick, D. S., Okey, A. B., Michnovicz, J. J. & Bradlow, H. L. (1993). Ah receptor binding properties of indole carbinols and induction of hepatic estradiol hydroxylation. *Biochemical Pharmacology* **45**, 1129–1136.

Jyonouchi, H., Hill, R. J., Tomita, Y. & Good, R. A. (1991). Studies of immunomodulating actions of carotenoids. I. Effects of β-carotene and astaxanthin on murine lymphocyte functions and cell surface marker expression in *in vitro* culture system. *Nutrition and Cancer* **16**, 93–105.

Jyonouchi, H., Zhang, L. & Tomita, Y. (1993). Studies of immunomodulating actions of carotenoids. II. Astaxanthin enhances *in vitro* antibody production to T-dependent antigens without facilitating polyclonal B-cell activation. *Nutrition and Cancer* **19**, 269–280.

Kasamaki, A., Takahashi, H., Tsumura, N., Niwa, J., Fujita, T. & Urasawa, S. (1982). Genotoxicity of flavoring agents. *Mutation Research* **105**, 387–392.

Kasamaki, A., Yasuhara, T. & Urasawa, S. (1987). Neoplastic transformation of Chinese hamster cells *in vitro* after treatment with flavoring agents. *Journal of Toxicological Sciences* **12**, 383–396.

Kato, K., Cox, A. D., Hisaka, M. M., Graham, S. M., Buss, J. E. & Der, C. J. (1992). Isoprenoid addition to Ras protein is the critical modification for its membrane association and transforming activity. *Proceedings of the National Academy of Sciences, USA* **89**, 6403–6407.

Kato, R., Nakadate, T., Yamamoto, S. & Sugimura, T. (1983). Inhibition of 12-O-tetradecanoylphorbol-13-acetate-induced tumor promotion and ornithine decarboxylase activity by quercetin: possible involvement of lipoxygenase inhibition. *Carcinogenesis* **4**, 1301–1305.

Kennedy, A. R. & Little, J. B. (1981). Effects of protease inhibitors on radiation transformation *in vitro*. *Cancer Research* **41**, 2103–2108.

Kilgore, K. S. & Lucchesi, B. R. (1993). Reperfusion injury after myocardial infarction: the role of free radicals and the inflammatory response. *Clinical Biochemistry* **26**, 359–370.

Klone, A., Hussnatter, R. & Sies, H. (1992). Cloning, sequencing and characterization of the human alpha-glutathione S-transferase gene corresponding to the cDNA clone pGTH2. *Biochemical Journal* **285**, 925–928.

Kore, A. M., Jeffery, E. H. & Wallig, M. A. (1993). Effects of 1-isothiocyanato-3-(methylsulfinyl)-propane on xenobiotic metabolizing enzymes in rats. *Food and Chemical Toxicology* **31**, 723–729.

Korpássy, B. & Mosonyi, M. (1950). The carcinogenic activity of tannic acid. Liver tumours induced in rats by prolonged subcutaneous administration of tannic acid solutions. *British Journal of Cancer* **4**, 411–420.

Krinsky, N. I. (1993). Actions of carotenoids in biological systems. *Annual Review of Nutrition* **13**, 561–587.

Ladenstein, R. (1984). Molecular enzymology of seleno-glutathione peroxidase. *Peptide and Protein Reviews* **4**, 173–214.

Landers, J. P. & Bunce, N. J. (1991). The Ah receptor and the mechanism of dioxin toxicity. *Biochemical Journal* **276**, 273–287.

Larocca, L. M., Teofili, L., Leone, G., Sica, S., Pierelli, L., Menichella, G., Scambia, G., Benedetti Panici, P., Ricci, R., Piantelli, M. & Ranelletti, F. O. (1991). Antiproliferative activity of quercetin on normal bone marrow and leukaemic progenitors. *British Journal of Haematology* **79**, 562–566.

Lee, H. P., Gourley, L., Duffy, S. W., Estève, J., Lee, J. & Day, N. E. (1991). Dietary effects on breast-cancer risk in Singapore. *Lancet* **337**, 1197–1200.

Lee, R. B. (1967). *What Hunters Do for a Living: A Comparative Study in Man the Hunter*, p. 41 [R. B. Lee and D. Vore, editors]. Chicago: Aldine.

Levine, L., Hinkle, P. M., Voelkel, E. F. & Tashjian, A. H. (1972). Prostaglandin production by mouse fibrosarcoma cells in culture: inhibition by indomethacin and aspirin. *Biochemical and Biophysical Research Communications* **47**, 888–896.

Li, Y. & Jaiswal, A. K. (1992). Regulation of human NAD(P)H: quinone oxidoreductase gene – role of AP1

binding site contained within human antioxidant response element. *Journal of Biological Chemistry* **267**, 15097–15104.
Lind, C., Hochstein, P. & Ernster, L. (1982). DT-diaphorase as a quinone reductase: a cellular control device against semiquinone and superoxide radical formation. *Archives of Biochemistry and Biophysics* **216**, 178–185.
Liu, J. & Mori, A. (1993). Antioxidant and pro-oxidant activities of *p*-hydroxybenzyl alcohol and vanillin: effects on free radicals, brain peroxidation and degradation of benzoate, deoxyribose, amino acids and DNA. *Neuropharmacology* **32**, 659–669.
Lotan, R. (1980). Effects of vitamin A and its analogs (retinoids) on normal and neoplastic cells. *Biochimica et Biophysica Acta* **605**, 33–91.
Lotan, R. (1992*a*). Evaluation of the results of clinical trials with retinoids in relation to their basic mechanism of action. In *Cancer Chemoprevention*, pp. 71–82 [L. Wattenberg, M. Lipkin, C. W. Boone & G. J. Kelloff, editors]. Boca Raton, FL: CRC Press.
Lotan, R. (1992*b*). Cell invasion as a target for chemoprevention. In *Cellular and Molecular Targets for Chemoprevention*, pp. 339–350 [V. E. Steele, G. D. Stoner, C. W. Boone & G. J. Kelloff, editors]. Boca Raton, FL: CRC Press.
Lucas, D., Menez, J. F., Berthou, F., Cauvin, J. M. & Deitrich, R. A. (1992). Differences in hepatic microsomal cytochrome P-450 isoenzyme induction by pyrazole, chronic ethanol, 3-methylcholanthrene, and phenobarbital in high alcohol sensitivity (HAS) and low alcohol sensitivity (LAS) rats. *Alcoholism – Clinical and Experimental Research* **16**, 916–921.
Luk, G. D. (1992). Clinical and biologic studies of DFMO in the colon. In *Cancer Chemoprevention*, pp. 515–530 [L. Wattenberg, M. Lipkin, C. W. Boone & G. J. Kelloff, editors]. Boca Raton, FL: CRC Press.
McCann, P. P., Bitonti, A. J. & Pegg, A. E. (1992). Inhibition of polyamine metabolism and the consequent effects on cell proliferation. In *Cancer Chemoprevention*, pp. 531–539 [L. Wattenberg, M. Lipkin, C. W. Boone & G. J. Kelloff, editors]. Boca Raton, FL: CRC Press.
McManus, M. E. & McKinnon, R. A. (1991). Measurement of cytochrome P450 activation of xenobiotics using the Ames Salmonella test. *Methods in Enzymology* **206**, 501–509.
Maeda, S. & Nakagawa, M. (1977). General chemical and physical analyses on various kinds of green tea. *Chagyo Kenkyu Hokoku* **45**, 85–92.
Maiorino, M., Zamburlini, A., Roveri, A. & Ursini, F. (1993). Prooxidant role of vitamin-E in copper induced lipid peroxidation. *FEBS Letters* **330**, 174–176.
Mallett, A. K., Bearne, C. A., Lake, B. G. & Rowland, I. R. (1989). Modified mutagen activation in hepatic fractions from rats fed dietary rutin. Interaction between gut flora and host metabolism. *Food and Chemical Toxicology* **27**, 607–611.
Mallett, A. K. & Rowland, I. R. (1990). Bacterial enzymes: their role in the formation of mutagens and carcinogens in the intestine. *Digestive Diseases* **8**, 71–79.
Mamont, P. S., Böhlen, P., McCann, P. P., Bey, P., Schuber, F. & Tardif, C. (1976). α-Methyl ornithine, a potent competitive inhibitor of ornithine decarboxylase, blocks proliferation of rat hepatoma cells in culture. *Proceedings of the National Academy of Sciences, USA* **73**, 1626–1630.
Mannervik, B. (1985). The isoenzymes of glutathione transferase. *Advances in Enzymology* **57**, 357–417.
Mannervik, B., Alin, P., Guthenberg, C., Jensson, H. & Warholm, M. (1985). Glutathione transferases and the detoxification of products of oxidative metabolism. In *Proceedings of the 6th International Symposium on Microsomes and Drug Oxidations*, pp. 221–228 [A. R. Boobis, J. Caldwell, F. De Mattheis & C. R. Elcombe, editors]. London: Taylor & Francis.
Manervik, B. & Danielson, U. H. (1988). Glutathione S-transferases – structure and catalytic activity. *CRC Critical Reviews in Biochemistry* **23**, 283–337.
Markaverich, B. M., Roberts, R. R., Alejandro, M. A., Johnson, G. A., Middleditch, B. S. & Clark, J. H. (1988). Bioflavonoid interaction with rat uterine type II binding sites and cell growth inhibition. *Journal of Steroid Biochemistry* **30**, 71–78.
Maruta, A., Enaka, K. & Umeda, M. (1979). Mutagenicity of quercetin and kaempferol on cultured mammalian cells. *GANN* **70**, 273–276.
Mathews-Roth, M. M. (1982). Antitumor activity of β-carotene, canthaxanthin and phytoene. *Oncology* **39**, 33–37.
Meister, A. (1991). Glutathione deficiency produced by inhibition of its synthesis, and its reversal; applications in research and therapy. *Pharmacology and Therapeutics* **51**, 155–194.
Michnovicz, J. J. & Bradlow, H. L. (1990). Induction of estradiol metabolism by dietary indole-3-carbinol in humans. *Journal of the National Cancer Institute* **82**, 947–949.
Michnovicz, J. J. & Bradlow, H. L. (1991). Altered estrogen metabolism and excretion in humans following consumption of indole-3-carbinol. *Nutrition and Cancer* **16**, 59–66.
Miners, J. O. & Mackenzie, P. I. (1991). Drug glucuronidation in humans. *Pharmacology and Therapeutics* **51**, 347–369.
Minotti, G., Di Gennaro, M., D'Ugo, D. & Granone, P. (1991). Possible sources of iron for lipid peroxidation. *Free Radical Research Communications* **12–13**, 99–106.
Moncada, S., Flower, R. J. & Vane, J. R. (1980). Prostaglandins, prostacyclin, and thromboxane A$_2$. In *Goodman and Gilman's The Pharmacological Basis of Therapeutics*, 6th edn, pp. 668–681 [A. G. Gilman, L. S. Goodman & A. Gilman, editors]. New York: MacMillan.

Moon, R. C., Rao, K. V. N., Detrisac, C. J. & Kelloff, G. J. (1992). Retinoid chemoprevention of lung cancer. In *Cancer Chemoprevention*, pp. 83–93 [L. Wattenberg, M. Lipkin, C. W. Boone & G. J. Kelloff, editors]. Boca Raton, FL: CRC Press.

Morel, I., Lescoat, G., Cogrel, P., Sergent, O., Pasdeloup, N., Brissot, P., Cillard, P. & Cillard, J. (1993). Antioxidant and iron-chelating activities of the flavonoids catechin, quercetin and diosmetin on iron-loaded rat hepatocyte cultures. *Biochemical Pharmacology* **45**, 13–19.

Mori, A., Nishino, C., Enoki, N. & Tawata, S. (1988). Cytotoxicity of plant flavonoids against HeLa cells. *Phytochemistry* **27**, 1017–1020.

Morse, M. A., Eklind, K. I., Amin, S. G., Hecht, S. S. & Chung, F. L. (1989). Effects of alkyl chain length on the inhibition of NNK-induced lung neoplasia in A/J mice by arylalkyl isothiocyanates. *Carcinogenesis* **10**, 1757–1759.

Morse, M. A. & Stoner, G. D. (1993). Cancer chemoprevention: principles and prospects. *Carcinogenesis* **14**, 1737–1746.

Mukai, K., Sawada, K., Kohno, Y. & Terao, J. (1993). Kinetic study of the prooxidant effect of tocopherol. Hydrogen abstraction from lipid hydroperoxides by tocopheroxyls in solution. *Lipids* **28**, 747–752.

Murphy, S. P., Subar A. F. & Block G. (1990). Vitamin E intakes and sources in the United States. *American Journal of Clinical Nutrition* **52**, 361–367.

Murray, A. W., Fournier, A. & Hardy, S. J. (1978). Proteolytic activation of protein kinase C: a physiological reaction? *Trends in Biochemical Science* **12**, 53–54.

Musk, S. R. R. & Johnson, I. T. (1993a). Allyl isothiocyanate is selectively toxic to transformed cells of the human colorectal tumour line HT29. *Carcinogenesis* **14**, 2079–2083.

Musk, S. R. R. & Johnson, I. T. (1993b). *In vitro* genetic toxicology testing of naturally occurring isothiocyanates. In *Food and Cancer Prevention: Chemical and Biological Aspects*, pp. 58–61 [K. W. Waldron, I. T. Johnson & G. R. Fenwick, editors]. Cambridge: Royal Society of Chemistry.

Nagasawa, H. & Konishi, R. (1988). Stimulation by caffeine of spontaneous mammary tumorigenesis in mice. *European Journal of Cancer and Clinical Oncology* **24**, 803–805.

Nakadate, T., Yamamoto, S., Aizu, E. & Kato, R. (1984). Effects of flavonoids and antioxidants on 12-O-tetradecanoylphorbol-13-acetate-caused epidermal ornithine decarboxylase induction and tumor promotion in relation to lipoxygenase inhibition by these compounds. *GANN* **75**, 214–222.

Nanno, M., Morotomi, M., Takayama, H., Kuroshima, T., Tanaka, R. & Mutai, M. (1986). Mutagenic activation of biliary metabolites of benzo(a)pyrene by β-glucuronidase-positive bacteria in human faeces. *Journal of Medical Microbiology* **22**, 351–355.

Nelson, D. R., Kamataki, T., Waxman, D. J., Guengerich, F. P., Estabrook, R. W., Feyereisen, R., Gonzalez, F. J., Coon, M. J., Gunsalus, I. C., Gotoh, O., Okuda, K. & Nebert, D. W. (1993). The P450 superfamily – update on new sequences, gene mapping, accession numbers, early trival names of enzymes, and nomenclature. *DNA and Cell Biology* **12**, 1–51.

Nguyen, T. & Pickett, C. B. (1992). Regulation of rat glutathione *S*-transferase Ya subunit gene expression – DNA-protein interaction at the antioxidant responsive element. *Journal of Biological Chemistry* **267**, 13535–13539.

Nijhoff, W. A., Groen, G. M. & Peters, W. H. M. (1993). Induction of rat hepatic and intestinal glutathione S-transferases and glutathione by dietary naturally occurring anticarcinogens. *International Journal of Oncology* **3**, 1131–1139.

Nishikawa, A., Furukawa, F., Imazawa, T., Yoshimura, H., Mitsumori, K. & Takahashi, M. (1992). Effects of caffeine, nicotine, ethanol and sodium selenite on pancreatic carcinogenesis in hamsters after initiation with *N*-nitrosobis(2-oxopropyl)amine. *Carcinogenesis* **13**, 1379–1382.

Nishino, H. (1992). Antitumor-promoting activity of glycyrrhetinic acid and its related compounds. In *Cancer Chemoprevention*, pp. 457–467 [L. Wattenberg, M. Lipkin, C. W. Boone & G. J. Kelloff, editors]. Boca Raton, FL: CRC Press.

Nomura, A., Henderson, B. E. & Lee, J. (1978). Breast cancer and diet among the Japanese in Hawai. *American Journal of Clinical Nutrition* **31**, 2020–2025.

O'Brian, C. A., Ward, N. E. & Vogel, V. G. (1990). Inhibition of protein kinase C by the 12-O-tetradecanoylphorbol-13-acetate antagonist glycyrrhetic acid. *Cancer Letters* **49**, 9–12.

Oguni, I., Nasu, K., Kanaya, S., Ota, Y., Yamamoto, S. & Nomura, T. (1989). Epidemiological and experimental studies on the antitumor activity by green tea extracts. *Japanese Journal of Nutrition* **47**, 43–48.

Ohmori, S., Misaizu, T., Nakamura, T., Takano, N., Kitagawa, H. & Kitada, M. (1993). Differential role in lipid peroxidation between rat P450 1A1 and P450 1A2. *Biochemical Pharmacology* **46**, 55–60.

Ohta, T., Watanabe, K., Moriya, M., Shirasu, Y. & Kada, T. (1983a). Anti-mutagenic effects of coumarin and umbelliferone on mutagenesis induced by 4-nitroquinoline 1-oxide or UV irradiation in *E. coli*. *Mutation Research* **117**, 135–138.

Ohta, T., Watanabe, K., Moriya, M., Shirasu, Y. & Kada, T. (1983b). Analysis of the antimutagenic effect of cinnamaldehyde on chemically induced mutagenesis in *Escherichia coli*. *Molecular and General Genetics* **192**, 309–315.

Ohta, T., Watanabe, M., Shirasu, Y. & Inoue, T. (1988). Post-replication repair and recombination in *uvrA umuC* strains of *Escherichia coli* are enhanced by vanillin, an antimutagenic compound. *Mutation Research* **201**, 107–112.

Ohta, T., Watanabe, M., Watanabe, K., Shirasu, Y. & Kada, T. (1986). Inhibitory effects of flavourings on mutagenesis induced by chemicals in bacteria. *Food and Chemical Toxicology* **24**, 51–54.

Okuzumi, J., Nishino, H., Murakoshi, M., Iwashima, A., Tanaka, Y., Yamane, T., Fujita, Y. & Takahashi, T. (1990). Inhibitory effects of fucoxanthin, a natural carotenoid, on N-*myc* expression and cell cycle progression in human malignant tumor cells. *Cancer Letters* **55**, 75–81.

Okuzumi, J., Takahashi, T., Yamane, T., Kitao, Y., Inagake, M., Ohya, K., Nishino, H. & Tanaka, Y. (1993). Inhibitory effects of fucoxanthin, a natural carotenoid, on N-ethyl-N'-nitro-N-nitrosoguanidine-induced mouse duodenal carcinogenesis. *Cancer Letters* **68**, 159–168.

Pacifici, R. E. & Davies, K. J. A. (1990). Protein degradation as an index of oxidative stress. *Methods in Enzymology* **186**, 485–502.

Packer, L. (1991). Protective role of vitamin E in biological systems. *American Journal of Clinical Nutrition* **53**, 1050S–1055S.

Packer, L. (1992). New horizons in vitamin E research – the vitamin E cycle, biochemistry, and clinical applications. In *Lipid-soluble Antioxidants: Biochemistry and Clinical Applications*, pp. 1–15 [A. S. H. Ong and L. Packer, editors]. Basel: Birkhauser Verlag.

Pegg, A. E. (1988). Polyamine metabolism and its importance in neoplastic growth and as a target for chemotherapy. *Cancer Research* **48**, 759–774.

Pence, B. C., Buddingh, F. & Yang, S. P. (1986). Multiple dietary factors in the enhancement of dimethylhydrazine carcinogenesis: main effect of indole-3-carbinol. *Journal of the National Cancer Institute* **77**, 269–276.

Powell, S. R. & Tortolani, A. J. (1992). Recent advances in the role of reactive oxygen intermediates in ischemic injury. 1. Evidence demonstrating presence of reactive oxygen intermediates. 2. Role of metals in site-specific formation of radicals. *Journal of Surgical Research* **53**, 417–429.

Pratt, D. E. (1992). Natural antioxidants from plant material. In *Phenolic Compounds in Food and their Effect on Health*, vol. 2, pp. 54–71. [M.-T. Huang, C.-T. Ho & C. Y. Lee, editors]. Washington DC: American Chemical Society.

Preston-Martin, S., Pike, M. C., Ross, R. K., Jones, P. A. & Henderson, B. E. (1990). Increased cell division as a cause of human cancer. *Cancer Research* **50**, 7415–7421.

Prochaska, H. J., De Long, M. J. & Talalay, P. (1985). On the mechanisms of induction of cancer-protective enzymes: a unifying proposal. *Proceedings of the National Academy of Sciences, USA* **82**, 8232–8236.

Prochaska, H. J. & Santamaria, A. B. (1988). Direct measurement of NAD(P)H:quinone reductase from cells cultured in microtiter wells: a screening assay for anticarcinogenic enzyme inducers. *Analytical Biochemistry* **169**, 328–336.

Pung, A., Rundhaug, J. E., Yoshizawa, C. N. & Bertram, J. S. (1988). β-Carotene and canthaxanthin inhibit chemically- and physically-induced neoplastic transformation in 10T1/2 cells. *Carcinogenesis* **9**, 1533–1539.

Rao, C. V., Simi, B. & Reddy, B. S. (1993). Inhibition by dietary curcumin of azoxymethane-induced ornithine decarboxylase, tyrosine protein kinase, arachidonic acid metabolism and aberrant crypt foci formation in the rat colon. *Carcinogenesis* **14**, 2219–2225.

Raveendran, M., Thanislass, J., Maheswari, G. U. & Devaraj, H. (1993). Induction of prooxidant state by the food flavor cinnamaldehyde in rat liver. *Journal of Nutritional Biochemistry* **4**, 181–183.

Reddy, B. S., Weisburger, J. H., Narisawa, T. & Wynder, E. L. (1974). Colon carcinogenesis in germ-free rats treated with 1,2-dimethylhydrazine and N-methyl-N'-nitro-N-nitrosoguanidine. *Cancer Research* **34**, 2368–2372.

Reif, D. W. (1992). Ferritin as a source of iron for oxidative damage. *Free Radical Biology and Medicine* **12**, 417–427.

Remacle, J., Lambert, D., Raes, M., Pigeolet, E., Michiels, C. & Toussaint, O. (1992). Importance of various antioxidant enzymes for cell stability – confrontation between theoretical and experimental data. *Biochemical Journal* **286**, 41–46.

Riley, R. J. & Workman, P. (1992). DT-diaphorase and cancer chemotherapy. *Biochemical Pharmacology* **43**, 1657–1669.

Rousseau, E. J., Davison, A. J. & Dunn, B. (1992). Protection by β-carotene and related compounds against oxygen-mediated cytotoxicity and genotoxicity – implications for carcinogenesis and anticarcinogenesis. *Free Radical Biology and Medicine* **13**, 407–433.

Rowland, I. R., Mallett, A. K. & Wise, A. (1985). The effect of diet on the mammalian gut flora and its metabolic activities. *CRC Critical Reviews in Toxicology* **16**, 31–103.

Rozen, F., Nguyen, T. & Pickett, C. B. (1992). Isolation and characterization of a human glutathione S-transferase Ha1 subunit gene. *Archives of Biochemistry and Biophysics* **292**, 589–593.

Ruch, R. J., Cheng, S.-J. & Klaunig, J. E. (1989). Prevention of cytotoxicity and inhibition of intercellular communication by antioxidant catechins isolated from Chinese green tea. *Carcinogenesis* **10**, 1003–1008.

Rueff, J., Laires, A., Gaspar, J., Borba, H. & Rodrigues, A. (1992). Oxygen species and the genotoxicity of quercetin. *Mutation Research* **265**, 75–81.

Rushmore, T. H., Morton, M. R. & Pickett, C. B. (1991). The antioxidant responsive element. Activation by oxidative stress and identification of the DNA consensus sequence required for functional activity. *Journal of Biological Chemistry* **266**, 11632–11639.

Rushmore, T. H. & Pickett, C. B. (1993). Glutathione S-transferases, structure, regulation, and therapeutic implications. *Journal of Biological Chemistry* **268**, 11475–11478.

Saatcioglu, F., Perry, D. J., Pasco, D. S. & Fagan, J. B. (1990). Aryl hydrocarbon (Ah) receptor DNA binding activity. Sequence specificity and Zn^{2+} requirement. *Journal of Biological Chemistry* **265**, 9251–9258.

Sasaki, Y. F., Imanishi, H., Ohta, T. & Shirasu, Y. (1987). Effects of vanillin on sister-chromatid exchanges and chromosome aberrations induced by mitomycin C in cultured Chinese hamster ovary cells. *Mutation Research* **191**, 193–200.

Sasaki, Y. F., Imanishi, H., Watanabe, M., Ohta, T. & Shirasu, Y. (1990a). Suppressing effect of antimutagenic flavorings on chromosome aberrations induced by UV-light or X-rays in cultured Chinese hamster cells. *Mutation Research* **229**, 1–10.

Sasaki, Y. F., Ohta, T., Imanishi, H., Watanabe, M., Matsumoto, K., Kato, T. & Shirasu, Y. (1990b). Suppressing effects of vanillin, cinnamaldehyde, and anisaldehyde on chromosome aberrations induced by X-rays in mice. *Mutation Research* **243**, 299–302.

Sato, M., Narisawa, T., Sano, M., Takahashi, T. & Goto, A. (1983). Growth inhibition of transplantable murine colon adenocarcinoma 38 by indomethacin. *Journal of Cancer Research and Clinical Oncology* **106**, 21–26.

Schelp, F.-P. & Pongpaew, P. (1988). Protection against cancer through nutritionally-induced increase of endogenous proteinase inhibitors – a hypothesis. *International Journal of Epidemiology* **17**, 287–292.

Schneider, J., Huh, M. M., Bradlow, H. L. & Fishman, J. (1984). Antiestrogen action of 2-hydroxyestrone on MCF-7 human breast cancer cells. *Journal of Biological Chemistry* **259**, 4840–4845.

Scholar, E. M., Wolterman, K., Birt, D. F. & Bresnick, E. (1989). The effect of diets enriched in cabbage and collards on murine pulmonary metastasis. *Nutrition and Cancer* **12**, 121–126.

Schrauzer, G. N. (1992). Mechanistic aspects of anticarcinogenic action. *Biological Trace Element Research* **33**, 51–62.

Schuckelt, R., Brigelius-Flohe, R., Maiorino, M., Roveri, A., Reumkens, J., Strassburger, W., Ursini, F., Wolf, B. & Flohe, L. (1991). Phospholipid hydroperoxide glutathione peroxidase is a selenoenzyme distinct from the classical glutathione peroxidase as evident from cDNA and amino acid sequencing. *Free Radical Research Communications* **14**, 343–361.

Scott, B. C., Butler, J., Halliwell, B. & Aruoma, O. I. (1993). Evaluation of the antioxidant actions of ferulic acid and catechins. *Free Radical Research Communications* **19**, 241–253.

Setchell, K. D. R. & Aldercreutz, H. (1988). Mammalian lignans and phyto-oestrogens. Recent studies on their formation, metabolism and biological role in health and disease. In *Role of the Gut Flora in Toxicity and Cancer*, pp. 315–345 [I. R. Rowland, editor]. London: Academic Press.

Shamsuddin, A. M. & Sakamoto, K. (1992). Antineoplastic action of inositol compounds. In *Cancer Chemoprevention*, pp. 285–308 [L. Wattenberg, M. Lipkin, C. W. Boone & G. J. Kelloff, editors]. Boca Raton, FL: CRC Press.

Shamsuddin, A. M. & Ullah, A. (1989). Inositol hexaphosphate inhibits large intestinal cancer in F344 rats 5 months following induction by azoxymethane. *Carcinogenesis* **10**, 625–626.

Shertzer, H. G. & Sainsbury, M. (1991). Chemoprotective and hepatic enzyme induction properties of indole and indenoindole antioxidants in rats. *Food and Chemical Toxicology* **29**, 391–400.

Shi, X., Dalal, N. S. & Jain, A. C. (1991). Antioxidant behaviour of caffeine: efficient scavenging of hydroxyl radicals. *Food and Chemical Toxicology* **29**, 1–6.

Shimoi, K., Nakamura, Y., Tomita, I., Hara, Y. & Kada, T. (1986). The pyrogallol related compounds reduce UV-induced mutations in *Escherichia coli* B/r WP2. *Mutation Research* **173**, 239–244.

Shimoi, K., Nakamura, Y., Tomita, I. & Kada, T. (1985). Bio-antimutagenic effects of tannic acid on UV and chemically induced mutagenesis in *Escherichia coli* B/r. *Mutation Research* **149**, 17–23.

Shklar, G. & Schwartz, J. (1988). Tumor necrosis factor in experimental cancer regression with alphatocopherol, beta-carotene, canthaxanthin and algae extract. *European Journal of Cancer and Clinical Oncology* **24**, 839–850.

Slaga, T. J. (1983). Overview of tumor promotion in animals. *Environmental Health Perspectives* **50**, 3–14.

Smith, T. J., Guo, Z., Li, C., Ning, S. M., Thomas, P. E. & Yang, C. S. (1993). Mechanisms of inhibition of 4-(methylnitrosamino)-1-(3-pyridyl)-1-butanone bioactivation in mouse by dietary phenethyl isothiocyanate. *Cancer Research* **53**, 3276–3282.

Soudamini, K. K. & Kuttan, R. (1989). Inhibition of chemical carcinogenesis by curcumin. *Journal of Ethnopharmacology* **27**, 227–233.

Stähelin, H. B., Gey, K. F., Eichholzer, M. & Lüdin, E. (1991). β-Carotene and cancer prevention: the Basel study. *American Journal of Clinical Nutrition* **53**, 265S–269S.

Sugie, S., Yoshimi, N., Okumura, A., Tanaka, T. & Mori, H. (1993). Modifying effects of benzyl isothiocyanate and benzyl thiocyanate on DNA synthesis in primary cultures of rat hepatocytes. *Carcinogenesis* **14**, 281–283.

Sutherland, L., Ebner, T. & Burchell, B. (1993). The expression of UDP-glucuronosyltransferases of the UGT1 family in human liver and kidney and in response to drugs. *Biochemical Pharmacology* **45**, 295–301.

Suzuki, S., Takada, T., Sugawara, Y., Muto, T. & Kominami, R. (1991). Quercetin induces recombinational mutations in cultured cells as detected by DNA fingerprinting. *Japanese Journal of Cancer Research* **82**, 1061–1064.

Taffet, S. M. & Russell, S. W. (1981). Macrophage-mediated tumor cell killing: regulation of expression of cytolytic activity by prostaglandin E. *Journal of Immunology* **126**, 424–427.

Takada, H., Hirooka, T., Hiramatsu, Y. & Yamamoto, M. (1982). Effects of β-glucuronidase inhibitor on azoxymethane-induced colonic carcinogenesis in rats. *Cancer Research* **42**, 331–334.

Takahashi, S., Hakoi, K., Yada, H., Hirose, M., Ito, N. & Fukushima, S. (1992). Enhancing effects of diallyl sulfide on hepatocarcinogenesis and inhibitory actions of the related diallyl disulfide on colon and renal carcinogenesis in rats. *Carcinogenesis* **13**, 1513–1518.

Talalay, P., De Long, M. J. & Prochaska, H. J. (1988). Identification of a common chemical signal regulating the induction of enzymes that protect against chemical carcinogenesis. *Proceedings of the National Academy of Sciences, USA* **85**, 8261–8265.

Talalay, P. & Prochaska, H. J. (1987). Mechanisms of induction of NAD(P)H:quinone reductase. *Chemica Scripta* **27A**, 61–66.

Tanaka, T., Morishita, Y., Suzui, M., Kojima, T., Okumura, A. & Mori, H. (1994). Chemoprevention of mouse urinary bladder carcinogenesis by the naturally occurring carotenoid astaxanthin. *Carcinogenesis* **15**, 15–19.

Tawfiq, N., Wanigatunga, S., Heaney, R. K., Musk, S. R. R., Williamson, G. & Fenwick, G. R. (1994). Induction of the anticarcinogenic enzyme quinone reductase by food extracts using murine hepatoma cells. *European Journal of Cancer Prevention*, in press.

Telang, N. T., Suto, A., Wong, G. Y., Osborne, M. P. & Bradlow, H. L. (1992). Induction by estrogen metabolite 16 α-hydroxyestrone of genotoxic damage and aberrant proliferation in mouse mammary epithelial cells. *Journal of the National Cancer Institute*, **84**, 634–638.

Testa, U., Pelosi, E. & Peschle, C. (1993). The transferrin receptor. *Critical Reviews in Oncogenesis* **4**, 241–276.

Tong, C., Fazio, M. & Williams, G. M. (1980). Cell cycle-specific mutagenesis at the hypoxanthine phosphoribosyltransferase locus in adult rat liver epithelial cells. *Proceedings of the National Academy of Sciences, USA* **77**, 7377–7379.

Troll, W., Frenkel, K. & Wiesner, R. (1984). Protease inhibitors as anticarcinogens. *Journal of the National Cancer Institute* **73**, 1245–1250.

Troll, W., Lim, J. S. & Belman, S. (1992). Protease inhibitors suppress carcinogenesis *in vivo* and *in vitro*. In *Cancer Chemoprevention*, pp. 503–512 [L. Wattenberg, M. Lipkin, C. W. Boone & G. J. Kelloff, editors]. Boca Raton, FL: CRC Press.

Troll, W., Wiesner, R. & Frenkel, K. (1987). Anticarcinogenic action of protease inhibitors. *Advances in Cancer Research* **49**, 265–283.

Trosko, J. E. & Chang, C.-C. (1984). Role of intercellular communication in tumor promotion. In *Mechanisms of Tumor Promotion*, vol. 4, *Cellular Responses to Tumor Promoters*, pp. 119–145 [T. J. Slaga, editor]. Boca Raton, FL: CRC Press.

Ursini, F. & Bindoli, A. (1987). The role of selenium peroxidases in the protection against oxidative damage of membranes. *Chemistry and Physics of Lipids* **44**, 255–276.

Ursini, F., Maiorino, M. & Gregolin, C. (1985). The selenoenzyme phospholipid hydroperoxide glutathione peroxidase. *Biochimica et Biophysica Acta* **839**, 62–70.

VanderPloeg, L. C. & Welsch, C. W. (1991). Inhibition by caffeine of ovarian hormone-induced mammary gland tumorigenesis in female GR mice. *Cancer Letters* **56**, 245–250.

Verhagen, H., van Poppel, G., Willems, M. I., Bogaards, J. J. P., Rompelberg, C. J. M. & van Bladeren, P. J. (1993). Cancer prevention by natural food constituents. *International Food Ingredients no. 1/2*, 22–29.

Wallig, M. A., Kuchan, M. J. & Milner, J. A. (1993). Differential effects of cyanohydroxybutene and selenium on normal and neoplastic canine mammary cells *in vitro*. *Toxicology Letters* **69**, 97–105.

Wang, Z. Y., Agarwal, R., Zhou, Z. C., Bickers, D. R. & Mukhtar, H. (1991). Inhibition of mutagenicity in *Salmonella typhimurium* and skin tumor initiating and tumor promoting activities in SENCAR mice by glycyrrhetinic acid: comparison of 18 α- and β-stereoisomers. *Carcinogenesis* **12**, 187–192.

Wattenberg, L. W. (1981). Inhibition of carcinogen-induced neoplasia by sodium cyanate, tert-butylisocyanate and benzyl isothiocyanate administered subsequent to carcinogen exposure. *Cancer Research* **41**, 2991–2994.

Wattenberg, L. W. (1983). Inhibition of neoplasia by minor dietary constituents. *Cancer Research* **43**, 2448S–2453S.

Wattenberg, L. W. (1985). Chemoprevention of cancer. *Cancer Research* **45**, 1–8.

Wattenberg, L. W. (1993). Inhibition of carcinogenesis by nonnutrient constituents of the diet. In *Food and Cancer Prevention: Chemical and Biological Aspects*, pp. 12–23 [K. W. Waldron, I. T. Johnson & G. R. Fenwick, editors]. Cambridge: Royal Society of Chemistry.

Wattenberg, L. W., Schafer, H. W., Waters, L. & Davis, D. W. (1989a). Inhibition of mammary tumor formation by broccoli and cabbage. *Proceedings of the American Association for Cancer Research* **30**, 181.

Wattenberg, L. W., Sparnins, V. L. & Barany, G. (1989b). Inhibition of N-nitrosodiethylamine carcinogenesis in mice by naturally occurring organosulfur compounds and monoterpenes. *Cancer Research* **49**, 2689–2692.

Weed, H. G., McGandy, R. B. & Kennedy, A. R. (1985). Protection against dimethylhydrazine-induced adenomatous tumors of the mouse colon by the dietary addition of an extract of soybeans containing the Bowman–Birk protease inhibitor. *Carcinogenesis* **6**, 1239–1241.

Wei, H., Tye, L., Bresnick, E. & Birt, D. F. (1990). Inhibitory effect of apigenin, a plant flavonoid, on epidermal ornithine decarboxylase and skin tumor promotion in mice. *Cancer Research* **50**, 499–502.

Wells, W. W., Xu, D. P., Yang, Y. & Rocque, P. A. (1990). Mammalian thioltransferase (glutaredoxin) and protein disulfide isomerase have dehydroascorbate reductase activity. *Journal of Biological Chemistry* **265**, 15361–15364.

Welsch, C. W., DeHoog, J. V. & O'Connor, D. H. (1988). Influence of caffeine consumption on carcinomatous and normal mammary gland development in mice. *Cancer Research* **48**, 2078–2082.

Williams, D. E., Carpenter, H. M., Buhler, D. R., Kelly, J. D. & Dutchuk, M. (1992). Alterations in lipid peroxidation, antioxidant enzymes, and carcinogen metabolism in liver microsomes of vitamin-E-deficient trout and rat. *Toxicology and Applied Pharmacology* **116**, 78–84.

Williams-Ashman, H. G. & Canellakis, Z. N. (1979). Polyamines in mammalian biology and medicine. *Perspectives in Biology and Medicine* **22**, 421–453.

Winterbourn, C. C., Vile, G. F. & Monteiro, H. P. (1991). Ferritin, lipid peroxidation and redox cycling xenobiotics. *Free Radical Research Communications* **12–13**, 107–114.

Witschi, H. & Kennedy, A. R. (1989). Modulation of lung tumor development in mice with the soybean-derived Bowman–Birk protease inhibitor. *Carcinogenesis* **10**, 2275–2277.

Wortelboer, H. M., Vanderlinden, E. C. M., Dekruif, C. A., Noordhoek, J., Blaauboer, B. J., Van Bladeren, P. J. & Falke, H. E. (1992). Effects of indole-3-carbinol on biotransformation enzymes in the rat. In vivo changes in liver and small intestinal mucosa in comparison with primary hepatocyte cultures. *Food and Chemical Toxicology* **30**, 589–599.

Wu, L. & Whitlock, J. P. (1993). Mechanism of dioxin action – receptor-enhancer interactions in intact cells. *Nucleic Acids Research* **21**, 119–125.

Xanthoudakis, S., Miao, G., Wang, F., Pan, Y. C. E. & Curran, T. (1992). Redox activation of Fos-Jun DNA binding activity is mediated by a DNA repair enzyme. *EMBO Journal* **11**, 3323–3335.

Xu, Y., Ho, C.-T., Amin, S. G., Han, C. & Chung, F.-L. (1992). Dietary inhibitors of chemical carcinogenesis. 17. Inhibition of tobacco-specific nitrosamine-induced lung tumorigenesis in A/J mice by green tea and its major polyphenol as antioxidants. *Cancer Research* **52**, 3875–3879.

Yamaguchi, T. (1980). Mutagenicity of isothiocyanates, isocyanates and thioureas on *Salmonella typhimurium*. *Agricultural and Biological Chemistry* **44**, 3017–3018.

Yamamoto, S. & Kato, R. (1992). Inhibitors of the arachidonic acid cascade and their chemoprevention of skin cancer. In *Cancer Chemoprevention*, pp. 141–151 [L. Wattenberg, M. Lipkin, C. W. Boone & G. J. Kelloff, editors]. Boca Raton, FL: CRC Press.

Yamasaki, H. (1984). Modulation of cell differentiation by tumor promoters. In *Mechanisms of Tumor Promotion. Cellular Responses to Tumor Promoters*, pp. 1–26 [T. J. Slaga, editor]. Boca Raton, FL: CRC Press.

Yang, C. S., Brady, J. F. & Hong, J. (1992). Dietary effects on cytochromes-P450, xenobiotic metabolism, and toxicity. *FASEB Journal* **6**, 737–744.

Yavelow, J., Collins, M., Birk, Y., Troll, W. & Kennedy, A. R. (1985). Nanomolar concentrations of Bowman–Birk soybean protease inhibitor suppress x-ray-induced transformation *in vitro*. *Proceedings of the National Academy of Sciences, USA* **82**, 5395–5399.

Yavelow, J., Finlay, T. H., Kennedy, A. R. & Troll, W. (1983). Bowman–Birk soybean protease inhibitor as an anticarcinogen. *Cancer Research* **43**, Suppl., 2454s–2459s.

Yoo, J. S. H., Hong, J. Y., Ning, S. M. & Yang, C. S. (1990). Roles of dietary corn oil in the regulation of cytochromes P450 and glutathione S-transferases in rat liver. *Journal of Nutrition* **120**, 1718–1726.

Zeigler, D. M. (1991). Mechanism, multiple forms and substrate specificities of flavin containing monooxygenases. In *N-oxidation of Drugs: Biochemistry, Pharmacology and Toxicology*, pp. 59–70 [P. Hlavica and L. A. Damani, editors]. London: Chapman and Hall.

Zhang, Y., Talalay, P., Cho, C.-G. & Posner, G. H. (1992). A major inducer of anticarcinogenic protective enzymes from broccoli: isolation and elucidation of structure. *Proceedings of the National Academy of Sciences, USA* **89**, 2399–2403.

Ziegler, R. G. (1991). Vegetables, fruits, and carotenoids and the risk of cancer. *American Journal of Clinical Nutrition* **53**, 251S–259S.

Printed in Great Britain

GLUCOSINOLATES AND GLUCOSINOLATE DERIVATIVES: IMPLICATIONS FOR PROTECTION AGAINST CHEMICAL CARCINOGENESIS

LIONELLE NUGON-BAUDON* AND SYLVIE RABOT

Unité d'Ecologie et de Physiologie du Système Digestif, Centre de Recherches de Jouy, Institut National de la Recherche Agronomique, 78352 Jouy-en-Josas Cédex, France

CONTENTS

GLUCOSINOLATES: OCCURRENCE AND METABOLIC FATE	205
GLUCOSINOLATES IN THE PLANT	205
GENESIS OF GLUCOSINOLATE DERIVATIVES	207
Enzymic hydrolysis and autolysis in cruciferous vegetables	207
Bacterial metabolism of glucosinolates	212
FROM THE PLANT TO THE DIET: INFLUENCE OF FOOD PROCESSING AND DIETARY HABITS	213
TOXICITY OF GLUCOSINOLATES AND GLUCOSINOLATE DERIVATIVES	215
GLUCOSINOLATES AND GLUCOSINOLATE DERIVATIVES: NEW CANDIDATES FOR PROTECTION AGAINST CHEMICAL CARCINOGENESIS	217
EPIDEMIOLOGICAL DATA: CRUCIFEROUS VEGETABLES AND CANCER INCIDENCE IN HUMAN POPULATIONS	217
EXPERIMENTAL DATA: CRUCIFEROUS VEGETABLES, GLUCOSINOLATES AND CHEMICAL CARCINOGENS IN ANIMAL MODELS	217
EXPERIMENTAL DATA: CRUCIFEROUS VEGETABLES, GLUCOSINOLATES AND XENOBIOTIC METABOLIZING ENZYMES	219
The xenobiotic metabolizing enzymes	219
The effects of cruciferous vegetables on the XME system	220
The effects of glucosinolates and glucosinolate derivatives on the XME system	222
CONCLUSIONS AND PENDING TOPICS	224
REFERENCES	225

GLUCOSINOLATES: OCCURRENCE AND METABOLIC FATE

GLUCOSINOLATES IN THE PLANT

Glucosinolates (GSL) are sulphur-containing molecules produced from amino acids by the secondary metabolism of plants. Their occurrence is limited to some families of dicotyledonous angiosperms. Considering edible plants only, they occur predominantly in *Cruciferae* and *Capparideae* and, sporadically, in *Caricaceae* and *Tropaeolaceae* (Table 1).

* Corresponding author.

Table 1. *Glucosinolate-containing edible plants*

Cruciferae	
Brassica oleracea L.	
gongylodes group	Kohlrabi
capitata group	Red/white cabbage
sabaüda group	Savoy cabbage
gemmifera group	Brussels sprouts
italica group	Broccoli
botrytis group	
var. *cauliflora* DC.	Cauliflower
var. *cymosa* Lam.	Calabrese (green sprouting broccoli)
acephala group	
var. *millecapitata* (Lev) Thell.	Thousand head kale
var. *medullosa* Thell.	Marrowstem kale
var. *selensia*	Curly kale
var. *sabellica*	Collard
Brassica alboglabra Bailey	Chinese kale
Brassica pekinensis (Lour.) Rupr.	Chinese cabbage (Pe-tsai)
Brassica chinensis L.	Chinese white cabbage (Pak-choi)
Brassica campestris L.	
ssp. *rapifera* (Metzg.) Sinsk.	Turnip
ssp. *oleifera* (Metzg.) Sinsk.	Turnip rape
Brassica napus L.	
var. *napobrassica* (L.) Peterm	Swede (Rutabaga)
or ssp. *rapifera* (Metzg.) Sinsk.	
var. *napus*	Winter, summer rape
Brassica nigra (L.L) Koch	Black mustard
Brassica juncea (L.) Czern et Coss	Brown mustard
Brassica carinata A. Br.	Abyssinian mustard (Ethiopian cabbage)
Sinapis alba L.	White mustard
Crambe maritima L.	Sea kale
Raphanus sativus L.	Radish
Armoracia lapathifolia Gilib	Horseradish
Wasabi japonica Matsum.	Wasabi (Japanese horseradish)
Eruca sativa (Miller) Thell.	Salad rocket
Lepidium sativum L.	Garden cress
Nasturtium officinalis R. Br.	Water cress
Capparaceae	
Capparis spinosa	Caper
Caricaceae	
Carica papaya L.	Papaya (Pawpaw)
Tropaeolaceae	
Tropaeolum majus L.	'Nasturtium' (Indian cress)

References are: Carlson *et al.* (1981), Fenwick *et al.* (1982), Carlson *et al.* (1987), Adams *et al.* (1989).

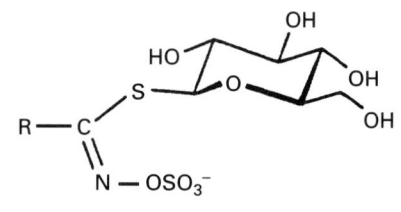

Fig. 1. The general structure of glucosinolates.

Species belonging to these families are widely consumed by humans as cooked or salad vegetables (cabbage, Brussels sprouts, cauliflower, turnip, radish, cress) or condiments (horseradish, mustard, caper); cruciferous forages (kale, rape, turnip) and oilseed meals (rape, turnip rape) are used as feedstuffs for animals (Fenwick *et al.* 1982).

More than 100 different GSL, which all share a common structure (Fig. 1), have been identified so far (Fenwick *et al.* 1982). GSL may be classified into several chemical families according to their side groups R (Fenwick *et al.* 1986; Quinsac, 1993), which include alkyl, alkenyl, hydroxyalkyl, hydroxyalkenyl, methylthioalkyl, methylsulphinylalkyl, methylsulphonylalkyl, arylalkyl and indolyl groups (Table 2). Furthermore, a new family of GSL, designated cinnamoylGSL, was recently identified (Linscheid *et al.* 1980; Bjerg & Sørensen, 1987). It differs from the usual pattern by the presence of cinnamic acid derivatives in the C(2) and/or C(6) positions on the glucose moiety.

Edible plants may contain up to fifteen different GSL. However, most of them synthesize between one and five of these compounds. Concern about the potential biological effects of GSL has in the last decade prompted various groups to examine the levels and profiles of these compounds in cruciferous vegetables. The reader interested in detailed information is referred to the extensive research performed at the Northern Regional Research Center of the US Department of Agriculture (Daxenbichler *et al.* 1979; Carlson *et al.* 1981, 1985, 1987) and at the Norwich Laboratory of the Institute of Food Research in Britain (Heaney & Fenwick, 1980*a, b*; Fenwick *et al.* 1982; Sones *et al.* 1984*a, b*; Lewis & Fenwick, 1987, 1988). Findings published by these and other workers are schematically summarized in Table 3. On the whole, great variations in the content as well as in the pattern of GSL occur according to the plant species. The wide range of GSL concentrations sometimes observed within an experiment and between different studies performed on the same vegetable indicates that further variations may occur according to the cultivar and the cultivation conditions. Carlson *et al.* (1985) have pointed out the remarkable differences in the GSL content between radishes originating from either the European–American or the Asian market. Analysis of Brussels sprouts and cauliflower cultivars grown at different sites in the UK shows great variations in the total GSL content (Heaney & Fenwick, 1980*a, b*; Sones *et al.* 1984*b*); however, the relative proportions of the individual GSL tend to remain fairly stable within a cultivar. Climate, soil type and agronomic practices, especially fertilizer applications and harvest date, are cited as causative factors for such variations (Josefsson, 1970; Heaney & Fenwick, 1980*a, b*; Fenwick *et al.* 1982; Lehrmann, 1989; Booth *et al.* 1990).

Another factor of tremendous importance is the part of the plant examined. Major quantitative and qualitative differences in the GSL accumulated by different organs (seeds, leaves, roots) and different tissues of the same organ (root peelings, cortex and medulla) occur in the same plant (Heaney & Fenwick, 1980*a, b*; Sang *et al.* 1984; Carlson *et al.* 1987; Adams *et al.* 1989). Such findings highlight the point that GSL biosynthesis in the plant is probably ruled by complex control mechanisms and that one cannot extrapolate data available for one part of the plant to another tissue.

GENESIS OF GLUCOSINOLATE DERIVATIVES

Enzymic hydrolysis and autolysis in cruciferous vegetables

The breakdown of GSL by myrosinase, a specific plant hydrolytic enzyme (thioglucoside glucohydrolase *EC* 3.2.3.1), has been extensively studied and reviewed (Duncan & Milne, 1989).

In intact cruciferous tissues, the enzyme is stored separately from the GSL substrates in specific cells named idioblasts. Contact between the two will result from mechanical injury

Table 2. *Glucosinolates occurring in edible plants*

Side chain	Glucosinolate	Trivial name
CH₃-	methyl-	glucocapparin
CH₃-CH₂-	ethyl-	glucolépidiin
CH₃-CH(CH₃)-	iso-propyl-	glucoputranjivin
CH₃-CH₂-CH(CH₃)-	1-methylpropyl-	glucocochlearin
CH₂=CH-CH₂-	prop-2-enyl-	sinigrin
CH₂=CH-CH₂-CH₂-	but-3-enyl-	gluconapin
CH₂=CH-CH₂-CH₂-CH₂-	pent-4-enyl-	glucobrassicanapin
CH₂=CH-CH(OH)-CH₂-	(R)-2-hydroxybut-3-enyl-	progoitrin
	(S)-2-hydroxybut-3-enyl-	epiprogoitrin
CH₂=CH-CH₂-CH(OH)-CH₂-	(R)-2-hydroxypent-4-enyl-	gluconapoleiferin
CH₃-S-CH₂-CH₂-CH₂-	3-methylthiopropyl-	glucoiberverin
CH₃-S-CH₂-CH₂-CH₂-CH₂-	4-methylthiobutyl-	glucoerucin
CH₃-S-CH=CH-CH₂-CH₂-	4-methylthiobut-3-enyl-	glucoraphasatin
CH₃-S-CH₂-CH₂-CH₂-CH₂-CH₂-	5-methylthiopentyl-	glucoberteroin
CH₃-SO-CH₂-CH₂-CH₂-	(R)-3-methylsulphinylpropyl-	glucoiberin
CH₃-SO-CH₂-CH₂-CH₂-CH₂-	(R)-4-methylsulphinylbutyl-	glucoraphanin
CH₃-SO-CH=CH-CH₂-CH₂-	(R)-4-methylsulphinylbut-3-enyl-	glucoraphenin
CH₃-SO-CH₂-CH₂-CH₂-CH₂-CH₂-	(R)-5-methylsulphinylpentyl-	glucoalyssin
CH₃-SO₂-CH₂-CH₂-CH₂-	3-methylsulphonylpropyl-	glucocheirolin
CH₃-SO₂-CH₂-CH₂-CH₂-CH₂-	4-methylsulphonylbutyl-	glucoerysolin
Ph-CH₂-	benzyl-	glucotropaeolin
Ph-CH₂-CH₂-	2-phenylethyl-	gluconasturtiin
Ph-CH(OH)-CH₂-	(R)-2-hydroxy-2-phenylethyl-	glucobarbarin
	(S)-2-hydroxy-2-phenylethyl-	glucosibarin
3-HO-C₆H₄-CH₂-	3-hydroxybenzyl-	glucolepigramin
4-HO-C₆H₄-CH₂-	4-hydroxybenzyl-	sinalbin
indolyl-CH₂- (R4)	indol-3-ylmethyl- (R1=R4=H)	glucobrassicin
	1-methoxyindol-3-ylmethyl- (R1=OCH₃ ; R4=H)	neoglucobrassicin
	1-sulphoindol-3-ylmethyl- (R1=SO₃⁻ ; R4=H)	sulphoglucobrassicin
	4-hydroxyindol-3-ylmethyl- (R1=H ; R4=OH)	4-hydroxyglucobrassicin
	4-methoxyindol-3-ylmethyl- (R1=H ; R4=OCH₃)	4-methoxyglucobrassicin

References are: Fenwick *et al.* (1982), Quinsac (1993).

Table 3. *Glucosinolate (GSL) content of the edible part of some cruciferous vegetables*

Vegetable	Reference	GSL content (mg/100 g fresh weight) Mean	Range	SIN	GNA	GBN	PRO	GIV	GER	GRH	GIB	GRA	GRE	GAL	GST	GBS	NGBS	4-OHGBS	4-OMGBS
Kohlrabi	Carlson et al. (1987)	39·4	—	—	—	—	—	—	—	—	—	—	—	—	—	✓	—	—	—
Cabbage	Sang et al. (1984)	—	—	—	—	—	—	✓	✓	—	—	—	—	—	—	✓	—	—	✓
	Sones et al. (1984a)	108·9	36·0–275·4	✓	—	—	✓	—	—	—	—	—	—	—	—	✓	—	—	—
Brussels sprouts	Slominski & Campbell (1989)	51·0	—	✓	—	—	✓	—	—	—	✓	—	—	—	—	✓	—	—	—
	Heaney & Fenwick (1980b)	199·0	90·0–390·0	✓	—	—	✓	—	—	—	—	—	—	—	—	✓	—	—	—
	Sones et al. (1984a)	226·2	145·5–393·9	✓	—	—	✓	—	—	—	—	—	—	—	—	✓	—	—	—
	McMillan et al. (1986)	247·0	—	—	—	—	✓	—	—	—	—	—	—	—	—	✓	—	—	—
	Carlson et al. (1987)	252·7	212·8–274·5	✓	—	—	✓	—	—	—	—	—	—	—	—	✓	—	—	—
	Slominski & Campbell (1989)	134·2	—	✓	—	—	✓	—	—	—	✓	—	—	—	—	✓	—	✓	—
Broccoli	Carlson et al. (1987)	86·0	46·7–121·0	—	—	—	—	—	—	—	—	—	—	—	—	✓	—	—	—
	Slominski & Campbell (1989)	27·6	—	—	—	—	—	—	—	✓	—	—	—	—	—	✓	—	✓	—
Cauliflower	Sones et al. (1984a)	62·0	13·8–208·3	✓	—	—	✓	—	—	✓	—	—	—	—	—	✓	—	—	—
	Sones et al. (1984b)	78·3	26·0–214·2	✓	—	—	✓	—	—	✓	—	—	—	—	—	✓	—	—	—
	Carlson et al. (1987)	43·2	18·8–73·4	✓	—	—	✓	—	—	✓	—	—	—	—	—	✓	—	—	—
	Slominsky & Campbell (1989)	12·2	—	—	—	—	—	—	—	✓	—	—	—	—	—	✓	—	✓	—
Calabrese	Lewis & Fenwick (1987)	62·3	42·1–94·5	—	—	—	✓	—	—	✓	—	—	—	—	—	✓	—	✓	—
Kale	Carlson et al. (1987)	144·0	29·4–317·1	✓	—	—	✓	—	—	—	—	—	—	—	—	✓	—	—	—
Collard	Carlson et al. (1987)	200·7	144·5–274·2	✓	—	—	✓	—	—	✓	—	—	—	—	—	✓	—	—	—
Chinese cabbage (pe-tsai)	Daxenbichler et al. (1979)	54·1	174·4–135·7	—	—	✓	✓	—	—	—	—	—	—	—	—	✓	—	✓	✓
Chinese cabbage (pak-choi)	Lewis & Fenwick (1988)	19·8	9·7–33·7	—	—	✓	✓	—	—	—	—	—	—	—	—	✓	—	—	—
	Lewis & Fenwick (1988)	53·4	39·0–70·4	—	✓	✓	—	—	—	—	—	—	—	✓	—	—	—	—	—
Turnip	Carlson et al. (1981)	125·1(b)	98·7–230·8(b)	—	—	—	✓	—	—	✓	—	—	—	—	—	—	—	✓	✓
	Sones et al. (1984a)	56·0(c)	39·2–165·7(c)	—	—	—	✓	—	—	✓	—	—	—	—	✓	—	—	✓	✓
Swede	Carlson et al. (1981)	162·4(b)	114·7–234·9(b)	—	—	—	✓	—	—	✓	—	—	—	—	✓	—	—	✓	✓
	Sang et al. (1984)	—	—	—	—	—	✓	—	—	✓	—	—	✓	—	—	—	—	—	—
	Adams et al. (1989)	—	—	—	—	—	✓	—	—	✓	—	—	✓	—	—	—	—	—	—
Mustard	Sang et al. (1984)	70·4(d)	44·0–116·4(d)	✓	—	—	—	—	—	—	—	—	✓	—	—	—	—	✓	✓
Radish	Sang et al. (1984)	108·8(e)	69·9–164·1(e)	—	—	—	—	—	—	✓	—	—	✓	—	—	—	—	✓	✓
	Carlson et al. (1985)	138·0(f)	65·8–251·8(f)	—	—	—	—	—	—	✓	—	—	✓	—	—	—	—	✓	✓

(a) SIN = sinigrin, GNA = gluconapin, GBN = glucobrassicanapin, PRO = progoitrin, GIV = glucoiberverin, GER = glucoerucin, GRH = glucoiberin, GRA = glucoraphanin, GRE = glucoraphenin, GAL = glucoalyssin, GST = gluconasturtin, GBS = glucobrassicin, NGBS = neoglucobrassicin, 4-OHGBS = 4-hydroxyglucobrassicin, 4-OMGBS = 4-methoxyglucobrassicin. (b) peeled roots. (c) turnip/swede. (d) European-American market. (e) Korean market. (f) Japanese market.

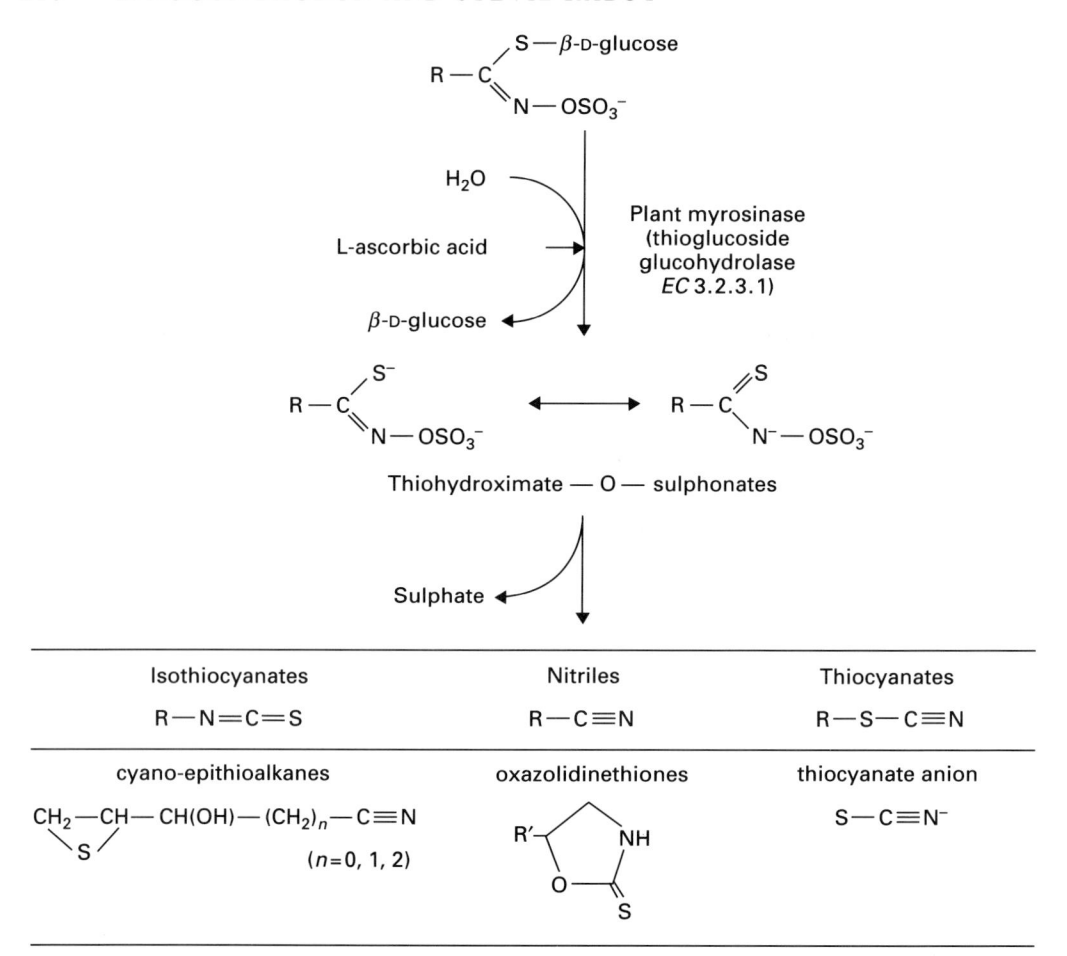

Fig. 2. The breakdown of glucosinolates by plant myrosinase.

of the plant tissue by, for example, cutting or chewing. Various isoenzymic forms of myrosinase have been isolated from different species and tissues (Fenwick et al. 1982; Buchwaldt et al. 1986). All of them hydrolyse the thioglucoside bond to release glucose and an unstable thiohydroximate-O-sulphonate, which is spontaneously further transformed (Lossen rearrangement; Ettlinger & Lundeen, 1956) to yield sulphate and a wide range of aglucones including isothiocyanates, nitriles, epithioalkanes, oxazolidinethiones, thiocyanate anions and, occasionally, organic thiocyanates (Fig. 2).

The enzymic step of the breakdown is usually enhanced by ascorbic acid, which acts as a specific coenzyme (Ettlinger et al. 1961; Ohtsuru & Hata, 1979). The structure of the aglucone eventually obtained is highly dependent on the structure of the side group R and on environmental factors such as pH, metallic ions (Fe^{2+}, Fe^{3+}, Cu^+ or Cu^{2+}) (Tookey & Wolff, 1970; Searle et al. 1984; Uda et al. 1986), and to a lesser extent temperature and moisture content (Tookey, 1973). For instance, low pH, low temperature or metallic ions will favour nitrile production, whereas neutral pH or high temperature will push the reaction toward isothiocyanate release (VanEtten et al. 1966; Gil & McLeod, 1980; Uda et al. 1986); the latter compound will tend to rearrange into oxazolidinethiones in an alkaline medium provided a hydroxyl group is present in the C(2) or C(3) position on the

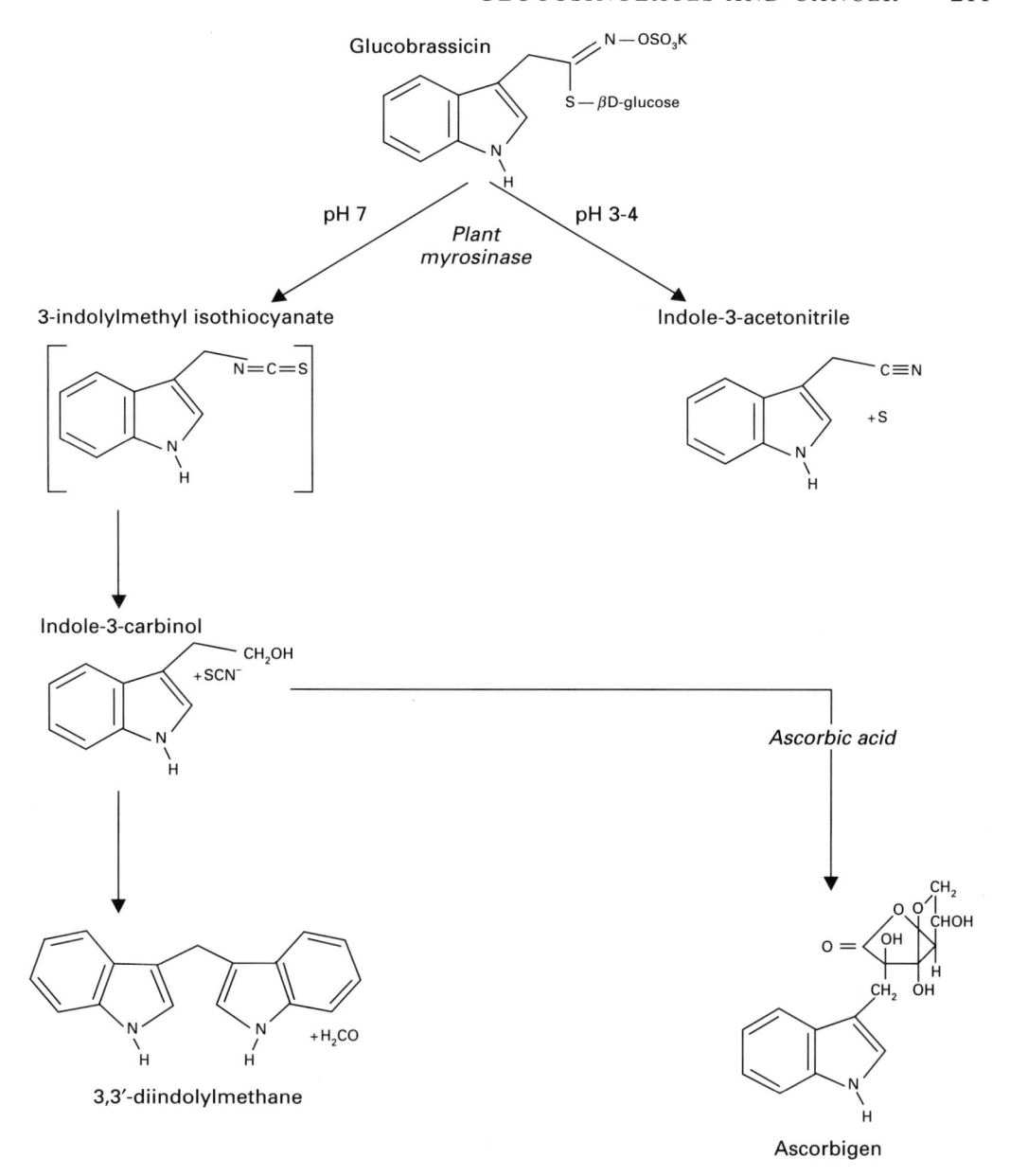

Fig. 3. The products of plant myrosinase hydrolysis of glucobrassicin.

side group R. Under the specific influence of ferrous ions autolysis can occur (Youngs & Perlin, 1967; Austin et al. 1968).

IndolylGSL may follow different patterns of enzymic breakdown. Fig. 3 summarizes the hydrolysis of glucobrassicin. Depending on the environmental conditions, glucobrassicin hydrolysis leads to the formation of indole-3-acetonitrile (IAN) (acidic pH, metallic ions) or of a putative unstable isothiocyanate derivative (neutral pH) which splits immediately to yield a thiocyanate anion and indole-3-carbinol (I3C). In the absence of ascorbic acid, two molecules of I3C condense to yield 3,3'-diindolylmethane. Should ascorbic acid be present in the medium, it will react with I3C to product ascorbigen (Searle *et al.* 1982). The glucobrassicin derivatives resulting from non-enzymic breakdown are different and vary with the pH: under acidic conditions, the first derivative would be IAN, thereafter transformed into indole-3-acetamide, indole-3-acetic acid and eventually skatole (minor product); a second pathway, which is less likely to occur, yields molecules such as indole-3-carboxaldehyde, indole-3-carboxylic acid and indole (McDanell *et al.* 1988). On the whole, the variety of compounds arising from the indolylGSL breakdown is greater than that of the aglucones derived from other GSL molecules.

Bacterial metabolism of glucosinolates

The observation that GSL derivatives could occur *in vivo* without prior ingestion of myrosinase prompted Greer and coworkers to look for myrosinase-like activity in body tissues and fluids. The exciting story of the investigations that led them to postulate that the human intestinal microflora was able to hydrolyse GSL *in vivo* is recorded in a review by Greer (1962).

An *in vitro* myrosinase-like activity in rat (Greer, 1962) and fowl (Marangos & Hill, 1974) faecal microflora was then demonstrated. Subsequently, different groups succeeded in isolating from human (Oginsky *et al.* 1965; Tani *et al.* 1974) and fowl (Miguchi *et al.* 1974) faecal microflora bacterial strains that were able to metabolize progoitrin or sinigrin *in vitro*.

Recent experiments performed with gnotobiotic animals in our laboratory at the Jouy-en-Josas Research Centre of the National Institute for Agricultural Research (INRA) in France definitely demonstrated that the myrosinase-like activity of the intestinal microflora was physiologically relevant since biological effects of cruciferous vegetables never occurred in germ free rodents and chickens given a GSL-rich but myrosinase free feed (Nugon-Baudon *et al.* 1988).

So far bacterial myrosinase-like activity has been considerably underinvestigated. Evidence for 5-vinyloxazolidine-2-thione (a progoitrin derivative) was assessed by Greer & Deeney (1959) in the urine of human volunteers after the ingestion of pure progoitrin, and by Oginsky *et al.* (1965) in the culture media of human Enterobacteriaceae strains incubated with the progoitrin. However, the breakdown of GSL by microflora is likely to be more complex than hydrolysis performed by the plant myrosinases. Experiments with gnotobiotic animals support the hypothesis that bacteria yield specific toxic GSL derivatives (Rabot *et al.* 1993a). Further conversion of several GSL derivatives into unknown compounds has been demonstrated in sheep rumen fluid *in vitro* (Lanzani *et al.* 1974; Duncan & Milne, 1992). There is little other information: the salient points in the studies of Oginsky *et al.* (1965) and Tani *et al.* (1974) are the influence of pH and the lack of influence, or even inhibitory effect, of ascorbic acid on bacterial myrosinase-like activities *in vitro*. In vivo, manipulation of the mineral (Vermorel & Evrard, 1987) or carbohydrate (Rabot *et al.* 1991) fraction of the diet helped to reduce the biological effects of cruciferous vegetables, implying that the bacterial myrosinase-like activities were altered in some way.

Improved and extended information on the myrosinase-like activities of the intestinal microflora would be of tremendous importance since plant myrosinase can be inactivated during processing of cruciferous vegetables; the implication is that a significant proportion of GSL must be actually metabolized by the intestinal microflora.

FROM THE PLANT TO THE DIET: INFLUENCE OF FOOD PROCESSING AND DIETARY HABITS

When one knows the basic GSL content of cruciferous edible plants, it does not mean that one has reached the end of the story. Before being consumed, cruciferous vegetables usually undergo processing operations that may influence the GSL content. De Vos & Blijleven (1988) have extensively reviewed this subject and we report here only the main points relevant to the discussion in subsequent sections of the present review.

Basic processes such as dicing, slicing, or shredding raw vegetables initiate the breakdown of GSL by myrosinase, since rupture of tissues puts the enzyme into contact with its substrates. However, some intact GSL may remain, depending on the degree of crushing (de Vos & Blijleven, 1988). Pulping might of course be expected to result in a high degree of GSL breakdown. Indeed no intact GSL can be recovered from homogenized cabbage (de Vos & Blijleven, 1988) and Brussels sprouts (Bradfield & Bjeldanes, 1987) after 30 min and 24 h respectively. A preponderance of nitrile derivatives and, from indolylGSL, of ascorbigen and I3C have been identified, although the latter compound is not particularly stable and tends to undergo conversion into other products, mainly ascorbigen. Thus McDanell *et al.* (1987) have shown that the concentration of ascorbigen in Savoy cabbage homogenized to a thick slurry prior to deep-freezing and freeze-drying was 1·5 g/kg dry matter; this level represents 75% of the theoretical total of breakdown products, based on the glucobrassicin content.

Cooking, steaming and blanching usually reduce GSL concentrations by 30–60%, depending on the vegetable and on the type of GSL (Sones *et al.* 1984a); the loss is due partly to enzymic hydrolysis and partly to leaching of the intact GSL and their derivatives into the cooking liquid (Srisangnam *et al.* 1980b; Slominski & Campbell, 1989). The pattern of intact GSL and breakdown products recovered after cooking is influenced by the thermal stability of the molecules: sinigrin, for instance, is more thermostable than progoitrin or glucoiberverin, allyl isothiocyanate (from sinigrin) totally disappears upon boiling, while 5-vinyloxazolidine-2-thione (from progoitrin) and 3-methyl-sulphinylpropylisothiocyanate (from glucoiberin) may partly escape decomposition (de Vos & Blijleven, 1988). Once again, among glucobrassicin derivatives, ascorbigen appears to be the major compound recovered after cooking (McDanell *et al.* 1987).

Fermented cruciferous products (sauerkraut, salt fermented vegetables) contain no intact GSL since this kind of process favours their quick and complete enzymic hydrolysis. In a study by Daxenbichler *et al.* (1980), reported by de Vos & Blijleven (1988), the main GSL derivatives identified in sauerkraut after a 2 week fermentation were the thiocyanate anion and 1-cyano-3-methylsulphinylpropane (a nitrile from glucoiberin). However McDanell *et al.* (1987) found that fermentation (18 h, 25 °C) of white or Savoy cabbage was less detrimental to intact GSL than cooking; as far as GSL derivatives were concerned, the content of IAN was seldom modified by fermentation whereas there was an important decrease of ascorbigen compared with the fresh material.

Storage processes such as freezing, dehydrating or irradiating have received much less attention. From the few and often contradictory studies reported, one can conclude only that whereas dehydrating preserves intact GSL (de Vos & Blijleven, 1988), irradiation with u.v. or ionizing radiation tends to favour their breakdown (Michajlovskij, 1968 cited in McDanell *et al.* 1988; Nugon-Baudon *et al.* 1988; de Vos & Blijleven, 1988).

Table 4 reports the average consumption of cruciferous vegetables in several countries for which nutritional survey data are available. While cruciferous vegetables are consumed worldwide, this table highlights the fact that quantitative and qualitative differences occur between the geographical regions and/or the dietary habits characteristic of each country.

Table 4. *Average weekly intake of some cruciferous vegetables in the UK, USA, Canada and Japan (g/person)*

Vegetables	UK (1980)	USA (1978)	Canada (1978)	Japan (1975)
Cabbage	123·2	77·0	34·2	136·5
Brussels sprouts	67·9	2·1	5·2	—
Broccoli	—	23·1	14·1	—
Cauliflower	71·4	14·0	9·4	—
Chinese cabbage	—	—	0·4	159·6
Turnip/Swede	38·5	—	24·6	—
Mustard	—	—	8·3	—
Radish	—	—	7·2	232·4
Coleslaw	—	—	14·3	—
Sauerkraut	—	11·2	6·7	—
Salt fermented vegetables	—	—	—	260·4

References are: Benns *et al.* (1978), Fenwick *et al.* (1982), Sones *et al.* (1984a).

Table 5. *Average weekly intake of glucosinolates from fresh vegetables in the UK and Canada*

	UK (1980) Glucosinolate content (mg/100 g fresh weight)	UK (1980) Glucosinolate intake (mg/person)	Canada (1978) Glucosinolate content (mg/100 g fresh weight)	Canada (1978) Glucosinolate intake (mg/person)
Cabbage	108·9	135·8	23·7	11·5 (Including coleslaw)
Brussels sprouts	226·2	120·4	122·4	6·4
Broccoli	—	—	29·3	4·2
Cauliflower	62·0	44·8	32·09	3·0
Turnip/Swede	56·0	21·7	122·6	30·2
Radish	—	—	11·8	0·9
		322·7		56·2

References are: Mullin & Sahasrabudhe (1978), Sones *et al.* (1984a).

Living standards may also account for variations in cruciferous vegetable consumption; as the income increases, there is an increase in total fresh green vegetable consumption (Sones *et al.* 1984a) and, among them, mild flavoured vegetables such as cauliflower or calabrese are preferred to cabbage or kale (Crisp, 1976 cited in Lewis & Fenwick, 1987). Sones *et al.* (1984a) and Mullin & Sahasrabudhe (1978) have estimated, from the British and Canadian consumption data reported in Table 4, the average intake of GSL in British and Canadian populations respectively. Assuming that the vegetables were eaten raw, the mean daily intakes were calculated to be 8·0 and 46·1 mg respectively in Canada and the UK (Table 5). Figures for individual GSL or GSL derivatives have occasionally been reported by some authors; the intake of progoitrin in the average UK diet is approximately 7 mg/day (Fenwick *et al.* 1983) and an average level of 28 μmol of glucoiberin is reported to be ingested daily by US citizens (Kore *et al.* 1993). Although this kind of information is very useful to draw a picture of levels of GSL ingested by humans, it must be treated with extreme caution since the final GSL and GSL derivative content of a dietary cruciferous vegetable depends on a tremendous number of factors. The researchers were of course aware of this uncertainty; indeed Sones *et al.* (1984*a*) have estimated that the amount of GSL ingested by certain individuals could exceed 300 mg/day.

On the whole, the findings reported here on GSL content of cruciferous vegetables and subsequent GSL consumption demonstrate that investigations about the biological effects of GSL should include measurement of, at the very least, the total GSL content and, ideally, of the content of individual GSL and hydrolysis products of the cruciferous vegetable included in the experimental diets. In addition, detailed information on how the cruciferous material and food are processed should also be provided; this should help nutritionists and toxicologists to obtain more valuable information from studies in which the experimental diets are inevitably different. Furthermore, if estimated figures for GSL consumption are helpful tools for the design of experimental diets, one should keep in mind that tremendous quantitative and qualitative variations occur in GSL consumption by humans.

TOXICITY OF GLUCOSINOLATES AND GLUCOSINOLATE DERIVATIVES

GSL derivatives are now known to be the toxic principles of cruciferous vegetables, and their toxic effects are well documented, especially in the case of rapeseed meal. Summarizing the vast literature published on this issue would be far too long; we have therefore stressed only the most striking points.

Experimentally, the general phenomenon observed is impaired performance of animals consuming GSL-rich feeds. The gross toxic effects can be described as reduced feed intake, growth depression, enlargement of target organs (liver, kidneys, thyroid gland) and reproductive disorders such as embryo mortality in mammals and decreased egg production in birds. The intensity of these effects varies with the animal species and, of course, the amount of GSL in their food (Bourdon *et al.* 1981; Butler *et al.* 1982; Bell, 1984; Vermorel *et al.* 1987; Etienne & Dourmad, 1987). In humans, reduced iodine uptake by the thyroid gland was reported after daily ingestion of 500 g cabbage for 2 weeks (Langer *et al.* 1971) or after a single meal of 300 to 500 g swede or turnip (Greer & Astwood, 1948). However, a more recent study by McMillan *et al.* (1986) did not lead to hypothyroidism in human volunteers consuming 150 g Brussels sprouts daily for 4 weeks. Nevertheless these contradictory findings are not too puzzling, since the thyroid function indices that were examined and the cruciferous vegetables and their GSL that were ingested were not the same.

Several attempts have been made to ascertain precisely which GSL or GSL derivatives are responsible for the different components of GSL toxicity. Addition of pure sinigrin or gluconapin to the diet led to liver hypertrophy in rats. Progoitrin seems to have a greater toxic potential; it has been shown to induce enlargement of the liver, kidneys and thyroid in rats (Bille *et al.* 1983; Vermorel *et al.* 1986). Goitrin (5-vinyloxazolidine-2-thione), one of the major derivatives of progoitrin, has been the most extensively studied GSL derivative, as far as toxicity is concerned. This goitrogen, very potent even at low doses (Krusius & Peltola, 1966; Langer & Michajlovskij, 1969; Akiba & Matsumoto, 1976), can induce decreased uptake of iodine by the thyroid gland in humans (Astwood *et al.* 1949) and in rats, modify the triiodothyronine:thyroxine ratio and alter the histological pattern of the thyroid in rats (Lo & Hill, 1971; Bell *et al.* 1972; Lo & Bell, 1972). It seems that goitrin interferes with organic iodination of thyroxine precursors in the gland, thus leading to compensatory goitre (Akiba & Matsumoto, 1976; Elfving, 1980). Isothiocyanates and thiocyanates were held responsible for similar thyroid disorders (Langer, 1964 cited in

Duncan & Milne, 1989; Langer & Štolc, 1965). The former prevent the iodination of tyrosine, as does goitrin, whereas the latter are known competitively to inhibit iodine uptake by thyroid cells (Langer & Greer, 1968; Muztar *et al.* 1979). Sinigrin and glucoiberin isothiocyanate derivatives were also shown to induce embryo death in the rat but the mechanism is still unknown (Nishie & Daxenbichler, 1980). Preferential target organs of the nitrile derivatives seem to be the liver and kidneys (VanEtten *et al.* 1969; Srivastava *et al.* 1975). The mechanism that underlies their toxicity seems to be their ability to interact with reduced glutathione, thus leading to substantial alterations in tissue glutathione levels as observed by Szabo *et al.* (1977) in the liver, kidneys, adrenals and lungs of rats after chronic ingestion or a single injection of acrylonitrile. The toxic effect of nitriles manifests itself as hypertrophy of the target organs, disruption of the normal lobular structure of the liver and irregular proliferation of the bile duct (VanEtten *et al.* 1969). As far as kidneys are concerned, enlarged nuclei of the epithelial cells lining the convoluted tubules have been observed (VanEtten *et al.* 1969). Gould *et al.* (1985) observed rapid production of kidney lesions, along with elevated plasma levels of nitrogen, urea and creatinine, which could suggest functional alterations of the kidneys.

Investigating the nature and the underlying mechanisms of toxic effects induced by GSL derivatives released by plant myrosinase gives very valuable information. However, it does not take into account the ability of the intestinal microflora to break down intact GSL or their derivatives into metabolites of which the nature and specific toxic potential are so far largely unknown. Experiments with gnotobiotic animals have proved to be an invaluable tool for addressing this topic. Those performed in our laboratory suggest that the different toxic patterns usually observed in different animal species are more likely to be due to differences in the autochthonous digestive microflora than to intrinsic host sensitivity toward GSL. Indeed, when given a diet based on rapeseed meal, conventional rats exhibit GSL-linked symptoms different from those of gnotobiotic rats harbouring either chicken or human microflora (Nugon-Baudon *et al.* 1988; Rabot *et al.* 1993*a*). The inoculation of germ free rats with single strains of fowl or human origin provided further information about the role of intestinal microflora in the production of toxic GSL derivatives. The toxic effects, observed in rats associated with a whole human microflora, namely reduced feed intake and weight gain, enlargement of the liver and thyroid and a decrease in both thyroxine (T4) and triiodothyronine (T3) plasma levels, could be reproduced in gnotobiotic rats harbouring a single human strain of *Bacteroides vulgatus* (Rabot *et al.* 1993*a*). Eventually, such simplified gnotobiotic models enabled our group to split the toxicity observed with complex intestinal microflora into different patterns (Table 6). A *Lactobacillus* strain isolated from a chicken crop was shown to induce goitre in gnotobiotic rats given a diet based on rapeseed meal (Nugon-Baudon *et al.* 1990*b*) whereas human strains of *Clostridium butyricum* and *Escherichia coli*, each isolated from healthy individuals, were responsible for liver hypertrophy and goitre associated with reduced T4 and T3 plasma levels respectively (Rabot *et al.* 1991, 1993*a*).

These findings reinforce the idea that, should the plant myrosinase in the diet be totally inactivated, the toxicity of GSL would depend strictly on the equilibrium between bacterial species possessing specific myrosinase-like activities. There exists an overall similarity in the nature of toxic effects observed in conventional rats given either pure GSL derivatives produced by plant myrosinase or a GSL-rich but myrosinase free diet. Nevertheless, one cannot exclude the possibility that extra metabolites, toxic or non-toxic, may be produced either from intact GSL or from previously released derivatives. This would of course enhance the difficulty that one encounters when trying to infer the potential toxicity of a diet containing cruciferous vegetables from a knowledge of its GSL and GSL derivative content.

Table 6. *Effects of a diet with rapeseed meal on weight gain, organ weight and thyroid hormones in gnotobiotic rats according to their bacterial status*

(Results are expressed as % of the mean values obtained with counterpart rats given a diet with soyabean meal)

Bacterial strain...	*Lactobacillus* (a) (LEM 220 strain)	*Bacteroides vulgatus* (b) (BV8H1 strain)	*Clostridium butyricum* (b) (CB1002 strain)	*Escherichia coli* (b) (EM0 strain)
Reference...	Nugon-Baudon *et al.* (1990*b*)	Rabot *et al.* (1993*a*)	Rabot *et al.* (1990)	Rabot *et al.* (1993*a*)
Duration of the trial (weeks)...	5	7	7	7
No. of animals...	7	6	6	6
Cumulative weight gain	109	28***	105	90
Liver	106	114***	115**	101
Kidneys	99	113	108**	95
Thyroid	148***	672***	111	302***
Tetraiodothyronine	112	57**	95	56***
Triiodothyronine	ND	71	ND	71**

Mean values were significantly different from those for counterpart animals given a soyabean meal diet: ** $P < 0.01$, *** $P < 0.001$.
(a) Isolated from a chicken crop.
(b) Isolated from the faecal flora of adult healthy humans.
ND: not determined.

GLUCOSINOLATES AND GLUCOSINOLATE DERIVATIVES: NEW CANDIDATES FOR PROTECTION AGAINST CHEMICAL CARCINOGENESIS

EPIDEMIOLOGICAL DATA: CRUCIFEROUS VEGETABLES AND CANCER INCIDENCE IN HUMAN POPULATIONS

Toxic effects of GSL and their derivatives in humans have seldom been described; in animals they are now less dramatic since new varieties of rape containing very low amounts of GSL have been bred. Nevertheless an ever increasing number of publications suggest a new potential of GSL-containing vegetables, namely that they may be serious candidates for protection against chemically induced cancer.

Different epidemiological studies (Graham *et al.* 1972, 1978; Haenzsel *et al.* 1980) seem to support the hypothesis that the consumption of cruciferous vegetables is associated with a lower risk of tumour formation in the human digestive tract (stomach, colon, rectum). Such observations led the (American) Committee on Diet, Nutrition and Cancer (1982) to suggest that the consumption of cruciferous vegetables "was associated with a reduction in the incidence of cancer at several sites in humans".

EXPERIMENTAL DATA: CRUCIFEROUS VEGETABLES, GLUCOSINOLATES AND CHEMICAL CARCINOGENS IN ANIMAL MODELS

The remarkable work carried out by Stoewsand and his team was a determining step in the experimental demonstration of the potentially beneficial effects of GSL consumption on chemically induced cancers. An initial experiment by these authors showed that giving rats a diet with 20% freeze-dried cauliflower reduced the toxic effects of aflatoxin B1 given orally (Stoewsand et al. 1978). This was subsequently supported by other studies by Boyd et al. (1982) using a diet with 25% freeze-dried cabbage, and by Salbe & Bjeldanes (1989) using a diet with 25% chopped and freeze-dried Brussels sprouts. The latter authors showed that aflatoxin B1 binding to hepatic DNA was much decreased when rats had been given Brussels sprouts for 2 weeks prior to the intraperitoneal or intragastric administration of the toxin. Female rats were also significantly protected against the carcinogenic properties of 7,12-dimethylbenz(a)anthracene administered by oral intubation when they received a feed containing 20% freeze-dried Brussels sprouts during the initiation period of carcinogenesis; in this 2-week experiment, the incidence of mammary tumours induced by 7,12-dimethylbenz(a)anthracene dropped from 77% in the control animals to 13% in the animals consuming the cruciferous vegetable (Stoewsand et al. 1988). Other studies have highlighted the protective effect of GSL-rich diets against chemically induced tumours (Wattenberg & Loub, 1978; Wattenberg et al. 1986).

However, these results showing the anticarcinogenic properties of cruciferous vegetables are counterbalanced by another series of experiments. Diets with 10% dried cabbage, for instance, have been shown to increase the incidence of pancreatic ductular carcinomas induced by N-nitroso-bis(2-oxopropyl)amine (Birt et al. 1987) in mice. A study carried out by Srisangnam et al. (1980a) is even more equivocal; the authors concluded that diets containing 10–20% sliced dehydrated cabbage enhanced the tumorigenicity of 1,2-dimethylhydrazine in mice whereas 40% cabbage in the feed has a protective effect. Although these results are very interesting, it is important to emphasize that in these cases 1,2-dimethylhydrazine and N-nitroso-bis(2-oxopropyl)amine were injected subcutaneously and that the greatest incidence of tumours was obtained with a high fat diet (22%; Birt et al. 1987).

The discrepancies observed between the findings can probably be explained partly by the tremendous variations in experimental design with respect to variables such as animal species, strain, sex, age, etc., the nature of the cruciferous vegetable and/or of the carcinogenic agent, the route of administration and/or the duration of the experiment. On the whole, these findings, albeit inconsistent, give definite evidence of the influence of cruciferous vegetables on chemical carcinogenesis.

In elucidating the anticarcinogenic properties of cruciferous vegetables, much of the work has focused on the effects of purified indolylGSL and derivatives. The main compound tested in these studies has been glucobrassicin and, more precisely, the derivatives obtained via its hydrolysis by plant myrosinase. When orally intubated into female rats before the administration of 7,12-dimethylbenz(a)anthracene, I3C (0·10 mmol/rat) and 3,3'-diindolylmethane (0·05 mmol/rat), but not IAN (0·10 mmol/rat), significantly reduced the incidence of mammary tumours (Wattenberg & Loub, 1978). Furthermore, mice given orally a 12 mg dose of the parent compound, glucobrassicin, a few days or even a few hours (4 h) before oral administration of benzo(a)pyrene (BaP), developed fewer forestomach and lung tumours (Wattenberg et al. 1986). In rats, I3C (1 g/kg diet) was shown to inhibit the hepatocarcinogenesis induced by diethylnitrosamine (40 mg/l drinking water) when it was administered concurrently with the carcinogen (Tanaka et al. 1990). Shertzer (1983, 1984) studied the change in binding to DNA of BaP or N-nitrosodimethylamine (NDMA) metabolites after mice were given I3C by gavage (163 mg/kg body weight); in both cases, there was evidence of a dramatic decrease in covalent binding. In contrast I3C proved unable to decrease the binding of aflatoxin B1 to hepatic DNA, whether it was administered via the intraperitoneal route or by gavage (Salbe & Bjeldanes, 1989). Pence et al. (1986) even demonstrated that I3C incorporated into the diet at a level of 1 g/kg dry matter enhanced 1,2-dimethylhydrazine induced tumorigenicity in rats; in this experiment, 1,2-dimethylhydrazine (10 mg/kg body weight) was injected intraperitoneally weekly for 16 weeks and the enhancing effect of I3C was significantly increased when the animals were given a high fat (20%) diet.

Apart from sinigrin which was shown to exhibit a protective effect similar to that of I3C against diethylnitrosamine induced hepatocarcinogenesis (Tanaka et al. 1990), other GSL have received little or even no attention, so that a great deal of uncertainty remains about the extent to which GSL can impede the carcinogenic process. Nevertheless, as was observed with diets based on cruciferous vegetables, the effects of glucobrassicin or of some of its derivatives are not always protective. One of the most likely explanations for these inconsistent results, i.e. enhancement versus reduction of the incidence of cancers in experimental animal models, is that GSL and/or GSL derivatives may modify the endogenous system of xenobiotic metabolizing enzymes (XME).

EXPERIMENTAL DATA: CRUCIFEROUS VEGETABLES, GLUCOSINOLATES AND XENOBIOTIC METABOLIZING ENZYMES

The xenobiotic metabolizing enzymes

We do not explain in detail how the XME system, which is very complex, works, since excellent reviews have been published (Burke & Orrenius, 1979; Kato, 1979; Caldwell, 1980). We give only a few examples which indicate the ways in which GSL can interfere with this biotransformation–detoxification system and help our understanding, at least in part, of their deleterious or protective effects.

The XME system is ubiquitous (skin, intestine, lungs, kidneys) with some exceptions, but is present mainly in the liver (Beaune, 1982, 1986). The reactions catalysed by the XME confer hydrophilic properties on endogenous compounds or molecules entering the organism that would otherwise be hard to eliminate due to their rather hydrophobic nature.

This system is usually described as having two phases, although a compound may be metabolized by either one or both phases (Jakoby, 1980). Phase one is represented by different enzymes such as flavin-containing monooxygenase, alcohol and aldehyde dehydrogenases, etc. The most widely studied enzymes belonging to phase I are undoubtedly the cytochrome P450 family (*EC* 1.14.14.1), probably because they metabolize a tremendous number of substances (Jakoby, 1980). We do not go into details of the biochemistry of the reactions catalysed by P450; schematically, these microsomal monooxygenases incorporate one atom of molecular oxygen into an organic substrate while using reducing equivalents (NADPH/H$^+$) to reduce the remaining oxygen atom to water. Since the discovery of cytochrome P450 by G. R. Williams in B. Chance's laboratory in 1955 (Conney, 1982), it has become obvious that it plays a key role in the metabolism of many xenobiotic or endogenous substances. So far approximately twenty P450 isoenzymes in the liver of the rat have been described (Nebert et al. 1989). As far as human P450 are concerned, results are of course less straightforward, due to the wide differences that may exist between individuals (genetic background, xenobiotic exposure, dietary habits, etc.; Wrighton et al. 1986; Guengerich, 1989; Sesardic et al. 1990). Individual forms of P450 may exhibit different degrees of specificity toward multiple substrates, i.e. high K_m activities toward some substrates and low K_m activities toward others.

The reactive products released by phase I can be further metabolized by phase II enzymes. Phase II catalyses the conjugation of phase I intermediates with endogenous ligands such as amino acids, glucuronic acid, sulphate or glutathione. As for P450, phase II is represented by large families of isoenzymes with overlapping substrate specificities (Habig et al. 1974; Jakoby, 1978; Wishart, 1978; Bock et al. 1979). UDPglucuronosyl-transferases (GT, *EC* 2.4.1.17) are microsomal enzymes that catalyse conjugation with

UDPglucuronic acid (Bock et al. 1987; Burchell et al. 1987). It seems that glucuronidation is the most important form of conjugation. Three of the isoenzymes identified so far in the rat are involved in the glucuronidation of endogenous substrates such as bilirubin and steroid hormones. Hepatic glucuronides are usually excreted *via* the bile. Most of the compounds that can be glucuronidated can also be sulphated by sulphotransferases (*EC* 2.8.2.1 etc.). These enzymes are located in the cytosol and catalyse the formation of sulphate monoesters with 3′-phosphoadenosine-5′-phosphosulphate. The result of the competition for a substrate between sulphotransferases and GT is usually in favour of the former, at least when the substrate concentration is low. With the exception of one microsomal form, glutathione *S*-transferases (GST, *EC* 2.5.1.18) are cytosolic proteins which conjugate glutathione on the sulphur atom of cysteine to various electrophiles (Mannervik, 1985; Pickett & Lu, 1989; Coles & Ketterer, 1990). GST also play a key role in the transport of hormones to the cell nucleus. Epoxide hydrolases (EH, *EC* 3.3.2.3) are found in both the cytosol and the endoplasmic reticulum. Their action is important since they degrade reactive epoxides by the addition of water, thus generally leading to the less reactive diols. However EH can sometimes contribute to the genesis of potent carcinogens as seen with the transformation of BaP: the EH mediated 7,8-dihydrodiol metabolite is less reactive than the parent molecule but cytochrome P450 can convert it into an extremely reactive epoxide responsible for the well-known mutagenic and carcinogenic properties of BaP. As with all other enzymes so far described, EH is also involved in the biotransformation of endogenous intermediates such as oestrogen and androgen epoxides (Timms *et al*. 1987).

Depending on their molecular weight, structure and polarity, conjugated metabolites are eliminated *via* urine or bile. Before urinary excretion, glutathione conjugates are further metabolized into mercapturic acids. Metabolites excreted *via* the biliary route may be partly hydrolysed by intestinal microflora and reabsorbed. This last transformation constitutes the first step of an enterohepatic cycle (Rowland, 1988).

Although XME is usually considered a detoxification system, such is not always the case. A lot of examples are known where it enhances or generates toxicity (carcinogenicity). On the whole it seems that the role of P450 in the toxification *v*. detoxification balance is far more ambiguous than that of phase II transferases. It is usually accepted, with some exceptions such as morphine-6-glucuronide (Caldwell, 1979), that an increase in the specific activities of transferases enhances detoxification. The XME system is very versatile and many factors may modulate its capacity. Apart from genetic characteristics (species, gender, individual), inducers may specifically enhance some of its activities, thus orienting its detoxification or toxification potential (Conney, 1982; Guengerich *et al*. 1982; Ullrich & Bock, 1984). Consequently, the induction of an isoform of P450 by a xenobiotic can have grave consequences for the fate of another xenobiotic, particularly if the latter is activated into a reactive toxic or carcinogenic metabolite by the isoform.

The effects of cruciferous vegetables on the XME system

A lot of work has been done since epidemiological and experimental findings first supported the idea of a protective role of GSL-rich diets against cancer. Most researchers have tried to elucidate the mechanism by which GSL and/or their derivatives could alter the XME system, both in the liver and the intestine.

Historically, the first work on that topic was performed by Wattenberg (1971) who showed that BaP hydroxylation in the rat intestine was very much enhanced when the animals were given a cabbage based diet. This study was then extended to other cruciferous vegetables and other activities of the phase I XME. McDanell *et al*. (1989) described the enhancement of ethoxyresorufin deethylation activity in the small intestine (5-fold), in the colon (4-fold) and in the liver (2·5-fold) of rats given a diet with 25% freeze-dried Brussels sprouts for 6 d. Similarly, feeding rats for 2 weeks on a diet containing 25% chopped and freeze-dried Brussels sprouts led to the induction (2-fold) of intestinal aryl hydrocarbon hydroxylase and ethoxycoumarin O-deethylase activities (Salbe & Bjeldanes, 1989). However no induction was seen in the liver, as already reported by Hendrich & Bjeldanes (1983) in mice fed on diets containing 20% chopped and freeze-dried cabbage or Brussels sprouts. A single meal of a GSL-containing food (25% dried cabbage) is not enough to modify the ethoxyresorufin deethylation activity in the liver and colon but it succeeds in inducing a temporary enhancement of this activity in the small intestine, the peak occurring 4–6 h post ingestion (McDanell et al. 1989). In our laboratory, monoclonal antibodies were used to investigate the influence of a diet with 39% rapeseed meal on the isoenzyme pattern of P450 in the liver of male rats. After 4 weeks, an overall reduction in the total P450 (-25%) occurred resulting from a 66% decrease of the 2C11 (male constitutive) form whereas the 1A1/1A2 (polycyclic hydrocarbon inducible) form was enhanced by 61%; the 2B1/B2 (phenobarbital inducible), 2E (ethanol inducible) and 3A (steroid inducible) forms also measured were not significantly modified (Nugon-Baudon et al. 1991).

Concerning phase II enzymes, giving rats for 10 d a diet containing 25% freeze-dried Brussels sprouts was shown to induce hepatic and intestinal GST and intestinal EH (Bradfield & Bjeldanes, 1984). Such phenomena were also observed by Aspry & Bjeldanes (1983) using diets containing 10–25% chopped freeze-dried broccoli. We have reproduced the induction of hepatic GST (2·5-fold) in rats given a diet with 39% rapeseed meal for 4 weeks and extended the investigations to hepatic GT; the activity of this last conjugative enzyme was dramatically enhanced (4-fold; Nugon-Baudon et al. 1990a). As far as hepatic EH is concerned, a slight stimulation (1·4-fold) was observed in mice fed for 10 d on a diet containing 20% chopped and freeze-dried Brussels sprouts but it did not occur when Brussels sprouts were replaced with cabbage (Hendrich & Bjeldanes, 1983).

The effects of various cruciferous vegetables on the phase I system seem to vary. Discrepancies between the experimental designs could be held responsible; indeed, Miller & Stoewsand (1983) have clearly shown that the phase I system of different strains of rats responded in different ways to a cabbage-containing diet. In contrast, results obtained on phase II are less divergent and there is now strong evidence of an overall induction of transferases in the intestine as well as the liver, whatever the experimental design.

It is now well known that the content and pattern of intact GSL and GSL derivatives are very much affected by the processing operations undergone by cruciferous vegetables before consumption. As cooking is one of the most usual treatments the question is, do cooked cruciferous vegetables modify the intestinal and/or hepatic XME, and if so how? Very few studies have been published on this issue. Recently, Wortelboer et al. (1992) have addressed the topic, using Brussels sprouts cooked for 20 min in unsalted water; consequently, the total GSL concentration dropped from 7·3 to 4·9 mmol/kg dry matter. Rats were given semi-synthetic diets containing either 0, 2·5, 5 or 20% cooked Brussels sprouts on a dry matter basis. Animals of each dietary group were killed after 2, 7, 14 or 28 d in order to assess the effects of the different levels of Brussels sprouts on hepatic and intestinal phase I and phase II enzymes. GST activity was induced throughout the experiment, in the intestine only by the 20% diet, and in the liver by diets containing at least 5% Brussels sprouts. From 2 d treatment onwards, the 20% diet also induced hepatic NAD(P)H quinone reductase (EC 1.6.99.2) and GT1 activities but it decreased hepatic GT2 activity. As far as P450 isoenzymes are concerned, polycyclic hydrocarbon inducible forms, i.e. 1A2 in the liver and 2B1/B2 in the small intestine, were induced in a dose related manner by all diets containing Brussels sprouts throughout the experiment. Apart from the immunochemical detection of apoproteins, the authors have used marker substrates to try to correlate Western-blot results and enzyme activities. Some were possible: enhanced ethoxyresorufin deethylation activity in the liver is correlated with the induction of 1A2, and increased 16α- and 16β-hydroxylation of testosterone by intestinal microsomes is correlated with the induction of the intestinal 2B isoenzyme. These results are quite important, since they show that cruciferous vegetables processed in the way that they usually are in a human diet may alter very significantly the XME system.

The effects of glucosinolates and glucosinolate derivatives on the XME system

Consistent with the work performed on the anticarcinogenic properties of pure GSL and GSL derivatives, most studies investigating the GSL linked alterations of the XME have strongly focused on indolylGSL and their enzymic derivatives.

Of four pure GSL, sinigrin, progoitrin, glucotropaeolin and glucobrassicin, only the last compound has been shown to induce phase I enzymes significantly, at least in the rat small intestine (McDanell *et al.* 1989). Loub *et al.* (1975) repeated the original work of Wattenberg (1971) on BaP hydroxylation, using pure I3C, 3,3'-diindolylmethane, IAN and ascorbigen as potential XME inducers. Given to rats by gavage a few hours before they were killed, these glucobrassicin derivatives induced BaP hydroxylation in the liver and the small intestine. I3C was tremendously active: a single 0·1 mmol dose induced 56- and 31-fold enhancements of BaP hydroxylation in the liver and small intestine respectively. Related studies, in which three glucobrassicin derivatives were given to rats twice daily for 3 d, corroborate this result (Pantuck *et al.* 1976); 3,3'-diindolylmethane (175 mg/kg body weight), IAN (95 mg/kg) and, to an even more important extent, I3C (100 mg/kg) each increased the intestinal metabolism of phenacetin, 7-ethoxycoumarin, hexobarbitone and BaP. I3C can modify the metabolism of other chemicals as well. In rainbow trout, dietary I3C (2 g/kg diet on a dry matter basis) is involved in substantial changes in the distribution, metabolism and elimination of aflatoxin B1, leading to significantly reduced hepatic DNA damage (Goeger *et al.* 1986). In liver microsomes prepared from rats fed for 2 weeks on an I3C-containing diet (30 mmol/kg on a dry matter basis) α-hydroxylation of NDMA and 4-(methylnitrosamino)-1-(3-pyridyl)-1-butanone (NNK), which are environmentally prevalent nitrosamines, is enhanced; in this case, the inducing effect of I3C is particularly harmful since it enhances the release of reactive intermediates binding to DNA (Chung *et al.* 1985). Bradfield & Bjeldanes (1984) reported that a dose of I3C as low as 50 mg/kg synthetic diet, given to rats for 10 d, led to a 6-fold increase of BaP hydroxylation activity in the small intestine. Gradually increasing the dose up to 500 mg/kg led to a positively correlated level of induction. The same effect was seen with intestinal ethoxycoumarin O-deethylase. However no effect on the hepatic counterparts of these activities could be seen, whatever the diet and the I3C concentration, though a slight increase in total P450 concentration occurred when animals were given the 500 mg/kg I3C diet. Shertzer (1982) also found contrasting results when he administered comparable doses of IAN or I3C to mice, rats and rabbits, orally or *via* the intraperitoneal route, daily for 10 d: IAN had no effect on either hepatic cytochrome P450 or BaP hydroxylation activity in any of the three animal species; with I3C, a 2-fold induction of hepatic P450 and BaP hydroxylation could be seen in the liver of mice and rats but not rabbits. On the whole, the induction level was much weaker than those reported elsewhere, especially considering that the administration of the GSL derivatives lasted a rather long time. The controversy increased with the findings of Babish & Stoewsand (1978); using rats given dietary levels of I3C ranging from 50 to 7500 mg/kg diet for 3 weeks, these authors observed a significant induction of intestinal BaP hydroxylation activity only at a dose that would correspond to a daily intake of 1·5 g/kg body weight, which is totally unrealistic for a human diet! Therefore the authors concluded that I3C is not the major inducer of phase I activities.

Once again, it is regrettable that so few studies have been conducted on other GSL or GSL derivatives. Nevertheless, among them, goitrin has received particular attention. According to Chang & Bjeldanes (1985), goitrin given to rats does not alter the ethoxycoumarin O-deethylase activity, either in the liver or in the small intestine, even at the lowest dose tested (40 mg/kg diet for 14 d). Recently Ozierenski et al. (1993) concluded that dietary goitrin is able to modify phase I activities in the rat liver, though in a contrasting way; whereas no significant modification of the overall P450 concentration occurred, aminopyrine N-demethylation was reduced and aniline p-hydroxylation was enhanced in a dose dependent manner. Ozierenski et al. (1993) have extended their investigations to a series of isothiocyanate derivatives and to 1-cyano-3-butene, the nitrile derived from gluconapin; on the whole, the overall concentration of P450 is significantly reduced and is accompanied by a dose related decrease of several P450 dependent activities such as aminopyrine N-demethylation, aniline p-hydroxylation and p-nitroanisole O-demethylation. Such results support our own findings concerning the heterogeneous alterations of the isoenzyme profile of P450 in the liver induced by rapeseed meal (Nugon-Baudon et al. 1991). In vivo consequences of the alterations of P450 activities by isothiocyanates have been investigated by Chung et al. (1985). These authors showed that isothiocyanates such as allyl-, benzyl- and phenylethylisothiocyanate, derived from sinigrin, glucotropaeolin and gluconasturtiin respectively, were good inhibitors of NDMA and NNK α-hydroxylation in liver microsomes prepared from rats fed for 2 weeks on a diet containing one of these compounds (3 mmol/kg dry matter); similar treatment with sinigrin also caused a significant decrease in the α-hydroxylation of these nitrosamines. In view of their promising inhibitory activities, the effects of dietary phenylethylisothiocyanate and sinigrin on the in vivo methylation of DNA by NDMA (25 mg/kg body weight by intraperitoneal injection) and NNK (85 mg/kg body weight by intravenous injection) were evaluated. The results were parallel to those obtained in the in vitro assays, suggesting that these compounds might be potent inhibitors of NDMA and NNK carcinogenesis.

Compared with phase I activities, the influence of GSL and GSL derivatives on phase II XME has been less fully investigated. Sparnins et al. (1982) showed that a semi-purified diet containing 6 g/kg I3C induced intestinal and hepatic GST (3-fold) after a 10 d trial in mice. A comparable result was reported later on hepatic EH (Cha et al. 1985). In both studies, the levels of I3C were very high and not to be found in a human diet. Using a diet containing 0·5 g/kg I3C, Wortelboer (1991) found a slight induction of liver and intestinal GST after 2 d and an induction of GT after 7 d in the rat. Nevertheless, other authors have published results that tend to show that no induction of intestinal or hepatic GST or EH activities by I3C is possible at normal dietary levels (Bradfield & Bjeldanes, 1984; Salbe & Bjeldanes, 1989). Although a diet with Brussels sprouts given to rats for 10 d induces both GST and EH in the liver as well as in the small intestine, synthetic diets containing 50–500 mg/kg I3C do not alter these activities at all (Bradfield & Bjeldanes, 1984).

There is now firm evidence that glucobrassicin and its derivatives cannot exclusively account for the phase II alterations observed when feeding crucifer-containing diets, far from it. This point has prompted several groups to look for effects of other GSL and GSL derivatives. One of the inducing molecules for phase II enzymes was identified as goitrin (Chang & Bjeldanes, 1985); when given to rats (40 mg/kg diet) for 14 d, this progoitrin derivative was able to increase significantly hepatic GST and EH activities. Ozierenski et al. (1993) addressed the same point, comparing the effects of goitrin and the gluconapin derivatives, 1-cyano-3-butene and butenyl isothiocyanate, and various isothiocyanates on GST in rat liver. All compounds tested, other than 1-cyano-3-butene, caused an increase in GST activity. In another very recent study, Zhang et al. (1992) applied a glucoiberin derivative, 1-isothiocyanato-(3R)-(methylsulphinyl) propane (IMSP), and a glucoraphanin derivative, 1-isothiocyanato-(4R)-(methylsulphinyl) butane, to a Hepa 1c1c7 murine hepatoma cell culture and found that both these derivatives were potent inducers of GST and NAD(P)H quinone reductase. These results were of extreme importance and deserved to be confirmed and qualified *in vivo*, which was done by Kore *et al.* in 1993. IMSP doses of 1, 10 and 100 µmol/kg body weight were given by gavage to rats once daily for 7 d; the lowest dose was, according to the authors, comparable to what an average western diet would contain. No alterations, whatever the dose, could be seen in hepatic levels of cytochrome P450, ethoxycoumarin O-deethylase or aminopyrine N-demethylase activities or in hepatic quinone reductase, GST and GT. However an important induction of intestinal quinone reductase (8-fold) and a moderate induction of intestinal GST (2-fold) occurred, but only at the highest dose of IMSP. Thus it was concluded that the IMSP content occurring in an average human diet may have no significant influence on either phase I or phase II enzymes.

All these findings definitely show that the XME alterations mediated by cruciferous vegetables involve many kinds of GSL derivatives. The numerous dose related studies reported here highlight the fact that one must be extremely cautious in extrapolating alterations observed under experimental conditions to real nutritional conditions; it seems that some molecules, albeit undeniably active towards XME, are eventually not relevant from a nutritional point of view and should rather be considered as candidates for pharmacological investigation.

CONCLUSIONS AND PENDING TOPICS

One of the major points which remains to be addressed is of course how far it is possible to extrapolate to humans results established in laboratory rodents.

Only a few studies have been performed so far in humans, for obvious reasons. Nevertheless the pharmacological fates of some drugs which are known to be metabolized by the XME system have been examined by Pantuck and coworkers in volunteers consuming cruciferous vegetables. The metabolism of phenacetin and antipyrine and the glucuronidation of paracetamol are enhanced by the consumption of Brussels sprouts, cabbage and other cruciferous-containing diets (Pantuck *et al.* 1979, 1984). In a review on indolylGSL, McDanell *et al.* (1988) support the idea that GSL derivatives are likely to be as active on the XME system of humans as they are in laboratory animals.

Among the most striking points, when looking at the findings reported in the present review, are the conflicting results which appear to arise from the diversity of amounts of cruciferous vegetables and GSL or GSL derivatives incorporated in the rodent diets. Furthermore, where cruciferous vegetables have been used, the reader has sometimes been poorly informed on the GSL content resulting from the process undergone by the vegetable before its incorporation into the diet. Since cruciferous vegetables are usually eaten after treatments such as mashing, fermenting, cooking, etc., reports on the effects of such treatments on the GSL related XME alterations would be a crucial matter to develop in order to extend data provided by the original works of McDanell *et al.* (1987), Wortelboer (1991) and Wortelboer *et al.* (1992).

All the studies reported here deal with the impact of GSL on environmental procarcinogens and/or carcinogens biotransformed via the XME system. Very few studies have investigated the extent to which GSL may alter steroid metabolism, though this point could be important for hormone dependent cancers in the human. Michnovicz & Bradlow (1991) have shown that in twelve healthy men and women ingestion of 6–7 mg/day of I3C for 7 d increased the 2-hydroxylation of oestradiol by about 50%, thus enhancing the urinary excretion of 2-hydroxyoestrone relative to the excretion of oestriol. In a

Table 7. *Effects of a diet with rapeseed meal on three hepatic xenobiotic metabolizing enzymes in germ free and conventional rats*

(Results are expressed as % of the mean values obtained with counterpart rats given a diet with soyabean meal)

Bacterial status...	Conventional	Germ free
Reference...	Nugon-Baudon et al. (1990a)	Rabot et al. (1993b)
Duration of trial (weeks)...	4	3
No. of animals...	11	8
Cytochrome P450	75**	80
Glutathione-*S*-transferase	236**	105
UDP-glucuronosyltransferase	372**	102

Mean values were significantly different from those for counterpart animals given a soyabean meal diet: ** $P < 0.01$.

spontaneous mammary tumour mouse model, tumour incidence and multiplicity were significantly reduced after mice had received a diet containing 500 or 2000 mg/kg I3C for 8 months; in this model, I3C increased the level of oestradiol 2-hydroxylation up to 5-fold. The authors concluded that the protective effect may have resulted from increased 2-hydroxylation and inactivation of endogenous oestrogens (Bradlow et al. 1991). This could be a clue that GSL influence on carcinogenesis might result from alterations of the XME mediated biotransformation of exogenous compounds as well as endogenous molecules.

Finally, all pure GSL derivatives examined so far, in cell cultures and *in vivo*, originate from hydrolysis by plant myrosinase. Since GSL derivatives produced by myrosinase-like activities of the intestinal microflora are able to induce toxic effects, one wonders whether they are able to induce XME alterations as well. A first answer arises from experiments performed in our laboratory using conventional and germ free rats: the decrease in total P450 concentration as well as the induction of GST and GT observed in the liver of conventional rats given a GSL-rich but myrosinase free diet cannot be reproduced in germ free animals (Table 7; Nugon-Baudon et al. 1990a; Rabot et al. 1993b). These findings indicate that, should myrosinase be absent from the diet, bacterial metabolism would substitute for it and produce GSL derivatives capable of altering the XME system. Primary GSL derivatives produced by plant myrosinase or by the microflora may also undergo further transformations in the body. Whether or not these putative second metabolic steps are mediated by the intestinal microflora, one may think that the active GSL metabolites may not be exclusively the aglucones released by the plant myrosinase or the primary metabolites released by the microflora. We have been able to offer some support for this hypothesis (Nugon-Baudon *et al.* 1990a) by showing that a pretreatment with phenobarbital led to enhancement of several GSL linked toxic effects in conventional rats.

On the whole there is still a tremendous and varied scope for further research in the field of relationships between glucosinolates and cancer. The numerous studies already performed and the hypotheses already suggested demonstrate the challenge to the imagination and ingenuity of nutritionists, pharmacologists, chemists and bacteriologists posed by the puzzle.

REFERENCES

Adams, H., Vaughan, J. G. & Fenwick, G. R. (1989). The use of glucosinolates for cultivar identification in swede, *Brassica napus* L var *napobrassica* (L) Peterm. *Journal of the Science of Food and Agriculture* **46**, 319–324.

Akiba, Y. & Matsumoto, T. (1976). Antithyroid activity of goitrin in chicks. *Poultry Science* **55**, 716–719.
Aspry, K. E. & Bjeldanes, L. F. (1983). Effects of dietary broccoli and butylated hydroxyanisole on liver-mediated metabolism of benzo[a]pyrene. *Food and Chemical Toxicology* **21**, 133–142.
Astwood, E. B., Greer, M. A. & Ettlinger, M. G. (1949). l-5-vinyl-2-thiooxazolidone, an antithyroid compound from yellow turnip and from brassica seeds. *Journal of Biological Chemistry* **181**, 121–130.
Austin, F. L., Gent, C. A. & Wolff, I. A. (1968). Degradation of natural thioglucosides with ferrous salts. *Journal of Agricultural and Food Chemisty* **16**, 752–755.
Babish, J. G. & Stoewsand, G. S. (1978). Effect of dietary indole-3-carbinol on the induction of the mixed-function oxidases of rat tissue. *Food and Cosmetics Toxicology* **16**, 151–155.
Beaune, P. (1982). *Les cytochromes P450 des microsomes de foie humain: activités monooxygénasiques et purification partielle (Microsomal P450 Cytochromes in Human Liver: Monoxygenase Activities and Partial Purification)*, PhD thesis, University of Paris 6, 139 pp.
Beaune, P. (1986). [Liver P450 cytochromes in humans.] *Médecine/Sciences* **2**, 358–363.
Bell, J. M. (1984). Nutrients and toxicants in rapeseed meal: a review. *Journal of Animal Science* **58**, 996–1010.
Bell, J. M., Benjamin, B. R. & Giovannetti, P. M. (1972). Histopathology of thyroids and livers of rats and mice fed diets containing Brassica glucosinolates. *Canadian Journal of Animal Science* **52**, 395–406.
Benns, G. B., Hall, J. W. & Beare-Rogers, J. L. (1978). Intake of brassicaceous vegetables in Canada. *Canadian Journal of Public Health* **69**, 64–66.
Bille, N., Eggum, B. O., Jacobsen, I., Olsen, O. & Sorensen, H. (1983). Antinutritional and toxic effects in rats of individual glucosinolates ($\pm$myrosinases) added to a standard diet. 1. Effects on protein utilization and organ weights. *Zeitschrift für Tierphysiolgie, Tierernährung und Futtermittelkunde* **49**, 195–210.
Birt, D. F., Pelling, J. C., Pour, P. M., Tibbels, M. G., Schweickert, L. & Bresnick, E. (1987). Enhanced pancreatic and skin tumorigenesis in cabbage fed hamsters and mice. *Carcinogenesis* **8**, 913–917.
Bjerg, B. & Sørensen, H. (1987). Quantitative analysis of glucosinolates in oilseed rape based on HPLC of desulfoglucosinolates and HPLC of intact glucosinolates. *World Crops: Production, Utilization and Description* **13**, 125–150.
Bock, K. W., Josting, D., Lilienblum, W. & Pfeil, H. (1979). Purification of rat-liver microsomal UDP-glucuronyltransferase: separation of two enzyme forms inducible by 3-methylcholanthrene or phenobarbital. *European Journal of Biochemistry* **98**, 19–26.
Bock, K. W., Lilienblum, W., Fischer, G., Schirmer, G. & Bock-Hennig, B. S. (1987). Induction and inhibition of conjugating enzymes with emphasis on UDP-glucuronyltransferases. *Pharmacology and Therapeutics* **33**, 23–27.
Booth, E. J., Walker, K. C. & Griffiths, D. W. (1990). Effect of harvest date and pod position on glucosinolates in oilseed rape (*Brassica napus*). *Journal of the Science of Food and Agriculture* **53**, 43–61.
Bourdon, D., Perez, J.-M. & Baudet, J.-J. (1981). [New types of rapeseed meal fed to growing-finishing pigs: influence of glucosinolates and dehulling.] *Journées de la Recherche Porcine en France* **13**, 163–178.
Boyd, J. N., Babish, J. G. & Stoewsand, G. S. (1982). Modification by beet and cabbage diets of aflatoxin B_1-induced rat plasma α-foetoprotein elevation, hepatic tumorogenesis, and mutagenicity of urine. *Food and Chemical Toxicology* **20**, 47–52.
Bradfield, C. A. & Bjeldanes, L. F. (1984). Effect of dietary indole-3-carbinol on intestinal and hepatic monooxygenase, glutathione S-transferase and epoxide hydrolyase activities in the rat. *Food and Chemical Toxicology* **22**, 977–982.
Bradfield, C. A. & Bjeldanes, L. F. (1987). High-performance liquid chromatographic analysis of anticarcinogenic indoles in *Brassica oleracea*. *Journal of Agricultural and Food Chemistry* **35**, 46–49.
Bradlow, H. L., Michnovicz, J. J., Telang, N. T. & Osborne, M. P. (1991). Effects of dietary indole-3-carbinol on estradiol metabolism and spontaneous mammary tumors in mice. *Carcinogenesis* **12**, 1571–1574.
Buchwaldt, L., Larsen, L. M., Plöger, A. & Sørensen, H. (1986). Fast polymer liquid chromatography isolation and characterization of plant myrosinase, β-thioglucoside glucohydrolase, isoenzymes. *Journal of Chromatography* **363**, 71–80.
Burchell, B., Jackson, M. R., Coughtrie, M. W. H., Harding, D., Wilson, S. & Bend, J. R. (1987). Molecular characterization of hepatic UDP-glucuronyl transferases. In *Drug Metabolism: From Molecules to Man*, pp. 40–54 [D. Benford, J. W. Bridges and G. G. Gibson, editors]. London: Taylor and Francis.
Burke, M. D. & Orrenius, S. (1979). Isolation and comparison of endoplasmic reticulum membranes and their mixed function oxidase activities from mammalian extrahepatic tissues. *Pharmacology and Therapeutics* **7**, 549–599.
Butler, E. J., Pearson, A. W. & Fenwick, G. R. (1982). Problems which limit the use of rapeseed meal as a protein source in poultry diets. *Journal of the Science of Food and Agriculture* **33**, 866–875.
Caldwell, J. (1979). Minireview. The significance of phase II (conjugation) reactions in drug disposition and toxicity. *Life Sciences* **24**, 571–578.
Caldwell, J. (1980). Conjugation reactions. In *Concepts in Drug Metabolism*, vol. **10(A)**, pp. 211–217 [P. Jenner and B. Testa, editors]. Basel: Decker.
Carlson, D. G., Daxenbichler, M. E., VanEtten, C. H., Hill, C. B. & Williams, P. H. (1985). Glucosinolates in radish cultivars. *Journal of the American Society for Horticultural Science* **110**, 634–638.
Carlson, D. G., Daxenbichler, M. E., VanEtten, C. H., Kwolek, W. F. & Williams, P. H. (1987). Glucosinolates

in crucifer vegetables: broccoli, Brussels sprouts, cauliflower, collards, kale, mustard greens, and kohlrabi. *Journal of the American Society for Horticultural Science* **112**, 173–178.
Carlson, D. G., Daxenbichler, M. E., VanEtten, C. H., Tookey, H. L. & Williams, P. H. (1981). Glucosinolates in crucifer vegetables: turnips and rutabagas. *Journal of Agricultural and Food Chemistry* **29**, 1235–1239.
Cha, Y. N., Thompson, D. C., Heine, H. S. & Chung, J. H. (1985). Differential effects of indole, indole-3-carbinol and benzofuran on several microsomal and cytosolic enzyme activities in mouse liver. *Korean Journal of Pharmacology (Taehan Yakrihak Chapchi)* **21**, 1–11.
Chang, Y. & Bjeldanes, L. F. (1985). Effect of dietary *R*-goitrin on hepatic and intestinal glutathione *S*-transferase, microsomal epoxide hydratase and ethoxycoumarin *O*-deethylase activities in the rat. *Food and Chemical Toxicology* **23**, 905–909.
Chung, F.-L., Wang, M. & Hecht, S. S. (1985). Effects of dietary indoles and isothiocyanates on *N*-nitrosodimethylamine and 4-(methylnitrosamino)-1-(3-pyridyl)-1-butanone α-hydroxylation and DNA methylation in rat liver. *Carcinogenesis* **6**, 539–543.
Coles, B. & Ketterer, B. (1990). The role of glutathione and glutathione transferases in chemical carcinogenesis. *CRC Critical Reviews in Biochemistry and Molecular Biology* **25**, 47–70.
Committee on Diet, Nutrition and Cancer, National Research Council (1982). *Diet, Nutrition and Cancer.* Washington DC: National Academy Press.
Conney, A. H. (1982). Induction of microsomal enzymes by foreign chemicals and carcinogenesis by polycyclic aromatic hydrocarbons: G. H. A. Clowes Memorial Lecture. *Cancer Research* **42**, 4875–4917.
Daxenbichler, M. E., VanEtten, C. H. & Williams, P. H. (1979). Glucosinolates and derived products in cruciferous vegetables. Analysis of 14 varieties of Chinese cabbage. *Journal of Agricultural and Food Chemistry* **27**, 34–37.
de Vos, R. H. & Blijleven, W. G. H. (1988). The effect of processing conditions on glucosinolates in cruciferous vegetables. *Zeitschrift für Lebensmittel-Untersuchung und -Forschung* **187**, 525–529.
Duncan, A. J. & Milne, J. A. (1989). Glucosinolates. In *Anti-nutritional Factors, Potentially Toxic Substances in Plants (Aspects of Applied Biology* **19**), pp. 75–92.
Duncan, A. J. & Milne, J. A. (1992). Rumen microbial degradation of allyl cyanide as a possible explanation for the tolerance of sheep to brassica-derived glucosinolates. *Journal of the Science of Food and Agriculture* **58**, 15–19.
Elfving, S. (1980). Studies in the naturally occurring goitrogen 5-vinyl-2-thiooxazolidone. *Annals of Clinical Research* **12**, Suppl. 28, 7–47.
Etienne, M. & Dourmad, J.-Y. (1987). [Effects of high or low-glucosinolate varieties of rapeseed meal on reproduction in sow.] *Journées de la Recherche Porcine en France* **19**, 231–238.
Ettlinger, M. G., Dateo, G. P., Harrison, B. W., Mabry, T. J. & Thompson, C. P. (1961). Vitamin C as a coenzyme: the hydrolysis of mustard oil glucosides. *Proceedings of the National Academy of Sciences, USA* **47**, 1875–1880.
Ettlinger, M. G. & Lundeen, A. J. (1956). The structures of sinigrin and sinalbin; an enzymatic rearrangement. *Journal of the American Chemical Society* **78**, 4172–4173.
Fenwick, G. R., Butler, E. J. & Brewster, M. A. (1983). Are brassica vegetables aggravating factors in trimethylaminuria (fish odour syndrome)? *Lancet* **ii**, 916.
Fenwick, G. R., Heaney, R. K., Hanley, A. B. & Spinks, E. A. (1986). Glucosinolates in food plants. In *Food Research Institute, Norwich, Annual Report.*
Fenwick, G. R., Heaney, R. K. & Mullin, W. J. (1982). Glucosinolates and their breakdown products in food and food plants. *CRC Critical Reviews in Food Science and Nutrition* **18**, 123–201.
Gil, V. & MacLeod, A. J. (1980). The effects of pH on glucosinolate degradation by a thioglucoside glucohydrolase preparation. *Phytochemistry* **19**, 2547–2551.
Goeger, D. E., Shelton, D. W., Hendricks, J. D. & Bailey, G. S. (1986). Mechanisms of anti-carcinogenesis by indole-3-carbinol: effect on the distribution and metabolism of aflatoxin B_1 in rainbow trout. *Carcinogenesis* **7**, 2025–2031.
Gould, D. H., Fettman, M. J. Daxenbichler, M. E. & Bartuska, B. M. (1985). Functional and structural alterations of the rat kidney induced by the naturally occurring organonitrile, 2*S*-1-cyano-2-hydroxy-3,4-epithiobutane. *Toxicology and Applied Pharmacology* **78**, 190–201.
Graham, S., Dayal, H., Swanson, M., Mittelman, A. & Wilkinson, G. (1978). Diet in the epidemiology of cancer of the colon and rectum. *Journal of the National Cancer Institute* **61**, 709–714.
Graham, S., Schotz, W. & Martino, P. (1972). Alimentary factors in the epidemiology of gastric cancer. *Cancer* **30**, 927–938.
Greer, M. A. (1962). The natural occurrence of goitrogenic agents. *Recent Progress in Hormone Research* **18**, 187–219.
Greer, M. A. & Astwood, E. B. (1948). The antithyroid effect of certain foods in man as determined with radioactive iodine. *Endocrinology* **43**, 105–119.
Greer, M. A. & Deeney, J. M. (1959). Antithyroid activity elicited by the ingestion of pure progoitrin, a naturally occurring thioglycoside of the turnip family. *Journal of Clinical Investigation* **38**, 1465–1474.
Guengerich, F. P. (1989). Polymorphism of cytochrome P-450 in humans. *Trends in Pharmacological Sciences* **10**, 107–109.

Salbe, A. D. & Bjeldanes, L. F. (1989). Effect of diet and route of administration on the DNA binding of aflatoxin B$_1$ in the rat. *Carcinogenesis* **10**, 629–634.

Sang, J. P., Minchinton, I. R., Johnstone, P. K. & Truscott, R. J. W. (1984). Glucosinolate profiles in the seed, root and leaf tissue of cabbage, mustard, rapeseed, radish and swede. *Canadian Journal of Plant Science* **64**, 77–93.

Searle, L. M., Chamberlain, K. & Butcher, D. N. (1984). Preliminary studies on the effects of copper, iron and manganese ions on the degradation of 3-indolylmethyl-glucosinolate (a constituent of *Brassica* spp.) by myrosinase. *Journal of the Science of Food and Agriculture* **35**, 745–748.

Searle, L. M., Chamberlain, K., Rausch, T. & Butcher, D. N. (1982). The conversion of 3-indolylmethyl-glucosinolate to 3-indolylacetonitrile by myrosinase and its relevance to the clubroot disease of the Cruciferae. *Journal of Experimental Botany* **33**, 935–942.

Sesardic, D., Pasanen, M., Pelkonen, O. & Boobis, A. R. (1990). Differential expression and regulation of members of the cytochrome P450IA gene subfamily in human tissues. *Carcinogenesis* **11**, 1183–1188.

Shertzer, H. G. (1982). Indole-3-carbinol and indole-3-acetonitrile influence on hepatic microsomal metabolism. *Toxicology and Applied Pharmacology* **64**, 353–361.

Shertzer, H. G. (1983). Protection by indole-3-carbinol against covalent binding of benzo[*a*]pyrene metabolites to mouse liver DNA and protein. *Food and Chemical Toxicology* **21**, 31–35.

Shertzer, H. G. (1984). Indole-3-carbinol protects against covalent binding of benzo[*a*]pyrene and *N*-nitrosodimethylamine metabolites to mouse liver macromolecules. *Chemico-Biological Interactions* **48**, 81–90.

Slominski, B. A. & Campbell, L. D. (1989). Formation of indole glucosinolate breakdown products in autolyzed, steamed, and cooked *Brassica* vegetables. *Journal of Agricultural and Food Chemistry* **37**, 1297–1302.

Sones, K., Heaney, R. K. & Fenwick, G. R. (1984*a*). An estimate of the mean daily intake of glucosinolates from cruciferous vegetables in the UK. *Journal of the Science of Food and Agriculture* **35**, 712–720.

Sones, K., Heaney, R. K. & Fenwick, G. R. (1984*b*). Glucosinolates in *Brassica* vegetables. Analysis of twenty-seven cauliflower cultivars (*Brassica oleracea* L. var. *botrytis* subvar. *cauliflora* DC). *Journal of the Science of Food and Agriculture* **35**, 762–766.

Sparnins, V. L., Venegas, P. L. & Wattenberg, L. W. (1982). Glutathione S-transferase activity: enhancement by compounds inhibiting chemical carcinogenesis and by dietary constituents. *Journal of the National Cancer Institute* **68**, 493–496.

Srisangnam, C., Hendricks, D. G., Sharma, R. P., Salunkhe, D. K. & Mahoney, A. W. (1980*a*). Effects of dietary cabbage (*Brassica oleracea* L.) on the tumorigenicity of 1,2-dimethylhydrazine in mice. *Journal of Food Safety* **2**, 235–245.

Srisangnam, C., Salunkhe, D. K., Reddy, N. R. & Dull, G. G. (1980*b*). Quality of cabbage. II. Physical, chemical, and biochemical modification in processing treatments to improve flavor in blanched cabbage (*Brassica oleracea* L.). *Journal of Food Quality* **3**, 233–250.

Srivastava, V. K., Philbrick, D. J. & Hill, D. C. (1975). Response of rats and chicks to rapeseed meal subjected to different enzymatic treatments. *Canadian Journal of Animal Science* **55**, 331–335.

Stoewsand, G. S., Anderson, J. L. & Munson, L. (1988). Protective effect of dietary Brussels sprouts against mammary carcinogenesis in Sprague-Dawley rats. *Cancer Letters* **39**, 199–207.

Stoewsand, G. S., Babish, J. B. & Wimberly, H. C. (1978). Inhibition of hepatic toxicities from polybrominated biphenyls and aflatoxin B$_1$ in rats fed cauliflower. *Journal of Environmental Pathology and Toxicology* **2**, 399–406.

Szabo, S., Bailey, K. A., Boor, P. J. & Jaeger, R. J. (1977). Acrylonitrile and tissue glutathione: differential effect of acute and chronic interactions. *Biochemical and Biophysical Research Communications* **79**, 32–37.

Tanaka, T., Mori, Y., Morishita, Y., Hara, A., Ohno, T., Kojima, T. & Mori, H. (1990). Inhibitory effect of sinigrin and indole-3-carbinol on diethylnitrosamine-induced hepatocarcinogenesis in male ACI/N rats. *Carcinogenesis* **11**, 1403–1406.

Tani, N., Ohtsuru, M. & Hata, T. (1974). Isolation of myrosinase producing microorganism. *Agricultural and Biological Chemistry* **38**, 1617–1622.

Timms, C., Schladt, L., Robertson, L., Rauch, P., Schramm, H. & Oesch, F. (1987). The regulation of rat liver epoxide hydrolases in relation to that of other drug-metabolizing enzymes. In *Drug Metabolism: From Molecules to Man*, pp. 55–68 [D. Benford, J. W. Bridges and G. G. Gibson, editors]. London: Taylor and Francis.

Tookey, H. L. (1973). Crambe thioglucoside glucohydrolase (EC 3.2.3.1): separation of a protein required for epithiobutane formation. *Canadian Journal of Biochemistry* **51**, 1654–1660.

Tookey, H. L. & Wolff, I. A. (1970). Effect of organic reducing agents and ferrous ion on thioglucosidase activity of *Crambe abyssinica* seed. *Canadian Journal of Biochemistry* **48**, 1024–1028.

Uda, Y., Kurata, T. & Arakawa, N. (1986). Effects of pH and ferrous ion on the degradation of glucosinolates by myrosinase. *Agricultural and Biological Chemistry* **50**, 2735–2740.

Ullrich, D. & Bock, K. W. (1984). Glucuronide formation of various drugs in liver microsomes and in isolated hepatocytes from phenobarbital- and 3-methylcholanthrene-treated rats. *Biochemical Pharmacology* **33**, 97–101.

VanEtten, C. H., Daxenbichler, M. E., Peters, J. E. & Tookey, H. L. (1966). Variation in enzymatic degradation products from the major thioglucosides in *Crambe abyssinica* and *Brassica napus* seed meals. *Journal of Agricultural and Food Chemistry* **14**, 426–430.

VanEtten, C. H., Daxenbichler, M. E. & Wolff, I. A. (1969). Natural glucosinolates (thioglucosides) in foods and feeds. *Journal of Agricultural and Food Chemistry* **17**, 483–491.
Vermorel, M., Davicco, M.-J. & Evrard, J. (1987). Valorization of rapeseed meal. 3. Effects of glucosinolate content on food intake, weight gain, liver weight and plasma thyroid hormone levels in growing rats. *Reproduction, Nutrition, Développement* **27**, 57–66.
Vermorel, M. & Evrard, J. (1987). Valorization of rapeseed meal. 4. Effects of iodine, copper and ferrous salt supplementation in growing rats. *Reproduction, Nutrition, Développement* **27**, 769–779.
Vermorel, M., Heaney, R. K. & Fenwick, G. R. (1986). Nutritive value of rapeseed meal: effects of individual glucosinolates. *Journal of the Science of Food and Agriculture* **37**, 1197–1202.
Wattenberg, L. W. (1971). Studies of polycyclic hydrocarbon hydroxylases of the intestine possibly related to cancer. Effect of diet on benzpyrene hydroxylase activity. *Cancer* **28**, 99–102.
Wattenberg, L. W., Hanley, A. B., Barany, G., Sparnins, V. L., Lam, L. K. T. & Fenwick, G. R. (1986). Inhibition of carcinogenesis by some minor dietary components. In *Diet, Nutrition and Cancer*, pp. 13–21 [Y. Hayashi *et al.* editors]. Tokyo: VNU Science.
Wattenberg, L. W. & Loub, W. D. (1978). Inhibition of polycyclic aromatic hydrocarbon-induced neoplasia by naturally-occurring indoles. *Cancer Research* **38**, 1410–1413.
Wishart, G. J. (1978). Demonstration of functional heterogeneity of hepatic uridine diphosphate glucuronosyl-transferase activities after administration of 3-methylcholanthrene and phenobarbital to rats. *Biochemical Journal* **174**, 671–672.
Wortelboer, H. M. (1991). *Primary hepatocyte cultures as a model system for the determination of induction of biotransformation enzymes. Effects of glucosinolate hydrolysis products.* PhD Thesis, University of Utrecht, 145 pp.
Wortelboer, H. M., de Kruif, C. A., van Iersel, A. A. J., Noordhoek, J., Blaauboer, B. J., van Bladeren, P. J. & Falke, H. E. (1992). Effects of cooked Brussels sprouts on cytochrome P-450 profile and phase II enzymes in liver and small intestinal mucosa of the rat. *Food and Chemical Toxicology* **30**, 17–27.
Wrighton, S. A., Campanile, C., Thomas, P. E., Maines, S. L., Watkins, P. B., Parker, G., Mendez-Picon, G., Haniu, M., Shively, J. E., Levin, W. & Guzelian, P. S. (1986). Identification of a human liver cytochrome P-450 homologous to the major isosafrole-inducible cytochrome P-450 in the rat. *Molecular Pharmacology* **29**, 405–410.
Youngs, C. G. & Perlin, A. S. (1967). Fe(II)-catalyzed decomposition of sinigrin and related thioglycosides. *Canadian Journal of Chemistry* **45**, 1801–1804.
Zhang, Y., Talalay, P., Cho, C.-G. & Posner, G. H. (1992). A major inducer of anticarcinogenic protective enzymes from broccoli: isolation and elucidation of structure. *Proceedings of the National Academy of Sciences, USA* **89**, 2399–2403.

NUTRITIONAL INFLUENCES ON INTERACTIONS BETWEEN BACTERIA AND THE SMALL INTESTINAL MUCOSA

D. KELLY, R. BEGBIE AND T. P. KING

Rowett Research Institute, Bucksburn, Aberdeen, Scotland, AB2 9SB

CONTENTS

INTRODUCTION	233
COLONIZATION OF THE INTESTINAL TRACT	234
THE GLYCOCONJUGATE CHEMISTRY OF INTESTINAL SURFACES	234
MEMBRANE GLYCOCONJUGATES	235
MUCIN GLYCOCONJUGATES	235
BACTERIAL ADHESINS AND INTESTINAL RECEPTORS	235
FIMBRIAL ADHESINS	237
Type-1 fimbriae	237
Fimbrial serotypes associated with diarrhoeal disease	237
Fimbrial adhesins associated with Helicobacter pylori	240
OUTER MEMBRANE PROTEINS	240
LIPOTEICHOIC ACID	241
MUCIN AND BACTERIAL ADHESION	241
BACTERIAL GLYCOSIDASES	243
NUTRITIONAL INFLUENCES ON THE INTERACTIONS BETWEEN BACTERIA AND THE INTESTINAL MUCOSA	243
DIETARY MODULATION OF BACTERIAL RECEPTORS IN THE NEONATE	243
Probiosis	245
Milk oligosaccharides as probiotics	245
Breast milk and bifidobacteria	246
DIETARY MODULATION OF BACTERIAL RECEPTORS IN ADULTS	247
Dietary lectin toxicity and bacterial overgrowth	248
Malnutrition and bacterial translocation	249
Chemical probiosis	249
SUMMARY	251
REFERENCES	251

INTRODUCTION

Much recent research in intestinal bacteriology has been based on the tenet that the indigenous microflora plays an important role in preserving the health of individuals by preventing enteric colonization by opportunistic pathogens. This colonization resistance may be a consequence of competitive exclusion of pathogens by the indigenous flora. In other situations the control exerted by the indigenous flora is more subtle and may involve the synthesis of molecules which non-specifically stimulate the immune system or chemically interfere with the interaction between pathogens and intestinal epithelia. The concept is evolving that natural or synthetic antiadhesive compounds (soluble receptor analogues) might provide an alternative approach to rational drug design in antibacterial therapy and this has fuelled a widespread research interest in molecular interactions at the intestinal–bacterial interface. The role of nutrition in augmenting or inhibiting these interactions represents the major discussion point of this review.

COLONIZATION OF THE INTESTINAL TRACT

The intestinal tract and other mucous membranes are continually contaminated by micro-organisms from the environment. Through a combination of synergic and antagonistic interactions (Linton & Hinton, 1990) steady state is attained and the resultant bacterial population is variously referred to as the natural, indigenous, resident, normal or commensal microflora.

It has been estimated that several hundred microbial species inhabit the mammalian alimentary tract (Finegold *et al.* 1983). To provide a comprehensive description of the resident microbiota in the developing or adult intestine is a formidable task, and it is still only possible to indicate relative frequency and proportions. Study of the enteric flora is further complicated by the fact that within the small intestine there are several habitats including the villus surface, crypts, epithelial associated mucins and luminal mucus. Very different bacterial species colonize these diverse intestinal niches. The normal microbiota in a single community may also differ from that of another of the same species due to differences in diet, husbandry or climatic conditions. In spite of the technical difficulties several authors have attempted to describe, to varying degrees, the bacterial composition of the intestine of mammalian species (for reviews see Moughan *et al.* 1992; Maxwell & Stewart, 1994).

The selection and establishment of a stable intestinal microflora is dependent on several factors including local immune mechanisms, interbacterial interactions, the presence of bacterial receptors on mucosal surfaces, availability of nutrients from endogenous and dietary sources, digesta flow, conditions of pH, oxygen–reduction potentials and the availability of molecular oxygen within the intestine (Stewart *et al.* 1993).

THE GLYCOCONJUGATE CHEMISTRY OF INTESTINAL SURFACES

The differentiation processes of intestinal epithelial cells are associated with important changes in the synthesis and expression of membrane and secretory glycoproteins and glycolipids. This diversity is a feature not only of intestinal cell differentiation on the crypt–villus axis but also of membrane and mucin constituents in proximal and distal regions of the small intestine. The oligosaccharide structures of intestinal glycoconjugates are not themselves primary gene products but are constructed in a stepwise manner, monosaccharides being added to precursor oligosaccharides *via* several glycosyltransferases encoded by different genes (Neutra & Forstner, 1987; Ito & Hirota, 1992). The maturing glycoconjugate complexion of each intestinal cell is influenced by many different factors such as the intrinsic composition of glycosyltransferase species defined by the genotype of the individual, the relative activity or amount of these enzymes (repression or induction of the enzymes), competition between enzymes with overlapping substrate specificity, the organization of the enzymes in Golgi membranes, utilization of precursors and specific substrate sugars, and the activity level of degrading enzymes (Ito & Hirota, 1992). In spite of this considerable potential for diversity, glycosylation processes within the intestinal epithelium frequently proceed in a predictable fashion.

MEMBRANE GLYCOCONJUGATES

Temporal glycosylation changes have been extensively investigated in both rats and pigs. In these species a progressive change from $\alpha 2,6$ sialylation to $\alpha 1,2$ fucosylation of microvillar glycoconjugates occurs during postnatal development (Taatjes & Roth, 1990; King & Kelly, 1991; King et al. 1993). Cytochemical investigations indicate that intestinal membrane oligomannose-type *N*-linked oligosaccharides in the developing pig intestine are replaced by complex or hybrid *N*-linked structures (T. P. King, R. Begbie & D. Kelly, unpublished observations). In humans and many other mammalian species important differences in the glycosylation of intestinal membranes may be correlated with ABO histo-blood groups and are thus genetically mediated (Oriol, 1987; King & Kelly, 1990, 1991).

MUCIN GLYCOCONJUGATES

Intestinal goblet cells synthesize and secrete high molecular weight glycoproteins called mucins. Upon secretion, mucins hydrate and gel, generating a protective mucus blanket overlying the epithelial surface (Specian & Oliver, 1991). Non-mucin constituents of the mucus blanket include water, serum and cellular macromolecules, electrolytes, sloughed off epithelial cells, cell debris and other glycoproteins. A high percentage of the weight of mucin glycoproteins consists of oligosaccharides, *O*-linked to multiple serine or threonine residues in the polypeptide backbone. In addition, a growing number of mucins are reported to contain a small number of *N*-linked glycans (Strous & Dekker, 1992). In the human and pig small intestine immature goblet cells deep within the crypts produce neutral mucins containing little sialic acid (Specian & Oliver, 1991; King, 1994). As they mature and migrate to the villus tip, the mucins become increasingly sialylated; these sialic acid residues not only increase the acidity of the molecule but are also sites for further modification by *N*- and *O*-acylation (Filipe & Fenger, 1979). Age related changes in the glycosylation of goblet cell mucins are also a conspicuous feature of intestinal development in neonates (King & Kelly, 1990; King, 1994).

BACTERIAL ADHESINS AND INTESTINAL RECEPTORS

Many indigenous and pathogenic bacteria specifically adhere to complex carbohydrates of small intestinal membrane and mucin glycoconjugates. The diversity of these carbohydrate receptors plays an important role in host range, tissue tropism and the triggering of host responses (Hultgren et al. 1993). This is particularly noticeable in neonates where both beneficial and deleterious alterations in the microbial balance can accompany ontogenic epithelial glycosylation changes (Kelly et al. 1992; Stewart et al. 1993). The relationship may be passive and involve bacterial colonization mediated through binding to expressed glycoconjugates or may involve chemical modification of inhospitable sites through the

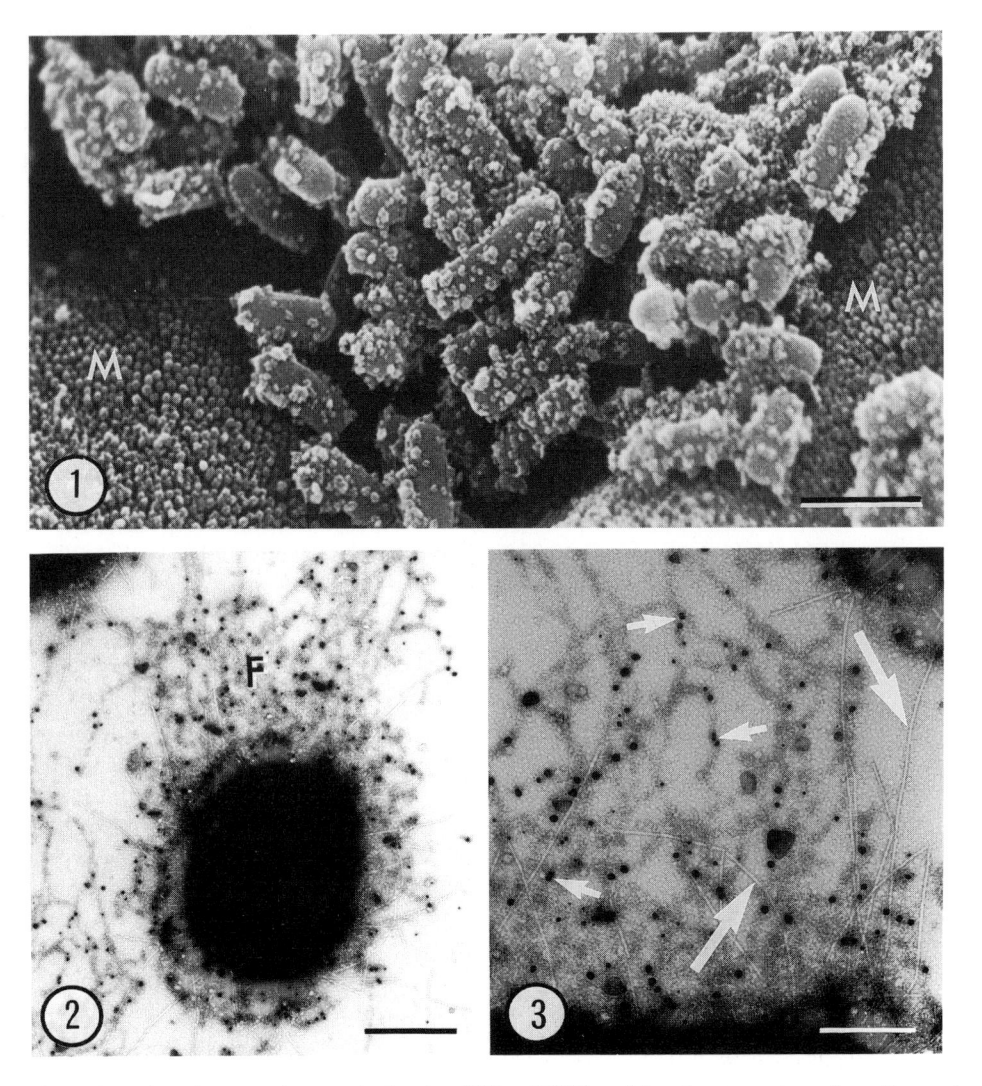

Fig. 1. Scanning electron micrograph showing K88ac ETEC on jejunal enterocyte surface of a newly weaned pig. K88 fimbriae are immunolabelled with an anti-K88a monoclonal antibody followed by immunogold cytochemistry and silver enhancement to produce a particulate label on the surfaces of the bacteria. M = tips of the intestinal microvilli, bar = 2 μm.

Fig. 2. Transmission electron micrograph of negatively stained whole mount of K88ab ETEC showing surface fimbriae (F). The presence of K88 fimbriae is revealed by immunogold/silver labelling, bar = 500 nm.

Fig. 3. Higher magnification of preparation in Fig. 2 showing type-1 fimbriae (large arrows) and aggregations of K88 fimbriae (small arrows and immunogold/silver labelling), bar = 250 nm.

actions of secreted exoglycosidase enzymes. Important bacterial surface factors involved in mediating adherence to intestinal surfaces include fimbriae (pili), outer membrane proteins and lipoteichoic acids.

FIMBRIAL ADHESINS

The pioneering studies of Duguid and his collaborators (Duguid & Old, 1980) demonstrated that many bacterial species, particularly those of the family Enterobacteriaceae, possess the ability to agglutinate red cells and that the bacterial agglutinins can be consigned to one of two groups designated mannose sensitive or mannose resistant depending on whether or not the haemagglutinating activity is mannose inhibitable. In many bacteria it has been determined that the mannose sensitive and mannose resistant agglutinins are located on surface fimbriae (Hultgren et al. 1993).

Type-1 fimbriae

Mannose sensitive fimbriae, usually termed type-1 fimbriae, are associated with several enteric species. Close serologic relationships exist among type-1 fimbriae of *Escherichia*, *Klebsiella* and *Shigella* spp., whereas fimbriae of *Salmonella* and *Citrobacter* species constitute a second group. In addition, the type-1 fimbriae of *Enterobacter*, *Edwardsiella*, *Hafnia*, *Serratia* and *Providencia* each comprise a distinct serologic group (Clegg & Gerlach, 1987). Type-1 fimbriae have the form of straight, tubular structures up to 2 μm in length and 7–8 nm in diameter. Structurally they are composed of polymers of FimA, a 17 kDa structural protein arranged in a right-handed helical array surrounding a hollow axillary core (Brinton, 1965). Three ancillary proteins, FimF, FimG and FimH, are also assembled as minor components of the filaments. FimH is thought to be primarily responsible for the lectin-like adhesive properties of the fimbriae, but association with FimG appears to be necessary for activity (reviewed by Sokurenko et al. 1992).

The adherence of *E. coli* O157:H7 strains to rabbit ileal brush borders is mediated by type-1 fimbriae binding to α-linked mannosyl residues present on surface glycoproteins (Durno et al. 1989). Rat intestinal mucin bears oligomannosyl receptors for the same *E. coli* type-1 fimbriae and it has been shown that these receptors are located on *N*-linked oligosaccharides of the 118 kDa link glycopeptide region of the mucin (Sajjan & Forstner, 1990a, b). *In vitro* investigations by Aslanzadeh & Paulissen (1990) indicated that type-1 fimbriae are involved in the adherence and pathogenicity of *Salmonella enteritidis* in mice. Conversely, Lockman & Curtiss (1992) have suggested that type-1 fimbriae are not virulence factors for colonization of the mouse small intestine by *Salmonella typhimurium*.

The role of type-1 fimbriae in enteric disease has been difficult to ascertain and remains the subject of some controversy (Holland, 1990; Krogfelt et al. 1991). These organelles are expressed by a large fraction of clinical isolates of the Enterobacteriaceae and it is perhaps this ubiquity that makes determination of their role in virulence more difficult (Sokurenko et al. 1992).

Fimbrial serotypes associated with diarrhoeal disease

Acute infectious enteritis caused by enterotoxigenic *E. coli* (ETEC) is a major cause of morbidity throughout the world. The disease is often mild and self limiting in healthy adults, but in the malnourished, in the aged and in young children and animals the symptoms may be severe (Gross, 1990). The virulence of ETEC is multifaceted and is attributable to both the production of fimbrial adhesins (Figs 1–3) and the elaboration of enterotoxins which bind to specific intestinal membrane receptors. The carbohydrate binding specificity of many ETEC strains responsible for the induction of diarrhoea in animals and man does not depend on the presence of mannose in the receptor structure.

Two important host specific adhesion fimbriae designated colonization factor antigens I and II (CFA/I and CFA/II) have been identified in some human ETEC strains (Evans & Evans, 1978). CFA/I fimbriae consist of only one type of adhesive subunit, of which only one at the tip is accessible to the receptor (Buhler et al. 1991). There is evidence that CFA/I fimbriae recognize the sialylated ganglioside GM2 (Faris et al. 1980). Three distinct antigens have been identified within the CFA/II group: coli surface antigen 1 (CS1), CS2 and CS3. CS1 and CS2 are 6–7 nm in diam., long rigid fimbriae similar to CFA/I, whereas CS3 fimbriae are thinner, flexible fimbrial structures, 2–3 nm in diam. (Knutton et al. 1985).

CFA/IV (PCF8775) expresses three coli surface antigens (CS4, CS5 and CS6) that appear on *E. coli* of specific serogroups and that seem to have individually distinct haemagglutination characteristics (Holland, 1990). There is little available information concerning the receptor characteristics of CS antigens although the CS3 antigen is believed to recognize oligosaccharide structures found on lactosaminoglycans, either on their backbone sequences or as subdeterminants of antigens related to histo-blood groups (Neeser et al. 1989).

K88 fimbrial adhesins associated with diarrhoea in piglets are plasmid encoded and are the major constituents of fine fimbriae of diam. 2–3 nm. K88 ETEC infections characteristically peak one week after birth and in the postweaning period although K88 receptors are present on the intestinal membranes of both neonates and adults (Harel et al. 1991; Mouricout, 1991). This suggests that susceptibility to K88 ETEC infection is complex and involves the interaction of luminal factors as well as receptor expression (Conway et al. 1990; Willemsen & de Graaf, 1992). K88 ETEC have been shown to adhere both *in vivo* and *in vitro* to enterocytes of K88 susceptible piglets but not to those of any K88 resistant pigs (Sellwood, 1979; Cox & Houvenaghel, 1993). Five porcine phenotypes can be distinguished with regard to brush border adhesiveness with K88ab, K88ac and K88ad serotypes (Bijlsma & Bouw, 1987). Differences in these five variants are likely to be due to the presence of highly specific and accessible brush border receptors (Bijlsma et al. 1982; Erickson et al. 1992). The precise molecular nature of these receptors has not yet been established. K88ab adhesins adhere to murine brush border membrane proteins with M_r of 57, 67 and 91 kDa (Laux et al. 1986). In the pig, glycoproteins of 67 kDa have been implicated in K88ab and ac adhesion while the K88ad receptor has been identified as a 40 kDa glycoprotein (Mouricout, 1991). Porcine brush border glycoproteins ranging in size from 40 to 70 kDa were recently shown to bind K88 fimbriae (Willemsen & de Graaf, 1992). Erickson et al. (1992) identified two porcine brush border glycoproteins (210 and 240 kDa) that bind K88ac fimbriae. The authors suggested that the presence of these glycoproteins on adhesive brush borders and absence on non-adhesive brush borders may be the basis for resistance and susceptibility of pigs to K88ac *E. coli* infections. Lower molecular weight proteins (37 and 48 kDa) were shown to bind biotinylated and ^{35}S-labelled K88ac adhesin but this binding was not specifically blocked by the presence of molar excess of unlabelled K88ac adhesin and the proteins were detected in both non-adhesive and adhesive phenotypes. Although the 37 and 48 kDa proteins do not fulfil the criteria for phenotypically important receptors, these proteins may be low affinity receptors that promote the initial interaction of K88ac *E. coli* with the phenotypically important 210 and 240 kDa brush border receptors (Erickson et al. 1992). Gibbons et al. (1975) concluded that an unsubstituted β-galactosyl residue was an important feature of the structural chemical requirements for binding the K88 adhesin. Competitive inhibition studies demonstrated that stachyose (Gal α1,6 Gal α 1,6 Glc α1, β2 Fru) and galactan (a polymer of D- and L-Gal) reduced binding of K88 to porcine brush border membranes by 33 and 50% respectively (Sellwood, 1984). Nilsson & Svensson (1983) found that K88 reacted with oligosaccharides containing galactosyl residues isolated from glycolipid fractions of the pig small intestine. Payne et al. (1993) recently suggested that β-linked galactose residues may form the molecular basis of both glycoprotein and glycolipid receptors for the K88 fimbrial adhesin in the porcine small intestine.

K99 ETEC strains which cause diarrhoea in newborn calves, lambs and pigs also express fine fimbriae of diam. 2–3 nm (Morris et al. 1980). Clinical studies indicate that with increasing host age the small intestine may develop resistance to adhesion mediated by K99 (Runnels et al. 1980). K99 *E. coli* express fimbrial adhesins which bind to sialylated glycoproteins and glycolipids (Mouricout, 1991). Two pig phenotypes have been identified, those expressing high levels of K99 receptors and those exhibiting low levels. Seignole *et al.* (1991) have determined that piglets susceptible to K99 adhesion express higher levels of sialylated glycolipids than those resistant to K99 attachment. Furthermore, the piglet intestinal mucosa contains a substantially higher content of acidic glycolipids than that of adult pigs (Teneberg *et al.* 1990). *N*-glycolylneuraminyl-lactosylceramide (NeuGc-GM3) has been identified in cattle as a major receptor for K99 (Lindahl *et al.* 1987; Teneberg *et al.* 1993). The membrane content of this ganglioside is maximal in newborn pigs and gradually decreases during development (Yuyama *et al.* 1993).

F41 is a chromosomally and/or plasmid encoded ETEC fimbrial adhesin which is often produced by strains which also elaborate K99 adhesins (Harel *et al.* 1991). There is clinical evidence to suggest that the polyfimbriated K99/F41 strains adhere in higher numbers to porcine intestinal villi than strains which only produce F41 or K99 fimbriae (Cox & Houvenhagel, 1993). F41 has a high affinity for terminal GalNac moieties (Lindahl & Wadstrom, 1986; Brooks *et al.* 1989). Cytochemical analysis of the membrane and mucin glycoconjugates in piglets reveals that these sugar moieties occur on immature glycoconjugates in newborn animals but are also present on histo-blood group A antigens in some weaned animals (King & Kelly, 1991).

ETEC strains carrying 987P fimbriae colonize the small intestine and cause diarrhoea in neonatal ($>$ 6 days post partum) pigs but not in older ($>$ 3 weeks) postweaning pigs in spite of the fact that intestinal receptors are present in both groups (Dean *et al.* 1989; Harel *et al.* 1991). A $<$ 17 kDa 987P binding component was present in the intestinal mucus of the older animals but not in neonates and Dean (1990) concluded that this receptor in the mucus may prevent *in vivo* adhesion of 987P$^+$ *E. coli* by competing with brush border 987P receptors.

Nagy *et al.* (1992) investigated the adhesion of ETEC strains which did not produce K88, K99, F41 or 987P fimbriae. Colonization of these so-called 4P$^-$ strains was characterized by adhesion to porcine intestinal villi overlying Peyer's patches, mediated by fimbriae. Small intestinal adhesion by these isolates was found to be dependent on receptors that develop progressively with age during the first 3 weeks after birth. The carbohydrate specificities of 4P$^-$ fimbriae are unknown.

The specific interaction of ETEC strains with the Peyer's patch epithelium may have particular relevance with regard to the delivery of bacterial antigens to the underlying gut associated lymphoid tissue. A wide range of pathogenic Gram-negative bacteria, including *E. coli* RDEC-1, *Vibrio cholerae*, *Salmonella typhimurium*, *Yersinia enterolytica* and *Campylobacter jejuni*, have been shown to adhere selectively or preferentially to M cells of experimental animals. A detailed description of the consequences of transepithelial transport of bacteria and bacterial antigens is beyond the scope of this review but has been recently extensively covered by Kraehenbuhl & Neutra (1992).

Salmonella typhimurium can cause severe diarrhoeal disease in pigs. Frequent reinoculation of the intestines by the faecal–oral route is considered an important mechanism for the persistent colonization of the porcine intestine, although evidence has also been presented that colonization is much enhanced in *Salmonella* strains which possess adhesive fimbriae (Isaacson & Kinsel, 1992). Thin aggregative fimbriae have been localized on the surfaces of several *Salmonella* species (Thorns *et al.* 1990; Collinson *et al.* 1993). The receptor specificities of the fimbrial adhesins have not been determined.

Mucosal adhesion by enteropathogenic *E. coli* (EPEC) involves two distinct stages: (1) initial attachment of bacteria promoted by plasmid encoded fimbrial adhesins and (2) effacement of microvilli, intimate EPEC attachment and cytoskeletal disruption (Knutton *et al.* 1987; Donnenberg & Kaper, 1992). *Vibrio cholerae* colonizes villus surfaces by adhering to peripheral components of the enterocyte glycocalyx but does not come into close contact with enterocyte microvillar membranes. However, on Peyer's patch M cells *Vibrio* forms tight membrane attachment sites that result in rapid phagocytosis and transport (Winner *et al.* 1991). Commensal Gram-positive segmented bacteria in the rodent ileum colonize mucosal surfaces by forming stable attachment sites on absorptive cells (Kraehenbuhl & Neutra, 1992).

Fimbrial adhesins associated with Helicobacter pylori

Helicobacter pylori is a microaerophilic bacterium found in the stomach and proximal small intestine of asymptomatic humans as well as patients with acid peptic disease and gastric adenocarcinoma (Falk *et al.* 1993; Ofek & Doyle, 1994). Electron microscope investigations have shown that *H. pylori* can adhere to apical membranes of epithelial cells in a manner reminiscent of the adherence pedestals of enteropathogenic *E. coli* (Ofek & Doyle, 1994). Surface fimbriae mediate the attachment of the bacteria to both gastric membranes and mucin secretions. *In vitro* adherence assays have shown the presence of *H. pylori* receptors on the surface mucous cells in the stomach and on the villus epithelium of the duodenum (Falk *et al.* 1993). *H. pylori* appears to be capable of expressing a number of adhesins with distinct specificities. Some *H. pylori* strains have been shown to cause haemagglutination of erythrocytes from only one species, whereas other strains cause haemagglutination of erythrocytes from several species including humans (Ofek & Doyle, 1994). It has been reported that different *H. pylori* strains recognize a heterogeneous class of sialoglycoconjugates (Lelwala-Guruge *et al.* 1992, 1993; Evans *et al.* 1993). There is an emerging body of data showing that *H. pylori* also bind to non-sialylated receptors. For example, using TLC immunostaining techniques, Saitoh *et al.* (1991) identified both sialylated and non-sialylated glycosphingolipid receptors in the human stomach. The relationship between histo-blood group antigen expression and colonization by *H. pylori* was investigated by Borén *et al.* (1993), who concluded that the Lewis (b) (Le(b))histo-blood group antigen mediates bacterial attachment of the organism to gastric mucosa. Bacteria did not bind to the Le(b) antigen substituted with terminal GalNAcα1,3 residue (histo-blood group A determinant), suggesting that *H. pylori* receptors might be reduced in individuals of blood group A phenotypes, as compared with blood group O individuals. *In vitro* adherence assays, employing blood group antigen specific monoclonal antibodies to block the attachment of *H. pylori* to histological sections of human stomach, showed that the attachment of *H. pylori* to fucosylated histo-blood group positive mucous cells was not dependent upon terminal non-substituted sialic acid residues (Falk *et al.* 1993).

OUTER MEMBRANE PROTEINS

Yersinia enterocolitica is an enteroinvasive bacterium that causes gastroenteritis. A 42 to 50 MDa virulence plasmid of *Y. enterocolitica* encodes a variety of proteins whose expression is regulated by both temperature and the availability of calcium (Mantle & Rombough, 1993). Several of these proteins are secreted, some of which associate with the outer membrane of the organism and are referred to (Michiels *et al.* 1990) as Yops (*Yersinia* outer membrane proteins). In addition the plasmid encodes for YadA, a fimbrial adhesin that is a true outer membrane protein (Kapperud *et al.* 1987). When expressed these plasmid encoded proteins alter the surface charge and hydrophobicity of the bacterium, promote auto-agglutination and mannose resistant haemagglutination and enhance adherence to epithelial membranes (Mantle & Rombough, 1993; Mantle & Husar, 1993). *Campylobacter jejuni* and *C. coli* are major causes of human enteritis. An outer membrane protein of approximately 27 kDa is believed to play a role in adherence of *C. jejuni* to intestinal cells (Kervella *et al.* 1993).

Lactobacilli represent a prevalent group of the indigenous flora of the intestine. For almost a century organisms of this genus have been alleged to augment the protective barrier of the intestine but their mode of action remains speculative. Pedersen & Tannock (1989) demonstrated that adherence of lactobacilli to epithelial surfaces of the digestive tract was a prerequisite for successful colonization and may also represent a key feature of their probiotic action. In a recent model Coconnier et al. (1992) proposed that an extracellular proteinaceous component, produced by adhering lactobacilli, provides a divalent bridge that links the bacteria to enterocyte surfaces. The extracellular bridging protein interacts with carbohydrate components of the bacterial cell and the intestinal epithelium. These proteins may also be derived from the host epithelium or introduced as lectin constituents in the diet (Pusztai et al. 1990; Tannock, 1992). Although the precise nature of the lectin receptors remains to be elucidated, there is evidence that specificity of different lactobacillus strains may reflect species and age related variations in the glycoconjugate complexion of intestinal surfaces (Tannock, 1992).

LIPOTEICHOIC ACID

The surfaces of Gram-positive bacteria are typically composed of two structurally related, negatively charged components. One is covalently linked to the peptidoglycan and is usually referred to as the teichoic acid or the secondary (or accessory) cell wall polymer (Poxton & Arbuthnott, 1990). The other is anchored in the cytoplasmic membrane and protrudes through the wall. This is the membrane or lipoteichoic acid (LTA) (Poxton & Arbuthnott, 1990). The chemical composition and possible modes of action of LTA were comprehensively reviewed by Christensen et al. (1985). LTA is a linear polyglycerophosphate containing a glycerophosphoryl–diglucosyl–acylglycerol moiety at its non-polar end. The amphipathic nature of LTA allows the molecule simultaneously to interact with charged (hydrophilic) substances and non-charged (hydrophobic) substances. LTA can therefore form molecular bridges. In *Streptococcus pyogenes* LTA is anchored to one or more proteins on the surface of the bacterial cells and interacts through its lipid moiety with the fatty acid binding sites of fibronectin molecules deposited on mucosal epithelial cells (Beachey & Courtney, 1989). Courtney et al. (1992) have recently demonstrated that two streptococcal adhesins, LTA and so-called M proteins, are involved in the adherence of group A streptococci to pharyngeal and buccal epithelial cells. Staphylococcal binding to mesothelial cells may involve the combined effects of LTA, teichoic acid and protein A (Haagen et al. 1990).

MUCIN AND BACTERIAL ADHESION

Many commensal and pathogenic bacteria have been found to colonize the intestinal mucus (Neutra & Forstner, 1987; Wanke et al. 1990). Whether such interactions favour the bacteria or the host continues to be the subject of much speculation. For many enteropathogens the initial contact by bacteria after entry into the small intestine from the stomach is with the unstirred mucus layer of the intestinal lumen (Wanke et al. 1990). In these situations the mucus may function as a permeability barrier or retention zone for bacteria, thereby restricting their free access to the underlying mucosa. Conversely, the ability of bacteria to interact with intestinal mucins can be an important step in facilitating colonization of the intestinal tract. If they are able to bind strongly to components of the intestinal mucus layer, their clearance by the motile and abrasive forces of digestion may be delayed and colonization of the intestinal tract may be favoured. Moreover, if the rate of bacterial growth in and penetration of the mucus barrier exceeds the rate at which this layer naturally turns over and is eliminated from the gut, then clearance will be further delayed, again favouring colonization of the gut (Pærregaard et al. 1991; Mantle & Husar, 1993; Mantle & Rombough, 1993).

One of the most extensively studied bacterial–mucus interactions concerns *Yersinia enterocolitica*, an enteroinvasive bacterium that causes gastroenteritis in man. Invading *Y. enterocolitica* are capable of binding to purified intestinal mucins from both humans and rabbits. The binding appears to be plasmid mediated and involves mucin Gal and GalNAc residues (Mantle & Husar, 1993). Data from the same laboratories have shown that *Y. enterocolitica* is capable of degrading small intestinal and colonic mucins. The major enzyme involved may be a plasmid encoded or plasmid regulated proteinase the activity of which may be important in the establishment of infection by 'dissolving' the normal protective mucus barrier and allowing the organism to gain access to the underlying mucosa (Mantle & Rombough, 1993).

Specific receptors for several ETEC fimbrial adhesins have been identified in intestinal mucus. Pig small intestinal mucus has receptors to K88 and 987P fimbriated *E. coli* (Dean et al. 1989; Conway et al. 1990; Metcalfe et al. 1991). Calf small intestinal mucus contains receptors for K99 and F41 fimbriae (Laux et al. 1986; Mouricout, 1991), while rabbit ileal mucus has been shown to contain receptors for specific *E. coli* RDEC-1 AF/R1 fimbriae (Drumm et al. 1988). The role of these receptors in infectious processes is unclear. As discussed, receptors for K88 ETEC fimbrial adhesins are present on the microvillar membranes of susceptible pig phenotypes irrespective of their age (Willemsen & de Graaf, 1992). Mucus receptors, on the other hand, are absent in adult animals (Willemsen & de Graaf, 1992). However, it has been demonstrated that K88 receptor levels transiently increase in the late preweaning period. Conway et al. (1990) suggested that these 5-week pigs were less susceptible to K88 infection because of the high concentration of K88 specific receptor present in their ileal mucus which restricted access of the bacteria to the underlying epithelial cell membranes.

Further research is needed to identify the chemical composition and true origins of mucin glycoproteins extracted from intestinal mucus samples. Although many of these proteins may originate from goblet cells it is equally possible that many are microvillar membrane glycoproteins that have been released into the intestinal lumen. Recent *in vitro* adherence studies employing K88 fimbriae as cytochemical probes on histological sections failed to show receptors in goblet cell mucins (T. P. King & D. Kelly, unpublished observations). It is possible that at least some of the mucus receptor glycoconjugates seen in high numbers in preweaned pigs (Conway et al. 1990) are derived from maternal milk or are released from sloughed epithelial cells. Blomberg & Conway (1989) observed that K88 ETEC adhered more poorly to ileal mucus from piglets in which the gastric tissue was densely colonized by lactobacilli compared with mucus from piglets with sparsely colonized gastric tissue. Recent *in vitro* investigations have shown that *Lactobacillus fermentum* releases into the culture supernatant one or more proteinaceous components which affect mucus so that adhesion of cells bearing K88ab and K88ac is reduced (Blomberg et al. 1993).

As discussed above, there is conflicting evidence concerning the role of intestinal mucus in the aetiology of enteric infections. Whatever the effect of mucus on the gut ecosystem, it is likely that dietary factors which either limit the production of mucin, alter its composition, or enhance its degradation have the potential to influence colonization of the intestine by commensal and pathogenic bacteria (Neutra & Forstner, 1987; Wanke et al. 1990).

BACTERIAL GLYCOSIDASES

After their synthesis intestinal glycoconjugates may also be altered by the action of the endo- and exo-glycosidases of the bacteria themselves. Although modification of oligosaccharides of the epithelial surface itself has not been reported, there is considerable evidence that bacterial enzymes may hydrolyse the same glycosidic configurations in mucins. Following occlusion of ileal reservoirs in experimental dogs, bacterial counts increased 3-fold and glycosidase activities 5-fold; chronic inflammation of the mucosae and the abolition of blood group epitopes were attributed to degradation of protective mucin by the agency of bacterial glycosidases (Ruseler-van Embden *et al.* 1992). *Shigella flexneri* was shown to possess blood group B degrading activity (Prizont, 1982), while strains of *Ruminococcus* were able to carry out extensive hydrolysis of A/H or B substances in hog gastric mucin and some bifidobacteria were equally active but lacked A degrading activity (Hoskins *et al.* 1985). Significantly, much of the activity was extracellular, implying that attack of glycosyl moieties located on membranes would be feasible with consequent creation or abolition of specific bacterial adhesion sites.

NUTRITIONAL INFLUENCES ON THE INTERACTIONS BETWEEN BACTERIA AND THE INTESTINAL MUCOSA

Components of the diet and the gut microflora are in intimate contact within the intestinal tract both in the lumen and at the absorptive surfaces. The effect of diet, however, may be direct or indirect. Dietary composition may influence the carbohydrate structures of the mucosal and mucin glycoconjugates with marked consequences for the adherent microflora. In the neonatal tract, exposure to components of maternal colostrum and milk may promote the selective growth of particular bacterial species. There is also good evidence for claims that the composition of adult diets may affect the bacterial ecology of the gut. Such effects offer potential routes whereby the gastrointestinal microflora may be manipulated for the benefit of the host.

DIETARY MODULATION OF BACTERIAL RECEPTORS IN THE NEONATE

As described in the previous sections, there is an increasing appreciation of the detailed carbohydrate structures constituting the receptor sites for adherent intestinal bacteria. The complex oligosaccharides involved are associated with proteins and/or lipids located in the mucosal membranes or with the secreted gastrointestinal glycoconjugates including mucins. It is conceivable that particular glycosyl structures appearing transiently represent 'windows of opportunity' for infection by enteropathogens whose adhesins exhibit the appropriate specificity, and such a rationale may explain, in part, why diarrhoea attributable to some species is particularly prevalent at certain stages of development.

During early development of the neonate, temporal expression of the glycosyl structures of the mucosal surfaces of the gut appears to be under endogenous control but may also be influenced markedly by diet (Kelly *et al.* 1992). Turck *et al.* (1993) found that the fucose, glucosamine and sulphate contents of mucin were increased in maternally fed piglets compared to their artificially fed counterparts. The authors suggested that breast fed animals may have a more effective mucus barrier against infection. Manipulation of the suckling regime has been shown (Figs 4 and 5) to alter the ontogenic expression of sialylated and fucosylated glycoconjugates (Kelly & King, 1991; Kelly *et al.* 1993). It is very

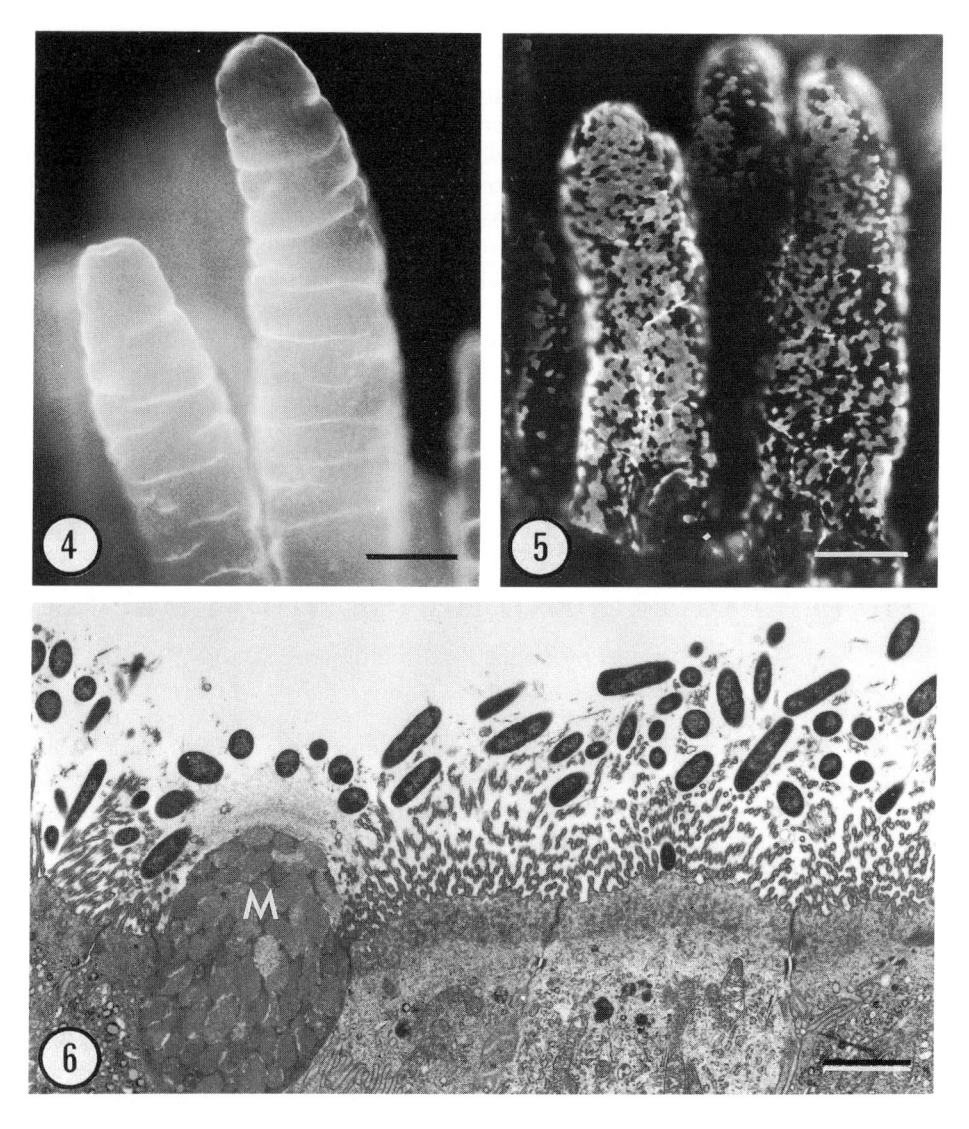

Figs 4 & 5. FITC-conjugated lectin from *Sambucus nigra* (SNA) used to label α2,6 NeuAc-Gal/GalNAc moieties on the villus surfaces of colostrum deprived (4) and colostrum fed (5) pigs at 1 week post partum. Postnatal reduction in epithelial sialylation, seen here as a mosaic of SNA+ and SNA− cells, is more extensive in the colostrum fed animals, bars = 50 μm.

Fig. 6. Transmission electron micrograph showing coliform bacterial overgrowth associated with extensively disrupted intestinal microvilli in a rat fed on a diet containing kidney bean lectin, M = goblet cell mucin, bar = 2 μm.

likely that the highly sialylated intestinal epithelial surface in the neonatal intestine during the first days of life influences the establishment of the enteric flora. Attachment sites for several organisms may be masked by sialic acids whereas other pathogenic and non-pathogenic bacterial strains secrete sialidase enzymes which enable them to overcome host defensive mechanisms and create novel binding sites for colonization (Corfield, 1992). As already discussed, other enteric bacteria such as K99 ETEC opportunistically bind to sialylated receptors and cause diarrhoea in neonatal but not adult pigs. The modulatory role of diet on temporal glycosylation changes in the neonatal intestine and its impact on bacterial colonization is a new and important subject for investigation.

Probiosis

Dietary inclusion of antimicrobials, particularly antibiotics, is established as an effective means of growth promotion in farm animals. Concern regarding the predicted emergence of resistant strains of pathogenic species has, however, provoked an increasing public rejection of this farming practice and has stimulated a demand for agents controlling growth and disease which are more 'natural' in concept. The rapid implantation of a commensal flora, coupled with the observation that germ free or antibiotic treated animals generally exhibit reduced resistance to invasion by enteropathogens, has given rise to the concept of protection by probiotics, i.e. 'a live microbial feed supplement which beneficially affects the host animal by improving its intestinal microbial balance' (Fuller, 1992). The probiotic principle was originally propounded by Metchnikoff (1907) when he claimed that the putrefactive actions of lower gut species could be alleviated by consumption of fermented milk. From the latter he was able to isolate a bacillus, probably *Lactobacillus delbrueckii* ssp. *bulgaricus* which, with *Streptococcus salivarius* ssp. *thermophilus*, is used in yoghurt production. More recently, the survival of these species at stomach pH has been questioned (Conway *et al.* 1987) and efforts to devise effective probiotic preparations have focused on the use of species or strains which are known to be acid resistant and active gut colonizers (Kleeman & Klaenhammer, 1982; Mayra-Makinen *et al.* 1983). Probiotic preparations for human use have generally taken the form of fermented dairy products (yogurt, bifidus milk, kefir, etc.) and claims have been made for their efficacy in the treatment of various disorders including intestinal infection, hepatic encephalopathy, hypercholesterolaemia and carcinogenesis. The species most frequently associated with positive responses is *L. acidophilus*, although a strain of *Lactobacillus* termed GG and a non-pathogenic yeast, *Saccharomyces boulardii*, have both been effective in controlling recurring diarrhoea caused by *Clostridium difficile* toxin after antibiotic therapy (Gorbach, 1990; Surawicz *et al.* 1989). The consumption of bifidobacteria in a variety of fermented milk products has recently been popularized, particularly in Japan, where the fermentation industries are well developed. Growth of bifids in the gut may be stimulated by direct inoculation of viable cells, by the incorporation of bifidus factors such as the synthetic disaccharide lactulose or, for maximum effect, a combination of these strategies. When lactulose (3 g/d) was administered to healthy adults, 6-fold increases in excreted bifidobacteria were observed in 1–2 weeks and this correlated with significant decreases in ammonia and putrefactive products and with a reduction in faecal pH (Terada *et al.* 1992).

A rather wider range of species has been evaluated for probiotic efficacy in animals and, indeed, many of the preparations marketed by commercial concerns contain a number of bacterial species together with other factors held to be beneficial. A comprehensive discussion of animal probiotics is outwith the scope of this review but the subject has been addressed for several agriculturally important species by Fuller (1992).

The mechanisms underlying probiotic action remain somewhat obscure but suggested actions include competitive exclusion of enteropathogens, production of bacterocins etc., the production of hydrogen peroxide, liberation of free bile acids and suppression of ammonia and toxic amines.

Milk oligosaccharides as probiotics

The beneficial effects of breast feeding in averting gastrointestinal infections are undisputed (Kovar *et al.* 1984) and have been attributed to the separate or synergic antibacterial action of factors which include secretory immunoglobulins, especially sIgA (immune exclusion), lactoferrin (iron sequestration), lysozyme (cell wall lysis), lymphocytes and phagocytes. This list must be extended to include the oligosaccharides of human breast milk which may fulfil a directly protective function in addition to their role as bifidus factors (see below). The non-immunoglobulin fraction of milk was shown to be an effective inhibitor of the adhesion of *Vibrio cholerae* biotypes to human or chick erythrocytes (Holmgren *et al.* 1983); inhibitory activity was attributed to both free oligosaccharides and glycoproteins. Haemagglutination by CFA/I- and CFA/II-fimbriated *E. coli* was inhibited by glycolipids in human milk (Holmgren *et al.* 1983). The same group have also shown that the trace amounts of free GM1 ganglioside present in human milk are sufficient to suppress the *in vivo* enterotoxic effects of cholera toxin and the heat labile *E. coli* toxin (Kolstø Otnæss *et al.* 1983), both of which are known to recognize GM1 receptors in the intestinal membranes (Schengrund & Ringler, 1989). Perhaps the most interesting observation to date is the inhibition of adherence of human strains of *E. coli* (CFA/I- and CFA/II-fimbriated) to guineapig small intestine by the non-immunoglobulin fraction from human milk; significantly, no inhibition of type-1 fimbriated *E. coli* was noted (Ashkenazi & Mirelman, 1987).

Breast milk and bifidobacteria

In the human infant, breast milk promotes the establishment of the genus *Bifidobacterium*, contrasting with the response to formula feeding where bifidobacteria were either absent or part of a more complex anaerobic flora including *Bacteroides* and *Plectridium* (Moreau *et al.* 1986; Balmer & Wharton, 1989). Notwithstanding the disagreement which exists in the literature regarding the relative effects of maternal milk and milk replacers both in the human (Simhon *et al.* 1982; Benno *et al.* 1984) and in the pig (Ducluzeau, 1983), bifidobacteria are a common constituent of the faecal microflora of the neonate one week after birth. Differences in the species composition of the bifidobacterial population in the human infant intestine have also been widely reported (comprehensively reviewed by Bezkorovainy, 1989), although it must be acknowledged that some of the discrepancy may be attributable to the shortcomings of the identification methods used.

Various arguments have been advanced to account for the prevalence of bifidobacteria in the large intestine of the breast fed infant. It has been suggested that bifidobacteria are able to sequester iron, thus depriving enteropathogens of an essential virulence factor (Bezkorovainy *et al.* 1986) or that the lower buffering capacity of human milk compared to that of formula preparations based on cows' milk permits a lower gut pH with consequent inhibitory action on enteropathogens (Willis *et al.* 1973); this hypothesis has been disputed (Rose, 1984). An alternative concept, based on the existence of growth factors present in human milk (György *et al.* 1954 *a, b*), has been more reliably validated.

In a more recent study (Beerens *et al.* 1980), specific *in vitro* growth factors for *B. bifidum* were found in human milk but were absent from the milk of other species including cow, sheep, horse and pig, although these latter did promote the growth of *B. infantis* and *B. longum*. Human blood group substances (Springer *et al.* 1954) and pig gastric mucin (György *et al.* 1954 *a*) were also found to be very active. Carbohydrate-rich glycoproteins from both human colostrum (Nichols *et al.* 1975) and milk are able to promote the growth of *B. bifidum* var. *pennsylvanicus* (Bezkorovainy & Nichols, 1976) and *B. infantis* (Azuma *et al.* 1984); fragments with similar composition may be released from human casein by tryptic and chymotryptic digestion (Bezkorovainy *et al.* 1979) or by β-elimination (Bezkorovainy & Topouzian, 1981). Digestion products from bovine casein also demonstrate some activity (Kehagias *et al.* 1977).

Much effort has been invested in attempting to determine the nature and mode of action of the human milk factors. A common feature of the growth promoting oligosaccharides present in human milk is the presence of an internal β-N-acetyl-D-glucosaminide residue and, indeed, it has been shown that methyl β-N-acetylglucosaminide (but not the α-methyl anomer) has some activity (Rose et al. 1984). In accord with this indicated anomeric specificity, N-acetyl-D-glucosamine is some 10-fold less active but 4-O-β-D-galactosyl-N-acetylglucosamine, isolated from pig gastric mucin (Zilliken et al. 1955), is 3–5 times more active, suggesting that more complex structural factors are important for activity.

DIETARY MODULATION OF BACTERIAL RECEPTORS IN THE ADULT

Generally, the most disruptive dietary change which the young animal experiences is that of weaning. This is a gradual process under natural conditions but it is now common practice to wean abruptly to comply with the demands of modern pig production (Fowler, 1985) and rapid adjustment to a radically different diet is required by both animal and microflora. The 'stress' incurred, including the withdrawal of maternal antibodies and other protective factors in sow's milk, may render the host vulnerable to infection by opportunistic and other pathogens. In addition, intestinal glycosylation changes associated with weaning may have an important influence on bacterial colonization. Diminished $\alpha 2,6$ sialylation and enhanced expression of histo-blood group O and A antigens on both mucin and membrane glycoconjugates have been observed in the postweaned porcine small intestine (T. P. King, R. Begbie & D. Kelly, unpublished observations). Similar post-weaning increases in fucosylated glycoconjugates have been observed in the rat intestine (Taatjes & Roth, 1990). Biol et al. (1992) concluded that an increase in the intestinal fucosyltransferase activity that follows weaning in rats is largely induced by the weaning process.

Intestinal glycosylation processes are sensitive to nutritional factors (Biol et al. 1992; Kelly et al. 1992). The synthesis of the internal core of N-linked glycans is regulated by diet-induced variations at the phosphoryldolichol level, whereas modulation of the biosynthesis of the external regions of N-glycans or the biosynthesis of O-linked moieties is controlled by variations in systems transferring fucose, galactose, sialic acid and hexosamines (Biol et al. 1992). The influence of dietary protein levels on intestinal glycosyltransferase activities in the rat intestine was investigated by Martin et al. (1989). Fucosyltransferase was increased when animals were given a high protein diet but was unaffected by a low protein diet. N-acetylgalactosaminyltransferase and sialyltransferase were not modified by the high protein diet, but were decreased by the low protein diet. There is some evidence that protein quality influences intestinal glycosylation (Biol et al. 1990), but this area requires further investigation. Decreased intestinal fucosyltransferase activity has been observed in rats fed on diets supplemented with saturated or unsaturated fats (Biol et al. 1981). Dietary fats do not modify the kinetic parameters or the isoenzymic pattern of the intestinal soluble fucosyltransferase but fatty acids derived from diets may exert a direct effect on its activity (Biol et al. 1992). It has been suggested that membrane fluidity and Golgi function may be altered by dietary fats (Biol et al. 1992). The ingestion of carbohydrate-rich diets was found to have no influence on intestinal galactosyltransferase, N-acetylgalactosyltransferase or sialyltransferase activities, but it slightly depressed fucosyltransferase activity in the intestine in rats (Biol et al. 1981).

The capacity of micronutrients such as retinoic acid, the active metabolite of vitamin A, to modulate growth and differentiation of normal and transformed cells is well established (Glass et al. 1991; Labarriere et al. 1993). In several cell types, changes in glycosyltransferase activities have been reported after retinoic acid treatment (Cummings & Mattox, 1988; Amos *et al.* 1990). Hypo- and hyper-vitaminosis A can significantly influence sugar-nucleotide availability and the initial stages of *N*-linked glycosylation of proteins (Biol *et al.* 1992). The influence of dietary vitamin levels on core glycosylation events and on the cellular systems transferring fucose, galactose, sialic acid, *N*-acetylglucosamine and *N*-acetylgalactosamine requires further study. There is evidence to suggest that there may be a relationship between dietary vitamin A and intestinal colonization by bacteria. In children deficient in vitamin A, increased rates of bacterial infection in the intestine have been observed (Gabriel *et al.* 1990). Vitamin A deficiency in rats is associated with the increased ability of *Salmonella typhimurium*, expressing type-1 mannose sensitive fimbriae, to adhere to proximal small intestinal enterocytes (Gabriel *et al.* 1990). Clearly, the role of minor dietary constituents as regulators of intestinal glycosylation and modulators of enteric health also deserves greater attention.

Dietary lectin toxicity and bacterial overgrowth

The relatively high protein and fibre contents of legume seeds make them attractive as feed and food components. The antinutritional effects associated with some seed components, however, have long been recognized by nutritionists and feed compounders, and it is customary to avoid problem seeds or to devise strategies such as heat treatment to inactivate the factors. Principal among these factors are lectins, which may be defined as 'proteins of non-immunoglobulin nature capable of specific recognition of, and reversible binding to, carbohydrate moieties of complex carbohydrates without altering the covalent structure of any of the recognised ligands' (Kocourek & Horejsi, 1983). Lectins are ubiquitous components of a wide range of commonly ingested foods (Nachbar & Oppenheim, 1980; Gibbons & Dankers, 1981) and may, in some circumstances, survive moderate processing treatments. Discussion of the wide range of physiological consequences of dietary lectin ingestion is outwith the scope of this review (for a recent review, see Pusztai, 1993); however, a particular feature, the induction of bacterial overgrowth, is relevant here and provides a useful model for discussion. Dramatic increases in the numbers of coliforms and other species have been consistently observed in the small intestines of conventional rats (Fig. 6; Wilson *et al.* 1980; Banwell *et al.* 1983, 1985), pigs (King *et al.* 1983), chicks (Untawale *et al.* 1978) and quail (Jayne-Williams & Hewitt, 1972) consuming diets containing kidney bean lectins (phytohaemagglutinins). The implication of bacterial overgrowth as a contributory factor in lectin mediated toxicity follows from the observation that germ free (Jayne-Williams & Hewitt, 1972; Rattray *et al.* 1974) or antibiotic treated animals (Banwell *et al.* 1983) were much less affected than their conventional counterparts. The nature of the involvement of phytohaemagglutinins in the dose dependent induction of a coliform-rich overgrowth which appears to be largely adherent to the mucosal surface (Banwell *et al.* 1985; Pusztai *et al.* 1993) has been the subject of much discussion (reviewed in Pusztai *et al.* 1990). The mucosal surface of the small intestine is normally sparsely populated with glycoconjugates carrying α-D-mannoside terminal residues (Pusztai *et al.* 1993; R. Begbie, unpublished) which are essential components of the adhesion sites for common type-1 fimbriated *E. coli* and other coliforms (Ofek & Sharon, 1990) and, to accommodate increased numbers of the latter, it is necessary to postulate the formation of novel sites. It has been suggested that these may be found in the form of the high mannose carbohydrate moieties of surface bound phytohaemagglutinins (King *et al.* 1983), but it has been claimed in a recent study in rats that the accelerated enterocyte turnover, stimulated by phytohaemagglutinin feeding, results in incomplete glycosylation of the brush border membrane glycoconjugates and consequent enrichment of polymannosylated glycan structures (Pusztai *et al.* 1993). Although proliferation of enterotoxigenic species has not been reported in this study it may be speculated that augmentation of mannose-rich receptor sites would predispose the intestine to colonization by pathogenic members of the Enterobacteriaceae such as *Salmonella*, *Shigella* and *Klebsiella* which also carry type-1 fimbriae (Duguid & Old, 1980; Ofek & Sharon, 1990). In this respect, a superficial analogy with the phytohaemagglutinin model may be found in reports of an increased *Salmonella typhimurium* population of the caeca of chickens infected with the coccidian parasite *Eimeria tenella* (Baba *et al.* 1993); increases in staining with the mannose specific lectins, concanavalin A and *Lens culinaris* agglutinin, suggested that the density of adhesion sites on the mucosal surface had somehow been increased by the coccidial infection.

Malnutrition and bacterial translocation

Although the gut can tolerate and indeed benefit from a threshold level of indigenous adhering bacteria (Fuller, 1992) it would be wrong to assume that the indigenous flora is always benign. Under conditions of nutritional or other physiological stress the indigenous population may exceed a safe threshold and constitute a pathological burden (King *et al.* 1983; Spitz *et al.* 1994).

Malnutrition and long term parenteral nutrition have been shown to alter intestinal morphology, impair immune function and induce bacterial overgrowth. In the extreme, these changes have been implicated in the pathophysiology of translocation of bacteria leading to sepsis (Langkamp-Henken *et al.* 1992; Deitch, 1994; Silk & Grimble, 1994; Spitz *et al.* 1994). Surface proteins (invasins, internalins) implicated in the process of cellular invasion and bacterial translocation are expressed by several enteric bacteria including *Shigella*, *Yersinia* and *Salmonella* (Wick *et al.* 1991). Furthermore, it has recently been demonstrated that ETEC which normally colonize the mucosal surface possess specialized features which facilitate cellular invasion (Elsinghorst & Kopecko, 1992). It has been suggested that under conditions of environmental stress such invasive gene products are expressed in a wide variety of normally non-invasive organisms. An important research goal of the future is to elucidate the mechanisms by which both adherent and non-adherent bacteria become invasive (Boedeker, 1994). The invasion processes are complex and involve the interaction of bacteria with extracellular matrix components and cytoskeletal elements and plagiarization of normal signal transduction mechanisms of the intestinal epithelium (Wick *et al.* 1991). Dietary management is important for the prevention of colonization of the gut by enteropathogens but also has a protective role in the prevention of transepithelial translocation of bacteria (Deitch, 1994).

Chemical probiosis

Many of the more successful enteric organisms have developed strategies to resist displacement from the gut milieu by the digestive flow and, *inter alia*, the development of anchoring adhesive fimbriae (pili) represents a common feature of both commensal and pathogenic gut microflora (Ofek & Sharon, 1990). In seeking to elucidate the carbohydrate specificity of particular adhesins authors have generally employed *in vitro* competition assays with authenticated oligosaccharide structures (Firon *et al.* 1983, 1984; Leffler & Svanborg-Edén, 1986; Lindahl *et al.* 1987; Van Driessche *et al.* 1989) or with lectins of known specificity (Sellwood, 1980; Neeser *et al.* 1989; Willemsen & de Graaf, 1992), and increasing appreciation of specific oligosaccharide structures targeted by enteropathogens has given rise to the notion of selective therapeutic or prophylactic exclusion of enteropathogens from the intestinal mucosae. Two related strategies have been encompassed in the concept of chemical probiosis (by analogy with microbial probiosis) promulgated by Pusztai *et al.* 1990: each is based on competition by molecular species.

In the first, a soluble, non-toxic lectin may be administered where its carbohydrate specificity is selected to compete for the glycosyl binding site of the fimbrial adhesin. This approach is exemplified by the use of GNA, the snowdrop lectin, which is highly specific for 1,3α-linked D-mannose units, to block the proliferation of type-1 fimbriated *E. coli* in rat small intestine induced by ingestion of phytohaemagglutinins (Pusztai et al. 1993). A daily dose of 40 mg GNA effectively eliminated the coliform overgrowth in rat small intestine. It is conceivable that other mannose specific lectins present in foodstuffs which are more regularly consumed (by humans) may also have beneficial effects and, in this context, lectins from garlic (Kaku et al. 1992), shallots (Mo et al. 1993) and BanLec-1 from bananas (Koshte et al. 1990) may hold particular interest. It may be appropriate to sound a cautionary note against the universal application of the concept in view of the reported enhancement of adherence of non-fimbriated *Salmonella typhimurium* to rat small intestine by concanavalin A (Abud et al. 1989); here the authors suggest that colonization may be promoted by the formation of lectin bridges between the bacterial and mucosal cell surfaces.

The alternative approach of masking the fimbrial adhesins with competing oligosaccharide structures is more widely documented. Early work on the specificity of the fimbrial adhesins served to establish that *in vitro* attachment of bacteria to cell surfaces could be effectively inhibited by mono- and oligo-saccharides of appropriate structure (reviewed in Duguid & Old, 1980; Sharon et al. 1981). Successful control of experimental (non-gut) infection by the *in vivo* administration of simple sugars or glycosides (Aronson et al. 1979; Andrade, 1980; Fader & Davis, 1980) has been reported; in these instances type-1, mannose sensitive fimbriae were involved. A similar strategy has been applied prophylactically with broiler chicks where administration of D-mannose (2·5% w/v) in the drinking water effectively reduced colonization by a *Salmonella typhimurium* challenging strain from 84% in controls to 30% in treated chicks and also depressed faecal counts by > 99% (Oyofo et al. 1988). High mannose glycopeptides from ovalbumin were potent inhibitors of adhesion of type-1 fimbriated *E. coli* to guineapig erythrocytes and human buccal cells (Neeser et al. 1986), while the activity of larger oligomannoside glycopeptides released from legume glycoproteins was increased after trimming by α-mannosidase. Such readily available structures may prove useful in inhibiting *in vivo* binding by mannose specific adhesins associated with type-1 fimbriated enteropathogens. It is a prerequisite that these competing analogues, and the more complex carbohydrate structures recognized by ETEC strains, resist gut degradation. Perhaps as a consequence of the importance of these structures in both the glycocalyx and intestinal mucin, the gut lumen appears not to be endowed with the necessary endogenous hydrolases and the prospects for dietary intervention are encouraging. Extrapolating from their observation that ileal mucus components protect against epithelial colonization by *E. coli* K88+, Conway et al. (1990) have suggested that it might be possible to extend the protection to an earlier age by incorporation of 'synthetic receptors' in the diet. In this respect, however, perhaps the most spectacular success claimed so far is the use of glycans derived by proteolysis of blood plasma glycoproteins to protect against lethal doses of K99 ETEC in colostrum deprived newborn calves (Mouricout et al. 1990). In this study, 3×250 mg doses of the glycan preparation administered orally at the first signs of diarrhoea were sufficient to cure an experimental infection induced by 10^{10} bacterial cells.

Several of the proteinaceous toxins associated with enteric pathogens such as *Vibrio cholerae* and *E. coli* also bind to intestinal carbohydrate sites and competition strategies have been envisaged as a means of combating their effects. The ganglioside, GM1, bound to medical charcoal, was shown to bind luminal *V. cholerae* fully and also to reduce purging in the early stages of the disease. Since the effect was transient, however, it was considered that oral GM1 might be useful mainly for prophylaxis in high risk groups (Stoll *et al.* 1980).

SUMMARY

In recent years there has been a major expansion in research on the molecular interactions between bacteria and the intestinal epithelium. Available data suggest that bacterial adhesins, in common with many well characterized plant lectins, recognize a diverse array of intestinal glycoconjugates. Serendipity and chemical inhibition studies have combined with good effect to identify receptor structures recognized by several fimbrial lectins. The chemistry of fimbrial lectins is in its infancy and the possibility should not be overlooked that simple chemical modifications may produce receptor molecules with even greater affinity for pathogens than the so-called native receptors (Mouricout, 1991). Molecular genetics and X-ray crystallography are proving to be invaluable in the identification of novel fimbrial and non-fimbrial adhesins, their specific interactions with host cell receptors and their role in pathogenesis. This information is essential for the design of effective disease prevention strategies. Nutritional intervention leading to altered receptor expression and/or the direct application of natural and artificial oligosaccharide structures as chemical probiotics may have considerable potential for the prophylaxis and therapy of intestinal infections.

REFERENCES

Abud, R. L., Lindquist, B. L., Ernst, R. K., Merick, J. M., Lebenthal, E. & Lee, P. C. (1989), Concanavalin A promotes adherence of *Salmonella typhimurium* to small intestinal mucosa of rats. *Proceedings of the Society for Experimental Biology and Medicine* **192**, 81–86.

Amos, B., Deutsch, V. & Lotan, R. (1990). Modulation by all-trans retinoic acid of glycoprotein glycosylation in murine melanoma cells: enhancement of fucosyl- and galactosyltransferase activities. *Cancer Biochemistry Biophysics* **11**, 31–43.

Andrade, J. R. C. (1980). [Role of fimbrial adhesiveness in experimental guinea-pig keratoconjunctivitis by *Shigella flexnerii*.] *Revista de Microbiología* **11**, 117–125.

Aronson, M., Medalia, O., Schori, L., Mirelman, D., Sharon, N. & Ofek, I. (1979). Prevention of colonization of the urinary tract of mice with *Escherichia coli* by blocking of bacterial adherence with methyl α-D-mannopyranoside. *Journal of Infectious Diseases* **139**, 329–332.

Ashkenazi, S. & Mirelman, D. (1987). Nonimmunoglobulin fraction of human milk inhibits the adherence of certain enterotoxigenic *Escherichia coli* strains to guinea pig intestinal tract. *Pediatric Research* **22**, 130–134.

Aslanzadeh, J. & Paulissen, L. J. (1990). Adherence and pathogenesis of *Salmonella enteritidis* in mice. *Microbiology and Immunology* **34**, 885–893.

Azuma, N., Yamauchi, K. & Mitsuoka, T. (1984). Bifidus growth-promoting activity of a glycomacropeptide derived from human κ-casein. *Agricultural and Biological Chemistry* **48**, 2159–2162.

Baba, E., Tsukamoto, Y., Fukata, T., Sasai, K. & Arakawa, A. (1993). Increase of mannose residues, as *Salmonella typhimurium*-adhering factor, on the cecal mucosa of germ-free chickens infected with *Eimeria tenella*. *American Journal of Veterinary Research* **54**, 1471–1475.

Balmer, S. E. & Wharton, B. A. (1989). Diet and faecal flora in the newborn: breast milk and infant formula. *Archives of Disease in Childhood* **64**, 1672–1677.

Banwell, J. G., Boldt, D. H., Meyers, J. & Weber, F. L. (1983). Phytohemagglutinin derived from red kidney bean (*Phaseolus vulgaris*): a cause for intestinal malabsorption associated with bacterial overgrowth in the rat. *Gastroenterology* **84**, 506–515.

Banwell, J. G., Howard, R., Cooper, D. & Costerton, J. W. (1985). Intestinal microbial flora after feeding phytohemagglutinin lectins (*Phaseolus vulgaris*) to rats. *Applied and Environmental Microbiology* **50**, 68–80.

Beachey, E. H. & Courtney, H. S. (1989). Bacterial adherence of group A streptococci to mucosal surfaces. *Respiration* **55**, Suppl. 1, 33–40.

Beerens, H., Romond, C. & Neut, C. (1980). Influence of breast-feeding on the bifid flora of the newborn intestine. *American Journal of Clinical Nutrition* **33**, 2434–2439.

Benno, Y., Sawada, K. & Mitsuoka, T. (1984). The intestinal microflora of infants: composition of fecal flora in breast-fed and bottle-fed infants. *Microbiology and Immunology* **28**, 975–986.

Bezkorovainy, A. (1989). Ecology of Bifidobacteria. In *Biochemistry and Physiology of Bifidobacteria*, pp. 29–72 [A. Bezkorovainy and R. Miller-Catchpole, editors]. Boca Raton, FL: CRC Press.

Bezkorovainy, A., Grohlich, D. & Nichols, J. H. (1979). Isolation of a glycopolypeptide fraction with *Lactobacillus bifidus* subspecies *pennsylvanicus* growth-promoting activity from whole human milk casein. *American Journal of Clinical Nutrition* **32**, 1428–1432.

Bezkorovainy, A. & Nichols, J. H. (1976). Glycoproteins from mature human milk whey. *Pediatric Research* **10**, 1–5.

Bezkorovainy, A. & Topouzian, N. (1981). *Bifidobacterium bifidum* var. *pennsylvanicus* growth promoting activity of human milk casein and its derivatives. *International Journal of Biochemistry* **13**, 585–590.

Bezkorovainy, A., Topouzian, N. & Miller-Catchpole, R. (1986). Mechanisms of ferric and ferrous iron uptake by *Bifidobacterium bifidum* var. *pennsylvanicus*. *Clinical Physiology and Biochemistry* **4**, 150–158.

Bijlsma, I. G. W. & Bouw, J. (1987). Inheritance of K88-mediated adhesion of *Escherichia coli* to jejunal brush borders in pigs: a genetic analysis. *Veterinary Research Communications* **11**, 509–518.

Bijlsma, I. G. W., De Nijs, A., Van Der Meer, C. & Frik, J. F. (1982). Different pig phenotypes affect adherence of *Escherichia coli* to jejunal brush borders by K88ab, K88ac, or K88ad antigen. *Infection and Immunity* **37**, 891–894.

Biol, M.-C., Martin, A., Gaertner, H., Puigserver, A., Richard, M. & Louisot, P. (1990). Intestinal glycosyltransferase activities. Nutritional regulation by a chemically modified protein: methionyl-casein. *Biochemistry International* **20**, 239–250.

Biol, M.-C., Martin, A. & Louisot, P. (1992). Nutritional and developmental regulation of glycosylation processes in digestive organs. *Biochimie* **74**, 13–24.

Biol, M.-C., Martin, A., Oehninger, C., Louisot, P. & Richard, M. (1981). Biosynthesis of glycoproteins in the intestinal mucosa. II. Influence of diets. *Annals of Nutrition and Metabolism* **25**, 269–280.

Blomberg, L. & Conway, P. L. (1989). An in vitro study of colonisation resistance to *Escherichia coli* Bd 1107/7508 (K88) in relation to indigenous squamous gastric colonisation in piglets of varying ages. *Microbial Ecology in Health and Disease* **2**, 285–291.

Blomberg, L., Henriksson, A. & Conway, P. L. (1993). Inhibition of adhesion of *Escherichia coli* K88 to piglet ileal mucus by *Lactobacillus* spp. *Applied and Environmental Microbiology* **59**, 34–39.

Boedeker, E. C. (1994). Adherent bacteria: breaching the mucosal barrier? *Gastroenterology* **106**, 255–257.

Borén, T., Falk, P., Roth, K. A., Larson, G. & Normark, S. (1993). Attachment of *Helicobacter pylori* to human gastric epithelium mediated by blood group antigens. *Science* **262**, 1892–1895.

Brinton, C. C. (1965). The structure, function, synthesis and genetic control of bacterial pili and a molecular model for DNA and RNA transport in Gram negative bacteria. *Transactions of the New York Academy of Sciences* **27**, 1003–1054.

Brooks, D. E., Cavanagh, J., Jayroe, D., Janzen, J., Snoek, R. & Trust, T. J. (1989). Involvement of the MN blood group antigen in shear-enhanced hemagglutination induced by the *Escherichia coli* F41 adhesin. *Infection and Immunity* **57**, 377–383.

Buhler, T., Hoschutzky, H. & Jann, K. (1991). Analysis of the colonization factor antigen (CFA/1) from enterotoxigenic *E. coli* O78:H11. *Bioforum* **14**, 55.

Christensen, G. D., Simpson, W. A. & Beachey, E. H. (1985). Adhesion of bacteria to animal tissues: complex mechanisms. In *Bacterial Adhesion*, pp. 279–305 [D. C. Savage and M. Fletcher, editors]. London: Plenum Press.

Clegg, S. & Gerlach, G. F. (1987). Enterobacterial fimbriae. *Journal of Bacteriology* **169**, 934–938.

Coconnier, M.-H., Klaenhammer, T. R., Kernéis, S., Bernet, M.-F. & Servin, A. L. (1992). Protein-mediated adhesion of *Lactobacillus acidophilus* BG2FO4 on human enterocyte and mucus-secreting cell lines in culture. *Applied and Environmental Microbiology* **58**, 2034–2039.

Collinson, S. K., Doig, P. C., Doran, J. L., Clouthier, S., Trust, T. J. & Kay, W. W. (1993). Thin, aggregative fimbriae mediate binding of *Salmonella enteritidis* to fibronectin. *Journal of Bacteriology* **175**, 12–18.

Conway, P. L., Gorbach, S. L. & Goldin, B. R. (1987). Survival of lactic acid bacteria in the human stomach and adhesion to intestinal cells. *Journal of Dairy Science* **70**, 1–12.

Conway, P. L., Welin, A. & Cohen, P. S. (1990). Presence of K88-specific receptors in porcine ileal mucus is age dependent. *Infection and Immunity* **58**, 3178–3182.

Corfield, T. (1992). Bacterial sialidases – roles in pathogenicity and nutrition. *Glycobiology* **2**, 509–521.

Courtney, H. S., Von Hunolstein, C., Dale, J. B., Bronze, M. S., Beachey, E. H. & Hasty, D. L. (1992). Lipoteichoic acid and M protein: dual adhesins of group A streptococci. *Microbial Pathogenesis* **12**, 199–208.

Cox, E. & Houvenaghel, A. (1993). Comparison of the in vitro adhesion of K88, K99, F41 and P987 positive *Escherichia coli* to intestinal villi of 4-week-old to 5-week-old pigs. *Veterinary Microbiology* **34**, 7–18.

Cummings, R. D. & Mattox, S. A. (1988). Retinoic acid-induced differentiation of the mouse teratocarcinoma cell line F9 is accompanied by an increase in the activity of UDP-galactose: β-D-galactosyl-α1,3-galactosyl-transferase. *Journal of Biological Chemistry* **263**, 511–519.

Dean, E. A. (1990). Comparison of receptors for 987P pili of enterotoxigenic *Escherichia coli* in the small intestines of neonatal and older pig. *Infection and Immunity* **58**, 4030–4035.

Dean, E. A., Whipp, S. C. & Moon, H. W. (1989). Age-specific colonization of porcine intestinal epithelium by 987P-piliated enterotoxigenic *Escherichia coli*. *Infection and Immunity* **57**, 82–87.

Deitch, E. A. (1994). Bacterial translocation: the influence of dietary variables. *Gut* Suppl. 1, S23–S27.

Donnenberg, M. S. & Kaper, J. B. (1992). Enteropathogenic *Escherichia coli*. *Infection and Immunity* **60**, 3953–3961.

Drumm, B., Roberton, A. M. & Sherman, P. M. (1988). Inhibition of attachment of *Escherichia coli* RDEC-1 to intestinal microvillus membranes by rabbit ileal mucus and mucin in vitro. *Infection and Immunity* **56**, 2437–2442.

Ducluzeau, R. (1983). Implantation and development of the gut flora in the newborn animal. *Annales de Recherches Vétérinaires* **14**, 354–359.

Duguid, J. P. & Old, D. C. (1980). Adhesive properties of Enterobacteriaceae. In *Bacterial Adherence*, pp. 185–217 [E. H. Beachey, editor]. London: Chapman & Hall.

Durno, C., Soni, R. & Sherman, P. (1989). Adherence of vero cytotoxin-producing *Escherichia coli* serotype O157:H7 to isolated epithelial cells and brush border membranes in vitro: role of type 1 fimbriae (pili) as a bacterial adhesin expressed by strain CL-49. *Clinical and Investigative Medicine* **12**, 194–200.

Elsinghorst, E. A. & Kopecko, D. J. (1992). Molecular cloning of epithelial cell invasion determinants from enterotoxigenic *Escherichia coli*. *Infection and Immunity* **60**, 2409–2417.

Erickson, A. K., Willgohs, J. A., McFarland, S. Y., Benfield, D. A. & Francis, D. H. (1992). Identification of two porcine brush border glycoproteins that bind the K88ac adhesin of *Escherichia coli* and correlation of these glycoproteins with the adhesive phenotype. *Infection and Immunity* **60**, 983–988.

Evans, D. G. & Evans, D. J. (1978). New surface-associated heat-labile colonization factor antigen (CFA/II) produced by enterotoxigenic *Escherichia coli* of serogroups 06 and 08. *Infection and Immunity* **21**, 638–647.

Evans, D. G., Karjalainen, T. K., Evans, D. J., Graham, D. Y. & Lee, C.-H. (1993). Cloning, nucleotide sequence, and expression of a gene encoding an adhesin subunit protein of *Helicobacter pylori*. *Journal of Bacteriology* **175**, 674–683.

Fader, R. C. & Davis, C. P. (1980). Effect of piliation on *Klebsiella pneumoniae* infection in rat bladders. *Infection and Immunity* **30**, 554–561.

Falk, P., Roth, K. A., Borén, T., Westblom, T. U., Gordon, J. I. & Normark, S. (1993). An in vitro adherence assay reveals that *Helicobacter pylori* exhibits cell lineage-specific tropism in the human gastric epithelium. *Proceedings of the National Academy of Sciences, USA* **90**, 2035–2039.

Faris, A., Lindahl, M. & Wadström, T. (1980). GM2-like glycoconjugate as possible erythrocyte receptor for the CFA/1 and K99 haemagglutinins of enterotoxigenic *Escherichia coli*. *FEMS Microbiology Letters* **7**, 265–269.

Filipe, M. I. & Fenger, C. (1979). Histochemical characteristics of mucins in the small intestine. A comparative study of normal mucosa, benign epithelial tumours and carcinoma. *Histochemical Journal* **11**, 277–287.

Finegold, S. M., Sutter, V. L. & Mathisen, G. E. (1983). Normal indigenous intestinal flora. In *Human Intestinal Microflora in Health and Disease*, pp. 3–31 [D. J. Hentges, editor]. New York: Academic Press.

Firon, N., Ofek, I. & Sharon, N. (1983). Carbohydrate specificity of the surface lectins of *Escherichia coli*, *Klebsiella pneumoniae* and *Salmonella typhimurium*. *Carbohydrate Research* **120**, 235–249.

Firon, N., Ofek, I. & Sharon, N. (1984). Carbohydrate-binding sites of the mannose-specific fimbrial lectins of Enterobacteria. *Infection and Immunity* **43**, 1088–1090.

Fowler, V. R. (1985). The nutrition of the piglet. In *Recent Developments in Pig Nutrition*, pp. 222–229 [D. J. A. Cole and W. Haresign, editors]. London: Butterworths.

Fuller, R. (1992). History and development of probiotics. In *Probiotics, the Scientific Basis*, pp. 1–8 [R. Fuller, editor]. London: Chapman & Hall.

Gabriel, E. P., Lindquist, B. L., Abud, R. L., Merrick, J. M. & Lebenthal, E. (1990). Effect of vitamin A deficiency on the adherence of fimbriated and nonfimbriated *Salmonella typhimurium* to isolated small intestinal enterocytes. *Journal of Pediatric Gastroenterology and Nutrition* **10**, 530–535.

Gibbons, R. J. & Dankers, I. (1981). Lectin-like constituents of foods which react with components of serum, saliva, and *Streptococcus mutans*. *Applied and Environmental Microbiology* **41**, 880–888.

Gibbons, R. A., Jones, G. W. & Sellwood, R. (1975). An attempt to identify the intestinal receptor for the K88 adhesin by means of a haemagglutination inhibition test using glycoproteins and fractions from sow colostrum. *Journal of General Microbiology* **86**, 228–240.

Glass, C. K., DiRenzo, J., Kurokawa, R. & Han, Z. (1991). Regulation of gene expression by retinoic acid receptors. *DNA and Cell Biology* **10**, 623–638.

Gorbach, S. L. (1990). Lactic acid bacteria and human health. *Annals of Medicine* **22**, 37–41.

Gross, R. J. (1990). *Escherichia coli* diarrhoea. In *Topley & Wilson's Principles of Bacteriology, Virology and Immunity*, 8th ed., vol. 3, *Bacterial Diseases*, pp. 469–487 [G. R. Smith and C. S. F. Easman, editors]. London: Edward Arnold.

György, P., Kuhn, R., Rose, C. S. & Zilliken, F. (1954a). Bifidus factor. II. Its occurrence in milk from different species and in other natural products. *Archives of Biochemistry and Biophysics* **48**, 202–208.

György, P., Norris, R. F. & Rose, C. S. (1954b). Bifidus factor. I. A variant of *Lactobacillus bifidus* requiring a special growth factor. *Archives of Biochemistry and Biophysics* **48**, 193–201.

Haagen, I. A., Heezius, H. C., Verkooyen, R. P., Verhoef, J. & Verbrugh, H. A. (1990). Adherence of peritonitis-causing staphylococci to human peritoneal mesothelial cell monolayers. *Journal of Infectious Diseases* **161**, 266–273.

Harel, J., Lapointe, H., Fallara, A., Lortie, L. A., Bigras-Poulin, M., Larivière, S. & Fairbrother, J. M. (1991). Detection of genes for fimbrial antigens and enterotoxins associated with *Escherichia coli* serogroups isolated from pigs with diarrhea. *Journal of Clinical Microbiology* **29**, 745–752.

Holland, R. E. (1990). Some infectious causes of diarrhea in young farm animals. *Clinical Microbiology Reviews* **3**, 345–375.

Holmgren, J., Svennerholm, A.-M. & Lindblad, M. (1983). Receptor-like glycocompounds in human milk that inhibit classical and El Tor *Vibrio cholerae* cell adherence (hemagglutination). *Infection and Immunity* **39**, 147–154.

Hoskins, L. C., Agustines, M., McKee, W. B., Boulding, E. T., Kriaris, M. & Niedermeyer, G. (1985). Mucin degradation in human colon ecosystems: isolation and properties of fecal strains that degrade ABH blood group antigens and oligosaccharides from mucin glycoproteins. *Journal of Clinical Investigation* **75**, 944–953.

Hultgren, S. J., Abraham, S., Caparon, M., Falk, P., Stgeme, J. W. & Normark, S. (1993). Pilus and nonpilus bacterial adhesins – assembly and function in cell recognition. *Cell* **73**, 887–901.

Isaacson, R. E. & Kinsel, M. (1992). Adhesion of *Salmonella typhimurium* to porcine epithelial surfaces: identification and characterization of two phenotypes. *Infection and Immunity* **60**, 3193–3200.

Ito, N. & Hirota, T. (1992). Histochemical and cytochemical localization of blood group antigens. *Progress in Histochemistry and Cytochemistry* **25** (2), 1–85.

Jayne-Williams, D. J. & Hewitt, D. (1972). The relationship between the intestinal microflora and the effects of diets containing raw navy beans (*Phaseolus vulgaris*) on the growth of Japanese quail (*Coturnix coturnix japonica*). *Journal of Applied Bacteriology* **35**, 331–344.

Kaku, H., Goldstein, I. J., Van Damme, E. J. M. & Peumans, W. J. (1992). New mannose-specific lectins from garlic (*Allium sativum*) and ramsons (*Allium ursinum*) bulbs. *Carbohydrate Research* **229**, 347–353.

Kapperud, G., Namork, E., Skurnik, M. & Nesbakken, T. (1987). Plasmid-mediated surface fibrillae of *Yersinia pseudotuberculosis* and *Yersinia enterocolitica*: relationship to the outer membrane protein YOP1 and possible importance for pathogenesis. *Infection and Immunity* **55**, 2247–2254.

Kehagias, C., Jao, Y. C., Mikolajcik, E. M. & Hansen, P. M. T. (1977). Growth response of *Bifidobacterium bifidum* to a hydrolytic product isolated from bovine casein. *Journal of Food Science* **42**, 146–150.

Kelly, D., Begbie, R. & King, T. P. (1992). Postnatal intestinal development. In *Neonatal Survival and Growth* (*BSAP Occasional Publication* no. 15), pp. 63–79 [M. A. Varley, P. E. V. Williams and T. L. J. Lawrence, editors]. Edinburgh: British Society of Animal Production.

Kelly, D. & King, T. P. (1991). The influence of lactation products on the temporal expression of histo-blood group antigens in the intestines of suckling pigs: lectin histochemical and immunohistochemical analysis. *Histochemical Journal* **23**, 55–60.

Kelly, D., King, T. P., McFadyen, M. & Coutts, A. G. P. (1993). Effect of preclosure colostrum intake on the development of the intestinal epithelium of artificially-reared piglets. *Biology of the Neonate* **64**, 235–244.

Kervella, M., Pages, J., Pei, Z., Grollier, G., Blaser, M. J. & Fauchere, J. (1993). Isolation and characterisation of two *Campylobacter* glycine-extracted proteins that bind to Hela cell membranes. *Infection and Immunity* **61**, 3440–3448.

King, T. P. (1994). Lectin cytochemistry and intestinal epithelial cell biology. In *Lectins: Biomedical Perspectives*, pp. 000–000 [A. Pusztai, editor] (in press).

King, T. P., Begbie, R. & Cadenhead, A. (1983). Nutritional toxicity of raw kidney beans in pigs. Immunocytochemical and cytopathological studies on the gut and the pancreas. *Journal of the Science of Food and Agriculture* **34**, 1404–1412.

King, T. P., Begbie, R., Spencer, R. & Kelly, D. (1993). Diversity of membrane sialo-glycoconjugates in the developing porcine small intestine. *Proceedings of the Nutrition Society* **52**, 195A.

King, T. P. & Kelly, D. (1990). Lectin and antibody affinity cytochemistry of intestinal goblet cells in suckling pigs. *Transactions of the Royal Microscopical Society* **1**, 649–652.

King, T. P. & Kelly, D. (1991). Ontogenic expression of histo-blood group antigens in the intestines of suckling pigs: lectin histochemical and immunohistochemical analysis. *Histochemical Journal* **23**, 43–54.

Kleeman, E. G. & Klaenhammer, T. R. (1982). Adherence of *Lactobacillus* species to human fetal intestinal cells. *Journal of Dairy Science* **65**, 2063–2069.

Knutton, S., Lloyd, D. R., Candy, D. C. A. & McNeish, A. S. (1985). Adhesion of enterotoxigenic *Escherichia coli* to human small intestinal enterocytes. *Infection and Immunity* **48**, 824–831.

Knutton, S., Lloyd, D. R. & McNeish, A. S. (1987). Adhesion of enteropathogenic *Escherichia coli* to human intestinal enterocytes and cultured human intestinal mucosa. *Infection and Immunity* **55**, 69–77.

Kocourek, J. & Horejsi, V. (1983). A note on the recent discussion on definition of the term "lectin". In *Lectins: Biology, Biochemistry, Clinical Biochemistry*, vol. 3, pp. 3–6 [T. C. Bog-Hansen and G. A. Spengler, editors]. Berlin: Walter de Gruyter.

Kolstø Otnæss, A.-B., Lægreid, A. & Ertresvåg, K. (1983). Inhibition of enterotoxin from *Escherichia coli* and *Vibrio cholerae* by gangliosides from human milk. *Infection and Immunity* **40**, 563–569.

Koshte, V. L., van Dijk, W., van der Stelt, M. E. & Aalberse, R. C. (1990). Isolation and characterization of BanLec-I, a mannoside-binding lectin from *Musa paradisiac* (banana). *Biochemical Journal* **272**, 721–726.

Kovar, M. G., Serdula, M. K., Marks, J. S. & Fraser, D. W. (1984). Review of the epidemiologic evidence for an association between infant feeding and infant health. *Pediatrics* **74** Suppl., 615–638.

Kraehenbuhl, J. P. & Neutra, M. R. (1992). Molecular and cellular basis of immune protection of mucosal surfaces. *Physiological Reviews* **72**, 853–879.

Krogfelt, K. A., McCormick, B. A., Burghoff, R. L., Laux, D. C. & Cohen, P. S. (1991). Expression of *Escherichia coli* F-18 type 1 fimbriae in the streptomycin-treated mouse large intestine. *Infection and Immunity* **59**, 1567–1568.

Labarriere, N., Piau, J. P., Zennadi, R., Blanchardie, P., Denis, M. & Lustenberger, P. (1993). Retinoic acid

modulation of α1,2-fucosyltransferase activity and sensitivity of tumor cells to LAK-mediated cytotoxicity. *In Vitro Cellular and Developmental Biology* **29**, 140–144.
Langkamp-Henken, B., Glezer, J. A. & Kudsk, K. A. (1992). Immunologic structure and function of the gastrointestinal tract. *Nutrition in Clinical Practice* **7**, 100–108.
Laux, D. C., McSweegan, E. F., Williams, T. J., Wadolkowski, E. A. & Cohen, P. S. (1986). Identification and characterization of mouse small intestine mucosal receptors for *Escherichia coli* K-12(K88ab). *Infection and Immunity* **52**, 18–25.
Leffler, H. & Svanborg-Edén, C. (1986). Glycolipids as receptors for *Escherichia coli* lectins or adhesins. In *Microbial Lectins and Agglutinins: Properties and Biological Activity*, pp. 83–111 [D. Mirelman, editor]. New York: Wiley.
Lelwala-Guruge, J., Ascencio, F., Ljungh, A. & Wadström, T. (1993). Rapid detection and characterization of sialic acid-specific lectins of *Helicobacter pylori*. *Acta Pathologica Microbiologica et Immunologica Scandinavica* **101**, 695–702.
Lelwala-Guruge, J., Ljungh, A. & Wadström, T. (1992). Haemagglutination patterns of *Helicobacter pylori*. Frequency of sialic acid-specific and non-sialic acid-specific haemagglutinins. *Acta Pathologica Microbiologica et Immunologica Scandinavica* **100**, 908–913.
Lindahl, M., Brossmer, R. & Wadström, T. (1987). Carbohydrate receptor specificity of K99 fimbriae of enterotoxigenic *Escherichia coli*. *Glycoconjugate Journal* **4**, 51–58.
Lindahl, M. & Wadström, T. (1986). Binding to erythrocyte membrane glycoproteins and carbohydrate specificity of F41 fimbriae of enterotoxigenic *Escherichia coli*. *FEMS Microbiology Letters* **34**, 297–300.
Linton, A. H. & Hinton, M. H. (1990). The normal microbiota of the body. In *Topley & Wilson's Principles of Bacteriology, Virology and Immunity*, 8th ed., vol. 1, *General Microbiology and Immunity*, pp. 311–329 [A. H. Linton and H. M. Dick, editors]. London: Edward Arnold.
Lockman, H. A. & Curtiss, R. (1992). Virulence of non-type 1-fimbriated and nonfimbriated nonflagellated *Salmonella typhimurium* mutants in murine typhoid fever. *Infection and Immunity* **60**, 491–496.
Mantle, M. & Husar, S. D. (1993). Adhesion of *Yersinia enterocolitica* to purified rabbit and human intestinal mucin. *Infection and Immunity* **61**, 2340–2346.
Mantle, M. & Rombough, C. (1993). Growth in and breakdown of purified rabbit small intestinal mucin by *Yersinia enterocolitica*. *Infection and Immunity* **61**, 4131–4138.
Martin, A., Biol, M. C. & Louisot, P. (1989). Intestinal glycosyltransferase activities: nutritional regulation by the quantity of dietary proteins. *Biochemical Archives* **5**, 297–308.
Maxwell, F. J. & Stewart, C. S. (1994). The microbiology of the gut and the role of probiotics. In *The Neonatal Pig: Development and Survival*, pp. 000–000 [M. Varley, editor]. Wallingford, Oxon: CAB International (in press).
Mayra-Makinen, A., Manninen, M. & Gyllenberg, H. (1983). The adherence of lactic acid bacteria to the columnar epithelial cells of pigs and calves. *Journal of Applied Bacteriology* **55**, 241–245.
Metcalfe, J. W., Krogfelt, K. A., Krivan, H. C., Cohen, P. S. & Laux, D. C. (1991). Characterization and identification of a porcine small intestine mucus receptor for the K88ab fimbrial adhesin. *Infection and Immunity* **59**, 91–96.
Metchnikoff, E. (1907). *The Prolongation of Life: Optimistic Studies*. London: Heinemann.
Michiels, T., Wattiau, P., Brasseur, R., Ruysschaert, J.-M. & Cornelis, G. R. (1990). Secretion of Yop proteins by *Yersinia*. *Infection and Immunity* **58**, 2840–2849.
Mo, H. Q., Van Damme, E. J. M., Peumans, W. J. & Goldstein, I. J. (1993). Purification and characterization of a mannose-specific lectin from shallot (*Allium ascalonicum*) bulbs. *Archives of Biochemistry* **306**, 431–438.
Moreau, M.-C., Thomasson, M., Ducluzeau, R. & Raibaud, P. (1986). [Kinetics of bacterial colonization in the human neonate in relation to the kind of milk.] *Reproduction, Nutrition, Développement* **26**, 745–753.
Morris, J. A., Thorns, C. J. & Sojka, W. J. (1980). Evidence for two adhesive antigens on the K99 reference strain *Escherichia coli* B41. *Journal of General Microbiology* **118**, 107–113.
Moughan, P. J., Birtles, M. J., Cranwell, P. D., Smith, W. C. & Pedraza, M. (1992). The piglet as a model animal for studying aspects of digestion and absorption in milk-fed human infants. *World Review of Nutrition and Dietetics* **67**, 40–113.
Mouricout, M. (1991). Swine and cattle enterotoxigenic *Escherichia coli*-mediated diarrhoea. Development of therapies based on inhibition of bacteria–host interactions. *European Journal of Epidemiology* **7**, 588–604.
Mouricout, M., Petit, J. M., Carias, J. R. & Julien, R. (1990). Glycoprotein glycans that inhibit adhesion of *Escherichia coli* mediated by K99 fimbriae: treatment of experimental colibacillosis. *Infection and Immunity* **58**, 98–106.
Nachbar, M. S. & Oppenheim, J. D. (1980). Lectins in the United States diet: a survey of lectins in commonly consumed foods and a review of the literature. *American Journal of Clinical Nutrition* **33**, 2338–2345.
Nagy, B., Arp, L. H., Moon, H. W. & Casey, T. A. (1992). Colonization of the small intestine of weaned pigs by enterotoxigenic *Escherichia coli* that lack known colonization factors. *Veterinary Pathology* **29**, 239–246.
Neeser, J.-R., Chambaz, A., Golliard, M., Link-Amster, H., Fryder, V. & Kolodziejczyk, E. (1989). Adhesion of colonization factor antigen II-positive enterotoxigenic *Escherichia coli* strains to human enterocytelike differentiated HT-29 cells: a basis for host–pathogen interactions in the gut. *Infection and Immunity* **57**, 3727–3734.

Neeser, J.-R., Koellreutter, B. & Wuersch, P. (1986). Oligomannoside-type glycopeptides inhibiting adhesion of *Escherichia coli* strains mediated by type 1 pili; preparation of potent inhibitors from plant glycoproteins. *Infection and Immunity* **52**, 428–436.

Neutra, M. R. & Forstner, J. F. (1987). Gastrointestinal mucus: synthesis, secretion, and function. In *Physiology of the Gastrointestinal Tract*, 2nd ed., pp. 975–1009 [L. R. Johnson, editor]. New York: Raven Press.

Nichols, J. H., Bezkorovainy, A. & Paque, R. (1975). Isolation and characterization of several glycoproteins from human colostrum whey. *Biochimica et Biophysica Acta* **412**, 99–108.

Nilsson, G. & Svensson, S. (1983). The role of the carbohydrate portion of glycolipids for the adherence of *Escherichia coli* K88+ to pig intestine. In *Proceedings of the 7th International Symposium on Glycoconjugates, Lund-Ronneby, Sweden*, pp. 637–638 [M. A. Chester, D. Heinegerd, A. Lundbald and S. Svensson, editors].

Ofek, I. & Doyle, R. J. (1994). *Bacterial Adhesion to Cells and Tissues*, pp. 397–400. London: Chapman and Hall.

Ofek, I. & Sharon, N. (1990). Adhesins as lectins: specificity and role in infection. *Current Topics in Microbiology and Immunology* **151**, 91–113.

Oriol, R. (1987). Tissular expression of ABH and Lewis antigens in humans and animals: expected value of different animal models in the study of ABO-incompatible organ transplants. *Transplantation Proceedings* **19**, 4416–4420.

Oyofo, B. A., De Loach, J. R., Corrier, D. E., Norman, J. O., Ziprin, R. L. & Mollenhauer, H. H. (1989). Prevention of *Salmonella typhimurium* colonization of broilers with D-mannose. *Poultry Science* **68**, 1357–1360.

Pærregaard A., Espersen, F., Jensen, O. M. & Skurnik, M. (1991). Interactions between *Yersinia enterocolitica* and rabbit ileal mucus: growth, adhesion, penetration, and subsequent changes in surface hydrophobicity and ability to adhere to ileal brush border membrane vesicles. *Infection and Immunity* **59**, 253–260.

Payne, D., O'Reilly, M. & Williamson, D. (1993). The K88 fimbrial adhesin of enterotoxigenic *Escherichia coli* binds to β1-linked galactosyl residues in glycosphingolipids. *Infection and Immunity* **61**, 3673–3677.

Pedersen, K. & Tannock, G. W. (1989). Colonization of the porcine gastrointestinal tract by lactobacilli. *Applied and Environmental Microbiology* **55**, 279–283.

Poxton, I. R. & Arbuthnott, J. P. (1990). Determinants of bacterial virulence. In *Topley & Wilson's Principles of Bacteriology, Virology and Immunity*, 8th ed., vol. 1, *General Microbiology and Immunity*, pp. 331–353 [A. H. Linton and H. M. Dick, editors]. London: Edward Arnold.

Prizont, R. (1982). Degradation of intestinal glycoproteins by pathogenic *Shigella flexneri*. *Infection and Immunity* **36**, 615–620.

Pusztai, A. (1993). Review. Dietary lectins are metabolic signals for the gut and modulate immune and hormone functions. *European Journal of Clinical Nutrition* **47**, 691–699.

Pusztai, A., Grant, G., King, T. P. & Clarke, E. M. W. (1990). Chemical probiosis. In *Recent Advances in Animal Nutrition*, pp. 47–60 [W. Haresign and D. J. A. Cole, editors]. London: Butterworths.

Pusztai, A., Grant, G., Spencer, R. J., Duguid, T. J., Brown, D. S., Ewen, S. W. B., Peumans, W. J., Van Damme, E. J. M. & Bardocz, S. (1993). Kidney bean lectin-induced *Escherichia coli* overgrowth in the small intestine is blocked by GNA, a mannose-specific lectin. *Journal of Applied Bacteriology* **75**, 360–368.

Rattray, E. A. S., Palmer, R. & Pusztai, A. (1974). Toxicity of kidney beans (*Phaseolus vulgaris* L.) to conventional and gnotobiotic rats. *Journal of the Science of Food and Agriculture* **25**, 1035–1040.

Rose, S. J. (1984). Bacterial flora of breast-fed infants. *Pediatrics* **74**, 563.

Runnels, P. L., Moon, H. W. & Schneider, R. A. (1980). Development of resistance with host age to adhesion of K99+ *Escherichia coli* to isolated intestinal epithelial cells. *Infection and Immunity* **28**, 298–300.

Ruseler-van Embden, J. G. H., Schouten, W. R., van Lieshout, L. M. C. & Auwerda, H. J. A. (1992). Changes in bacterial composition and enzymatic activity in ileostomy and ileal reservoir during intermittent occlusion: a study using dogs. *Applied and Environmental Microbiology* **58**, 111–118.

Saitoh, T., Natomi, H., Zhao, W., Okuzumi, K., Sugano, K., Iwamori, M. & Nagai, Y. (1991). Identification of glycolipid receptors for *Helicobacter pylori* by TLC-immunostaining. *FEBS Letters* **282**, 385–387.

Sajjan, S. U. & Forstner, J. F. (1990a). Characteristics of binding of *Escherichia coli* serotype O157:H7 strain CL-49 to purified intestinal mucin. *Infection and Immunity* **58**, 860–867.

Sajjan, S. U. & Forstner, J. F. (1990b). Role of the putative "link" glycopeptide of intestinal mucin in binding of piliated *Escherichia coli* serotype O157:H7 strain CL-49. *Infection and Immunity* **58**, 868–873.

Schengrund, C.-L. & Ringler, N. J. (1989). Binding of *Vibrio cholera* toxin and the heat-labile enterotoxin of *Escherichia coli* to G_{M1}, derivatives of G_{M1}, and non-lipid oligosaccharide polyvalent ligands. *Journal of Biological Chemistry* **264**, 13233–13237.

Seignole, D., Mouricout, M., Duval-Iflah, Y., Quintard, B. & Julien, R. (1991). Adhesion of K99 fimbriated *Escherichia coli* to pig intestinal epithelium: correlation of adhesive and non-adhesive phenotypes with the sialoglycolipid content. *Journal of General Microbiology* **137**, 1591–1601.

Sellwood, R. (1979). *Escherichia coli* diarrhoea in pigs with or without the K88 receptor. *Veterinary Record* **105**, 228–230.

Sellwood, R. (1980). The interaction of the K88 antigen with porcine intestinal epithelial cell brush borders. *Biochimica et Biophysica Acta* **632**, 326–335.

Sellwood, R. (1984). The K88 adherent system in swine. In *Attachment of Organisms to the Gut Mucosa*, vol. 1, pp. 21–29 [E. C. Boedeker, editor]. Boca Raton, FL: CRC Press.

Sharon, N., Eshdat, Y., Silverblatt, F. J. & Ofek, I. (1981). Bacterial adherence to cell surface sugars. In *Adhesion*

and *Micro-organisms Pathogenicity*, pp. 119–141 [K. Elliot, M. O'Connor and J. Whelan, editors]. London: Pitman Press.
Silk, D. B. A. & Grimble, G. K. (1994). Introduction. *Gut* Suppl. 1, V–VI.
Simhon, A., Douglas, J. R., Drasar, B. S. & Soothill, J. F. (1982). Effect of feeding on infants' faecal flora. *Archives of Disease in Childhood* **57**, 54–58.
Sokurenko, E. V., Courtney, H. S., Abraham, S. N., Klemm, P. & Hasty, D. L. (1992). Functional heterogeneity of type-1 fimbriae of *Escherichia coli*. *Infection and Immunity* **60**, 4709–4719.
Specian, R. D. & Oliver, M. G. (1991). Functional biology of intestinal goblet cells. *American Journal of Physiology* **260**, C183–C193.
Spitz, J., Hecht, G., Taveras, M., Aoys, E. & Alverdy, J. (1994). The effect of dexamethasone administration on rat intestinal permeability: the role of bacterial adherence. *Gastroenterology* **106**, 35–41.
Springer, G. F., Rose, C. S. & György, P. (1954). Blood-group mucoids: distribution and growth promoting properties for *Lactobacillus bifidus* var. *pennsylvanicus*. *Journal of Laboratory and Clinical Medicine* **43**, 532–542.
Stewart, C., Hillman, K., Maxwell, F., Kelly, D. & King, T. P. (1993). Recent advances in probiosis in pigs: observations on the microbiology of the pig gut. In *Recent Advances in Animal Nutrition*, pp. 197–219 [P. C. Garnsworthy and D. J. A. Cole, editors], Nottingham: Nottingham University Press.
Stoll, B. J., Holmgren, J., Bardhan, P. K., Huq, I., Greenhough, W. B., Fredman, P. & Svennerholm, L. (1980). Binding of intraluminal toxin in cholera: trial of GM1 ganglioside charcoal. *Lancet* **ii**, 888–891.
Strous, G. J. & Dekker, J. (1992). Mucin-type glycoproteins. *CRC Critical Reviews in Biochemistry and Molecular Biology* **27**, 57–92.
Surawicz, C. M., McFarland, L. V., Elmer, G. & Chinn, J. (1989). Treatment of recurrent *Clostridium difficile* colitis with vancomycin and *Saccharomyces boulardii*. *American Journal of Gastroenterology* **84**, 1285–1287.
Taatjes, D. J. & Roth, J. (1990). Selective loss of sialic acid from rat small intestinal epithelial cells during postnatal development: demonstration with lectin-gold techniques. *European Journal of Cell Biology* **53**, 255–266.
Tannock, G. W. (1992). The lactic microflora of pigs, mice and rats. In *The Lactic Acid Bacteria*, vol. 1, *The Lactic Acid Bacteria in Health and Disease*, pp. 21–48 [B. J. B. Wood, editor]. London: Elsevier Applied Science.
Teneberg, S., Willemsen, P. T. J., de Graaf, F. K. & Karlsson, K.-A. (1990). Receptor-active glycolipids of epithelial cells of the small intestine of young and adult pigs in relation to susceptibility to infection with *Escherichia coli* K99. *FEBS Letters* **263**, 10–14.
Teneberg, S., Willemsen, P. T. J., de Graaf, F. K. & Karlsson, K.-A. (1993). Calf small intestine receptors for K99 fimbriated enterotoxigenic *Escherichia coli*. *FEMS Microbiology Letters* **109**, 107–112.
Terada, A., Hara, H., Kataoka, M. & Mitsuoka, T. (1992). Effect of lactulose on the composition and metabolic activity of the human faecal flora. *Microbial Ecology in Health and Disease* **5**, 43–50.
Thorns, C. J., Sojka, M. G. & Chasey, D. (1990). Detection of a novel fimbrial structure on the surface of *Salmonella enteritidis* by using a monoclonal antibody. *Journal of Clinical Microbiology* **28**, 2409–2414.
Turck, D., Feste, A. S. & Lifschitz, C. H. (1993). Age and diet affect the composition of porcine colonic mucins. *Pediatric Research* **33**, 564–567.
Untawale, G. G., Pietraszek, A. & McGinnis, J. (1978). Effect of diet on adhesion and invasion of microflora in the intestinal mucosa of chicks. *Proceedings of the Society for Experimental Biology and Medicine* **159**, 276–280.
Van Driessche, E., Charlier, G., Schoup, J., Beeckmans, S., Pohl, P., Lintermans, P. & Kanarek, L. (1989). The effect of lectins and monosaccharides on the in vitro attachment of E. coli F17[+] to intestinal calf villi and immobilized glycoproteins. In *Recent Advances of Research in Antinutritional Factors in Legume Seeds*, pp. 43–48 [J. Huisman, T. F. B. van der Poel and I. E. Liener, editors]. Wageningen: Pudoc.
Wanke, C. A., Cronan, S., Goss, C., Chadee, K. & Guerrant, R. L. (1990). Characterization of binding of *Escherichia coli* strains which are enteropathogens to small-bowel mucin. *Infection and Immunity* **58**, 794–800.
Wick, M. J., Madara, J. L., Fields, B. N. & Normark, S. J. (1991). Molecular cross talk between epithelial cells and pathogenic microorganisms. *Cell* **67**, 651–659.
Willemsen, P. T. J. & de Graaf, F. K. (1992). Age and serotype dependent binding of K88 fimbriae to porcine intestinal receptors. *Microbial Pathogenesis* **12**, 367–375.
Willis, A. T., Bullen, C. L., Williams, K., Fagg, C. G., Bourne, A. & Vignon, M. (1973). Breast milk substitute: a bacteriological study. *British Medical Journal* **iv**, 67–72.
Wilson, A. B., King, T. P., Clarke, E. M. W. & Pusztai, A. (1980). Kidney bean (*Phaseolus vulgaris*) lectin-induced lesions in the rat small intestine. II. Microbiological studies. *Journal of Comparative Pathology* **90**, 597–602.
Winner, L. S., Mack, J., Weltzin, R. A., Mekalanos, J. J., Kraehenbuhl, J.-P. & Neutra, M. R. (1991). New model for analysis of mucosal immunity: intestinal secretion of specific monoclonal immunoglobulin A from hybridoma tumors protects against *Vibrio cholerae* infection. *Infection and Immunity* **59**, 977–982.
Yuyama, Y., Yoshimatsu, K., Ono, E., Saito, M. & Naiki, M. (1993). Postnatal change of pig intestinal ganglioside bound by *Escherichia coli* with K99 fimbriae. *Journal of Biochemistry* **113**, 488–492.
Zilliken, F., Smith, P. N., Tomarelli, R. M. & György, P. (1955). 4-*O*-β-D-Galactopyranosyl-*N*-acetyl-D-glucosamine in hog mucin. *Archives of Biochemistry and Biophysics* **54**, 398–405.

Printed in Great Britain

INDEX OF SUBJECTS

acrodermatitis enteropathica, zinc deficiency   164
adhesins, bacterial   235–240
adiposity *see* obesity
aflatoxins
   carcinogens   178, 182
   effect of glucosinolates on toxicity   217–218, 222
agglutinins, bacterial   237, 240, 246
aglucones, breakdown products of glucosinolates   210
alcohol dehydrogenases, xenobiotic metabolizing enzymes   219
alkaline phosphatase, assessment of zinc status   159
allergens *see* antigens
allergy
   definition   26
   infant, effects of human milk   33–34
   materno-fetal effects on development   25–38
amenorrhoea *see* menstruation
amino acids, effects on zinc uptake   142–145
anorexia *see* eating disorders
antibodies, materno-fetal transfer   27–28
antigens, materno-fetal transfer   31–32
antimicrobials, growth promotion in livestock   245
antioxidant responsive element   184
antioxidants, carcinogenesis   180–182, 184
antipyrine, effect of glucosinolate derivatives   224
appetite, effect of zinc deficiency   157
arachidonic acid in human milk and infant formulas   7–9
aryl hydrocarbon hydroxylase activity, effect of glucosinolates   221
ascorbic acid *see* vitamin C
ascorbigen, breakdown product of glucosinolates   212–213
asthma, infant, effect of maternal diet   35
atherosclerosis, prevention   43–60
athletes, women
   menstruation   67–86
   nutrition   70–73
atopic diseases *see* allergy

bacteria
   interactions with gut mucosa   233–251
   nutritional influences   243–250
beef in maternal elimination diets   35
benzo(*a*)pyrene
   carcinogen   178–179, 182, 184, 218
   effect of glucosinolates   220, 222
benzyl isothiocyanate, carcinogenesis   182, 186, 193
bifidobacteria
   establishment facilitated by breast milk   246
   glycosidase activity   243
   use as probiotics   245–247
biocytin *see* biotin
biotin in man   114–116
blood group antigens   235, 238–240
blood pressure, relation to atherosclerosis   51–54

bones
   effects of amenorrhoea   81–82
   effects of diet and physical exercise in women   80–84
   metabolism in women athletes   80–82
brain structures, requirement for fatty acids   8
broccoli, glucosinolates   206, 209, 214, 221
Brussels sprouts, glucosinolates   206–207, 209, 213–215
bulimia *see* eating disorders

cabbage, glucosinolates   206, 209, 213–215
caffeine, carcinogenesis   180–181, 185, 189
calabrese, glucosinolates   206, 209, 214
calciferol *see* vitamin D
calcium
   absorption from gut, relationship with zinc   139
   effect of elimination diets on intake in mothers   37
   effect on phytate inhibition of zinc absorption   154
   human milk and infant formulas   12–13
Campylobacter, enteritis in human beings   240
cancer *see also* carcinogenesis
   chemically induced, effects of glucosinolates   217–225
   metastases, prevention   192
   protective factors in plant foods   175–194
canthaxanthin *see* carotenoids
caper, glucosinolates   206–207
carbohydrates
   dietary, effect on glycosylation in gut   247
   human milk and infant formulas   11–12
   metabolism and insulin resistance   48–51
carbonic anhydrase, zinc constituent   130
carcinogenesis *see also* cancer
   blocking agents   178–184
   mechanisms   176–178
   suppressing agents   186–192
cardiovascular disease, prevention   43–60
β-carotene in human milk and infant formulas   14–15
carotenoids *see also* vitamin A
   anticarcinogens   176, 180, 191–192
   in man   96–98
catechin, carcinogenesis   181, 187, 191–192
cauliflower, glucosinolates   206–207, 209, 214
cell proliferation, inhibitors   186–190
cellular differentiation, prevention   191
children
   zinc deficiency   160–164
   zinc intake   156–157, 161–163
cholecalciferol *see* vitamin D
cholesterol
   infant nutrition   10–11
   plasma, factors affecting   44–48
chromium in human milk and infant formulas   14

Index of Subjects

cobalamin *see* vitamin B₁₂
colic in infants   31, 36
colostrum, effect on gut glycoconjugates   243
copper
    absorption from gut, relationship with zinc   139–140
    dietary, effect on zinc absorption   154
coumarin, carcinogenesis   182–183, 185, 193
cryptoxanthin *see* carotenoids
curcumin, carcinogenesis   183, 189, 193
cyanocobalamin *see* vitamin B₁₂
cytochrome P450
    carcinogen activators   178–180, 184
    xenobiotic metabolizing enzymes   219–225
cytotoxins, dietary   191

dairy products
    maternal elimination diets   35–36
    use as probiotics   245
diabetes, relation to atherosclerosis   48–49
diarrhoea
    effect of bacterial fimbriae   237–240, 243
    effect of dietary glycan   250
diet, prevention of cardiovascular disease   43–60
7,12-dimethylbenz(*a*)anthracene, carcinogenesis   186, 188–189
1,2-dimethylhydrazine   178, 184
diphenols, carcinogenesis   182, 190
DNA repair inducers, carcinogenesis   185
docosahexaenoic acid in human milk and infant formulas   7–9

eating disorders in women athletes   76–80
eczema in infants   31, 35–36
egg in maternal elimination diets   35–37
electrophile responsive element   184
embryo mortality, effects of dietary glucosinolates   215–216
energy
    balance, relation to atherosclerosis   54–58
    expenditure, women athletes   70
    intake, women athletes   70–76
    values for human milk   2–3
enteritis, effect of bacterial fimbriae   237–240
enzymes *see* alcohol dehydrogenases, aldehyde dehydrogenases, aryl hydrocarbon hydroxylase activity, cytochrome P450, epoxide hydrolases, ethoxycoumarin *O*-deethylase activity, ethoxyresorufin deethylation activity, glutathione *S*-transferases, monooxygenases, myrosinase, quinone reductase, sulphotransferases, UDPglucuronyltransferases, xenobiotic metabolizing enzymes
epithioalkanes, breakdown products of glucosinolates   210
epoxide hydrolases
    effect of glucosinolates   221, 223
    xenobiotic metabolizing enzymes   220
Escherichia coli
    effect of fimbriae on adherence to gut surfaces   236–240, 242, 250
    effect of milk glycoconjugates   246
ethoxycoumarin *O*-deethylase activity, effect of glucosinolates   221–224

ethoxyresorufin deethylation activity, effect of glucosinolates   220–222
exercise, prevention of cardiovascular disease   43–60

fats, dietary, effect on glycosylation in gut   247
fatty acids *see also* lipids
    polyunsaturated, in human milk and infant formulas   7–9
feedstuffs, toxicity due to dietary glucosinolates   215–217
fimbriae, bacterial   235–240
fish in maternal elimination diets   35–37
flavin nucleotides *see* riboflavin
flavonoids, carcinogenesis   180–182, 185, 189
folate, dietary, effect on zinc absorption   154
    in man   111–113
folic acid *see* folate
foods, toxicity due to dietary glucosinolates   215–217
free radicals, cell damage   178–181, 183, 188, 190–191
fruits, sources of anticarcinogens   175–194

germ free animals, toxicity of glucosinolate derivatives   216
gliadin
    breast milk   31
    materno-fetal transfer   29
glucobrassicin, glucosinolate   208, 211, 213, 218–219, 222–223
glucoiberverin, glucosinolate   213–214, 216
gluconapin, glucosinolate   208, 215, 223
glucosinolates   205–225
    bacterial metabolism   212, 225
    carcinogenesis, protection against   217–219
    food processing effects   213–215, 221, 224
    occurrence   205–207
    toxicity   215–217
β-glucuronidase, carcinogenesis   184–185
glutathione *S*-transferases
    carcinogenesis   181, 183–184
    effect of glucosinolates   221, 223–225
glycans *see* mucins
glycoconjugates
    dietary effects in gut   243–250
    gut surfaces   243–250
glycolipids *see* glycoconjugates
glycoproteins *see* glycoconjugates
β-glycosidase, carcinogenesis   184–185
glycosidases, bacterial, degradation of protective mucin   243
goitre, effect of glucosinolate derivates   215–217
goitrin, glucosinolate   215, 223
growth depression, effect of dietary glucosinolates   215–217
gut
    effect of glucosinolates on incidence of cancer   218
    mucosa, bacterial interactions   233–250
    sites of zinc absorption   137–139

haemagglutination *see Escherichia coli*, phytohaemagglutinins

Index of Subjects

hair, zinc content 158
Helicobacter pylori, adherence to gut surfaces 240
horseradish, glucosinolates 206–207
hydrocarbons, polycyclic aromatic, carcinogenesis 177
hypertension, relation to atherosclerosis 51–54

immune enhancement, anticarcinogenic 192
immune system, effect of zinc deficiency 157, 160
immunity, materno-fetal effects 25–38
immunoglobulin E, synthesis by fetus 33
immunoglobulin G, placental transfer 27–28
immunoglobulins
    breast milk, effects in infants 30–31
    human milk and infant formulas 6
indole-3-carbinol, carcinogenesis 180–181, 183, 189, 192–193
indoles, breakdown products of glucosinolates 211–212
infant formulas 1–16
inositol *see also* phytate
    human milk and infant formulas 15
insulin, resistance, relation to atherosclerosis 48–51
intrinsic factor, absorption of vitamin B_{12} 114
iron
    dietary, effect on zinc absorption 154
    effect on zinc absorption from the gut 139
    human milk and infant formulas 13
isothiocyanates
    aromatic, carcinogenesis 178, 182, 186, 193
    breakdown products of glucosinolates 210

kale, glucosinolates 206–207, 209, 214
kidney, effects of dietary glucosinolates 215, 217

lactation
    effect of maternal elimination diets 36–37
    zinc requirement 155, 166–167
lactobacilli
    adherence to gut mucosa 241–242
    use as probiotics 245
lactoferrin in human milk 5–6
β-lactoglobulin and antibodies, materno-fetal transfer 27–30
lactulose, use as probiotic 245
lectins
    bacterial adherence to gut surfaces 241
    dietary, effects on gut bacteria 244, 248–250
leucocytes, zinc content 158–159, 164–165
lipids *see also* fats, fatty acids, triacylglycerol
    blood, relation to atherosclerosis 44–48
    human milk and infant formulas 7–11
lipoproteins, relation to atherosclerosis 44–48
lipoteichoic acid, bacterial adherence to gut surfaces 241
liver, effects of dietary glucosinolates 215–217
lungs, effects of dietary glucosinolates 216
lutein *see* carotenoids
lycopene *see* carotenoids

maternal elimination diets and infant allergies 34–38
menadione, menaquinone *see* vitamin K

menstruation
    effects of diet and physical exercise 67–86
    effects of low energy intake 73–76
metallothionein
    assessment of zinc status 159
    gene expression 132–133
    zinc absorption in the gut 136–137
    zinc binding 131
milk
    effect on gut glycoconjugates 243
    human, model for infant formulas 1–16
        zinc content 156, 166–167
    maternal elimination diets 34, 36–37
    probiotic effect of oligosaccharides 245–246
    protein allergy 28–29
molybdenum in human milk and infant formulas 14
monooxygenases
    carcinogen activators 178–183
    xenobiotic metabolizing enzymes 219
mucins, gut surface 235
    bacterial adhesion 241–242
mucosa, gut, bacterial interactions 233–250
    nutritional influences 243–250
mustard, glucosinolates 206–207, 209, 214
myoinositol hexaphosphate *see* phytate
myrosinase, breakdown of glucosinolates 207, 210–211, 225

NAD(P)H dehydrogenase *see* quinone reductase
nervous system, requirement for fatty acids 7
niacin in man 109–111
nicotinamide, nicotinic acid *see* niacin
nitriles, breakdown products of glucosinolates 208, 213, 216
nitrosamines
    carcinogens 177
    effect of glucosinolate derivatives 223
non-protein nitrogen in human milk 7
nucleotides in human milk and infant formulas 15–16
nutrition
    effects on gut bacteria 243–250
    women athletes 67–86
nuts in maternal elimination diets 35–37

obesity, relation to atherosclerosis 55–58
oestrogens, cancer 189–190
oligosaccharides *see also* glycoconjugates, mucins
    dietary, use as probiotics 250
    human milk 11–12
ornithine decarboxylase, carcinogenesis 187–189, 191
ovalbumin
    antigen in breast milk 31
    materno-fetal transfer 29
oxazolidinethiones, breakdown products of glucosinolates 210

palmitic acid, availability to infants 9–10
pantothenic acid in man 116
paracetamol, effect of glucosinolate derivates 224
parakeratosis in pigs 130

Peyer's patch, bacterial adhesion 239–240
phenacetin, effect of glucosinolate derivatives 224
phenethyl isothiocyanate, carcinogenesis 180, 182
phorbol-12-O-tetradecanoate-13-acetate responsive element 184
phosphorus in human milk and infant formulas 13
phylloquinone *see* vitamin K
physical activity
    bone health in women 80–84
    menstruation 67–86
phytate *see also* inositol
    anticarcinogen 187
    zinc absorption 152–154, 156, 167
phytohaemagglutinins, dietary, effect on gut bacteria 248–250
phyto-oestrogens, anticarcinogens 190, 192–194
picolinic acid, effect on zinc absorption 142, 144
pig
    effect of zinc deficiency on growth 135
    zinc content 130–131
pili *see* fimbriae
polyamines, carcinogenesis 188–189
pregnancy
    effect of maternal elimination diets on infant allergies 36–37
    zinc deficiency 164–166
    zinc requirement 155, 157
probiosis, prevention of bacterial disease 245–247
progoitrin, glucosinolate 208, 212–214, 222
protein kinase C, carcinogenesis 187–188
proteinase inhibitors, carcinogenesis 187–188
proteins
    dietary, effects on glycosylation in gut 247
    human milk 3–7
proto-oncogenes 188
pteroylpolyglutamates *see* folate
pyridoxal, pyridoxamine, pyridoxine *see* vitamin B$_6$

quercetin, carcinogenesis 180–183, 189, 191, 193
quinone reductase
    carcinogenesis 181–183
    effect of glucosinolates 221, 224

radish, glucosinolates 206–207, 209, 214
rape, glucosinolates 206–207, 217
rapeseed meal, toxicity from glucosinolates 215–217, 221, 223, 225
receptors for bacterial adherence to gut surfaces 239, 242–250
retinal, retinoic acid *see* vitamin A
rhinitis, allergic, in infants, effect of maternal diet 35
riboflavin in man 106–107
Ruminococcus, glycosidase activity 243

Saccharomyces boulardii, control of diarrhoea 245
Salmonella enteritidis, effect of fimbriae on pathogenicity in mice 237
Salmonella typhimurium, diarrhoeal disease in pigs 239
salt *see* sodium chloride
sauerkraut, glucosinolates 213–214
selenium in human milk and infant formulas 14

Shigella flexneri, glycosidase activity 243
sialic acid *see also* mucins
    masking of bacterial adherence sites in gut 244
    requirement of infants 12
sinigrin, glucosinolate 208, 212–213, 215–216, 219, 222–223
sodium chloride, relation to hypertension 52–54
soya in maternal elimination diets 36
steroids, effects of glucosinolates 224
streptococci, group A, adherence to pharyngeal and buccal cells 241
sulphotransferases, xenobiotic metabolizing enzymes 220
sweed, glucosinolates 206, 209, 214–215

taste acuity, effect of zinc deficiency 157, 160
taurine in human milk and infant formulas 15
12-O-tetradecanoylphorbol-13-acetate 187–189
thiamin in man 104–106
thiocyanates, breakdown products of glucosinolates 210, 213, 215–216
thioglucoside glucohydrolase *see* myrosinase
thymulin, assessment of zinc status 159
thyroid, effects of dietary glucosinolates 215–217
tocopherol, tocotrienols *see* vitamin E
tolerance, immunological 29
    development in neonate 32–33
translocation of bacteria, effect of malnutrition 249
triacylglycerol structure and infant requirements 9–10
tryptophan
    conversion to niacin in man 110
    infant formulas 3–4
turnip, glucosinolates 206–207, 209, 214–215

UDPglucuronosyltransferase
    carcinogenesis 183–184
    xenobiotic metabolizing enzymes 219, 221, 224–225
urine, zinc content 158

vanillin, carcinogenesis 181, 185, 193
vegetables
    consumption data 213–214
    glucosinolates 205–225
    sources of anticarcinogens 175–194
vegetarianism, effects in women athletes 76
Vibrio cholerae
    colonization of villus surfaces 239
    probiotic effect of milk 246
visual function in infants, effects of fatty acids 7–9
vitamin A *see also* carotenoids
    dietary, effect on glycosylation in gut 247–248
    in man 96–98
vitamin B$_1$ *see* thiamin
vitamin B$_2$ *see* riboflavin
vitamin B$_6$ in man 107–109
vitamin B$_{12}$ in man 113–114
vitamin C
    carcinogenesis 179–181
    coenzyme for breakdown of glucosinolates 210
    in man 117–119
vitamin D in man 98–100

Index of Subjects

vitamin E
   carcinogenesis   176, 179–181
   in man   100–102
vitamin H *see* biotin
vitamin K in man   102–104
vitamins, availability in man   93–119

weaning, abrupt, effect on resistance to bacteria   247
whey proteins in human milk and infant formulas   4

xenobiotic metabolizing enzymes   219–224
   effects of glucosinolates   222–224
xenobiotic responsive element   180

Yersinia enterocolitica, gastroenteritis   240, 242
zeaxanthin *see* carotenoids
zinc
   absorption   136–140
   assessment of status   157–160
   bioavailability   141–142
   biology   130–145

zinc (*cont.*)
   body content   130
   catalyst   131
   complexes and chelates   140–145
   content in foods   152–153
   deficiency   133–135, 160–167
      aetiology   151–155
      prevention   161–163, 165, 167
      teratogenic effects   164
   homeostasis   130
   intake   153–154, 156
   intracellular   130
   nutrition in developing countries   151–167
   requirement   155–157
      infants   160–164
      lactation   166–167
      pregnancy   164–166
   sources in diet   151–153
   status assessment   133–135
   structural role in proteins   131